The Extended Organism

The
EXTENDED
ORGANISM

The Physiology of
Animal-Built Structures

J. SCOTT TURNER

HARVARD UNIVERSITY PRESS

Cambridge, Massachusetts, and London, England

Library of Congress Cataloging-in-Publication Data

Turner, J. Scott, 1951–
The extended organism : the physiology of animal-built structures /
J. Scott Turner.
    p.   cm.
Includes bibliographical references (p.).
ISBN 978-0-674-00151-0 (cloth)
ISBN 978-0-674-00985-1 (pbk.)
1. Animals—Habitations. 2. Animals—Physiology. I. Title.
QL756.T87   2000
591.56'4—dc21   99-057571

Designed by Gwen Nefsky Frankfeldt

For Debbie, Jackie, and Emma, who give me joy

Contents

Preface

This is a book about structures built by animals–sort of. I actually wish to explore an idea: that the edifices constructed by animals are properly external organs of physiology.

The subject of animal-built structures has traditionally been dominated by students of animal behavior and evolutionary biology. Consider the following. A collection of classic papers on animal architecture, *External Construction by Animals* (1976), compiled by Nicholas and Elsie Collias, is part of a series entitled *Benchmark Papers in Animal Behavior*. The authoritative work on the subject, *Animal Architecture* (1974), was written by Karl von Frisch, the Nobel Prize–winning behavioral biologist. A more recent book, by Michael Hansell, is entitled *Animal Architecture and Building Behavior* (1984): the title makes the point. Students of evolution, for their part, have long valued the fossilized remnants of animal-built structures as probes into the past, a view reflected in such books as S. K. Donovan's *Paleobiology of Trace Fossils* (1994) and Richard Crimes's *Trace Fossils: Biology and Taphonomy* (1990). Finally, of course, there is Richard Dawkins's eloquent book, *The Extended Phenotype* (1982).

For the most part, these treatments have been utilitarian in their approach to this subject—animal-built structures as examples of "frozen behavior," tools to probe the past, devices that genes use to project their influence beyond the organism's skin. I felt that a treatment of how these structures function for the animals that build them was lacking, and since I am a physiologist, I decided to try to fill the breach.

I have tried to make the book appealing to a broad audience. I hope students of animal behavior will enjoy a functional analysis of how such structures work. I hope physiologists will enjoy seeing the novel ways physiology can be applied to something besides organisms and cells. I hope evolutionary biologists will appreciate seeing the functional "flesh and bones" of the extended phenotype. Finally, I hope ecologists will enjoy delving into their discipline's historical roots as a physiological science.

Writing for breadth can be hazardous, though, for it is impossible to anticipate every reader's needs. Some passages may cover ground already familiar to some readers, but I hope all will find something new of interest. I have not shied away from quantitative and mathematical analyses, but all are straightforward and none require advanced training. I have used as many concrete examples as possible, but I have had to be very careful in making my selections: the hardest part of writing the book was in deciding what to leave out. I know many will wonder why I have left out their "favorite" structures—if so, please let me hear from you, just in case there is a second edition. For reasons more of personal interest than anything else, I have focused on invertebrates, so structures like birds' nests or beaver lodges have not been considered.

During the conception and writing of this book, I have been the beneficiary of numerous acts of generosity and good will. My editor at Harvard University Press, Ann Downer-Hazell, has been extraordinarily patient in helping me resolve difficult issues of style and presentation. She took a chance with me, an untested author, and I hope she now feels the gamble was worth it. Kate Schmit brought art and style to the final copy-editing, taking innumerable sentences and paragraphs that I thought I had tortured into a semblance of respectability and, with a brush and a flair, making them elegant. Three anonymous reviewers of earlier drafts of this book provided many thoughtful, perceptive, and critical comments. I hope they agree that they helped me make the book better. Bill Shields, my colleague at the SUNY College of Environmental Science and Forestry has been invaluable in challenging, openly, honestly and joyfully, some of my stranger thoughts—he is, in short, a *mensch*. Jim Nakas has been my resident expert on Greek and Latin etymology and phraseology. I have enjoyed talking with Charlie Hall about Howard Odum and maximum power. Earth-Watch, a remarkable organization, has been steadfast in supporting my work on termites; it was this work that got me thinking seriously about animal-built structures. The Trustees of the State University of New York generously granted me a sabbatical leave, which gave me the time I needed to complete writing the book. Finally, my wife Debbie and my two girls, Jackie and Emma, have had to put up with a husband and father who, while present physically, was often absent mentally. Their unwavering support and faith in me is more valuable to me than I can ever hope to express.

The Extended Organism

CHAPTER ONE

The Organism's Fuzzy Boundary

The building of structures by animals is widespread, ubiquitous even. Sometimes these structures are humble: simple tunnels in the ground or small piles of rubble. Sometimes they are grand—the nests built by some species of termites to house their colonies are, in their own way, magnificent, even sublime. Sometimes they are built unintentionally—the tracks left by an animal passing over or through sand or mud, for example. Sometimes they are built with seeming intention, like the "fog catchment trenches" the Namib desert beetle *Lepidochora* uses to capture water wafted inland from the Atlantic by coastal fogs. An animal-built structure may be ephemeral, disappearing with the next tide or gust of vigorous wind, or it may be relatively permanent, solidified by mucus or reinforced with silken threads. Some edifices are the work of masons, made from grains of sand laboriously glued together one after another. Others are the work of sculptors, like the holes carved from sandstone by sea urchins. Others still are the work of miners, like the tunnels dug into trees by bark beetles, or of weavers, like the woven nests of birds or the webs of spiders. Sometimes the workmanship seems sloppy, as in the rather haphazard piles of logs and twigs in a beaver's lodge, but occasionally it reflects an orderliness and precision of technique—as in the hexagonal wax cells of a honeycomb—that simply makes your jaw drop in wonder.

This book is about animal-built structures, but it also is about a question of broader interest in the fields of biology, evolution, and ecology. Are such structures best regarded as external to the animals that build them, or are they more properly considered parts of the animals themselves? I am an advocate for the latter interpretation, but the argument I present in this book is one with a twist: that animal-built structures are properly considered *organs of physiology*, in principle no different from, and just as much a part of the organism as, the more conventionally defined organs such as kidneys, hearts, lungs, or livers.

The idea that external structures are properly parts of the animals that build them really is not a new idea: the notion of an "extended phenotype," as Richard Dawkins has so aptly termed it, is well established and has become respectable, if not universally accepted, wisdom in biology. My goal in this book is to bring a physiological perspective to the idea, one which, I hope, complements that of a Darwinian like Richard Dawkins. An evolutionary biologist sees the extended phenotype as the extension of the action of genes beyond the outermost boundaries of an organism and asks how these extended phenotypes aid in the transmission of genes from one generation to the next. A physiologist, however, sees an extended phenotype in terms of mechanism and asks how it works, how it alters the flows of matter, energy, and information through the organism and between the organism and its environment. Although these two perspectives certainly complement one another, I hope to show that they also lead to somewhat different conclusions about the nature of life.

The Two Biologies

The crux of the problem, for both Darwinian and physiologist, is how one perceives the organism. On the face of it, this seems an absurd statement. One of the most obvious features of the living world is that it is composed of organisms, living things that we can hold in our hands, pin down under a microscope slide, give names to, feed and care for, catalog and place in a museum case, admire from afar. In organisms we recognize individuality, intention, purposefulness, function, *beingness*. To entertain *opinions* on what organisms are seems about as rational as disputing the value of π.

Yet that is where biology in the twentieth century has brought us, to a point where it is not at all clear just what organisms are.

We have arrived at this point via two intellectual journeys, resulting, if you will, in two biologies. On one hand, modern biology has relentlessly pursued an understanding of life as a mechanism, as a special case of chemistry, physics, and thermodynamics. For the most part, this mechanistic biology has played out in the study of how cells work, even down to their uttermost details of molecular action. An unintended consequence of this inward focus has been the fading of the organism from relevance: the organism itself has become, at best, an unwelcome distraction from the fascinating cellular and molecular business at hand. It's understandable, really—the mechanistic approach to biology is hardly worth pursuing if it doesn't look for unifying principles of life, fundamentals that do not depend upon whether a living thing is plant, animal, fungus, or bacterium.

At the same time, the twentieth century has seen the emergence of neo-Darwinism as a coherent philosophy of biology. Here, too, the organism has faded from prominence, but for different reasons. To the neo-Darwinist, the organism has become essentially an illusion, a wraith obscuring the "real" biology of the selfish genes that actually run the show. An organism, to the neo-Darwinian, is at most a transient coalition of genes, bound together in a conspiracy to promote the genetic interests of its members.

Although the two biologies have each prospered in their own rights, I think it also fair to say that they have proceeded more or less independently on their respective journeys. I do not mean to say that each of the two biologies has developed in ignorance of the other—far from it. Darwinians are comforted to know, no doubt, how genes really work, what is the chemical basis of heritability and phenotypic variability, and so forth. The molecular biologist may also rest easier knowing that the question of where all life came from is firmly in the good hands of biologists who, like them, are unwilling to admit the arbitrary power of gods as explanatory tools. Nevertheless, mechanistic biology and evolutionary biology are still, more or less, independent. Let's be honest, now—to what extent, for example, have the quantum-mechanical details of protein folding really informed or challenged the thinking of evolutionary biologists? From the other side, to what extent has, say, the evolution of song dialects in sparrows really changed the course of research in cellular signaling pathways? I think an honest answer to these questions would be "Not very much." This is a pity, really, because until the two biologies merge, until fundamental questions in one challenge the fundamental assumptions of the other, there can be no claim that we are even approaching a unified science of biology.

Beyond the Obvious Organism

The concept of the extended phenotype offers one way to bridge the divide between the two biologies. Using this bridge, however, requires that we think in yet another way about the organism, as illustrated by the question with which I opened this chapter: are animal-built structures properly things external to the animals that build them, or are they properly parts of the animals themselves? If we believe that animal-built structures are strictly external to their builders, we must posit an outer boundary to the organism, something which delimits it from (to use Boolean language) the NOT-organism. On the face of it, this seems an easy thing to do. The outer integument of the body—whether it be sheets of chitin, or woven socks of collagen, or shells of crystallized calcite or silica—seems to distinguish organisms quite obviously from their surroundings. But are we really justified in delimiting organisms in this way?

Well, yes and no. Certainly, the outer boundary of a living organism is a tangible, obvious thing, a wrapping of some material that keeps the organism contained nicely in a compact package. But it is worth remembering the etymology of the word obvious, which begins with the Latin *obvius* for "in the way" and ends with the modern, and really not very reassuring, usage of "evident without reasoning or observation." There

is always the possibility that a thing's "obviousness" is literally "in the way," a mask that prevents us from recognizing the thing for what it is. We must be willing to look past the obvious and ask what might be lurking beyond an organism's "outer" boundary.

Let us begin with a well-worn analogy, the supposed similarity between a turbulent eddy in a flowing stream and an organism. Eddies are familiar features of everyday life: we see one every time we pull the plug in a sink or bathtub. They are also to be found in the turbulent wakes of ships or stationary objects in moving streams, like bridge pilings in a river. Eddies develop when the inertia of a flowing fluid becomes just powerful enough to overcome the viscous forces that keep fluids flowing smoothly. Once an eddy develops, it dissipates this excess inertia as heat, giving rise to smaller eddies, which in turn give rise to still smaller eddies that merge finally into the surrounding fluid.[1]

Eddies are popular analogues for organisms because, on a superficial level, they seem to be so very

---

1. The tendency of turbulent eddies to dissipate their energy by spawning smaller eddies is summarized in a delightful bit of doggerel by L. F. Richardson (1922):

Big whorls have little whorls,
which feed on their velocity;
and little whorls have lesser whorls,
and so on to viscosity.

Richardson's verse itself is a parody of a poem that has become a favorite ditty of entomologists:

Big fleas have little fleas
upon their backs to bite 'em
And little fleas have lesser fleas,
And so, *ad infinitum*.

I am told by R. E. Lewis, the editor of *The Flea Newsletter*, that the original authorship of this poem has been lost. In all likelihood, though, it was inspired by Jonathan Swift, who published this rather more elegant poem in his *On Poetry: A Rhapsody* (1733):

So naturalist observe, a flea
Hath smaller fleas that on him prey;
and these have smaller still to bite 'em;
And so proceed *ad infinitum*.
Thus every poet, in his kind
is bit by him that comes behind.

similar. An eddy is a highly organized entity, seemingly self-contained, whose "purpose" is to dissipate excess inertia as heat. Likewise, an organism takes in energy in the form of light or chemical fuel and uses that energy to construct orderliness, eventually dissipating it as heat. An eddy is a transient phenomenon, persisting only as long as energy is fed into it. So too is an organism—take the fuel away, and a short time later the organism ceases to exist. Of course, one can only stretch an analogy so far before it breaks. One obvious weak point is the seemingly tangible boundary separating an organism from its surroundings. You can take a knife and cut open an organism: you cannot do that with a turbulent eddy.

Now, I confess I have laid a trap for you. If we assert that the boundary of an organism confers on it an identity, a beingness that makes it distinctive from the rest of the world, must we conclude that, because an eddy lacks a distinctive boundary, it cannot have identity?

The Identity of Permanent Eddies

Just downstream from Niagara Falls, there is a permanent eddy in the Niagara River, called simply The Whirlpool (Fig. 1.1). The Whirlpool is located at a dogleg turn of the river, at an old plunge pool formed when the Falls were located downstream from where they now are. As in all eddies, it is hard to see where The Whirlpool begins or where it ends. Nevertheless, The Whirlpool does seem to have an identity. It has a proper name. Its existence is obvious to anyone who looks over the cliff above the river. Its location can be found on maps. There is even a plaque on the walk along the riverbank that describes it. So, despite there being no obvious boundary separating The Whirlpool from the rest of the Niagara River, it does seem to have an identity. From where, then, does The Whirlpool's identity come?

The Whirlpool seems to derive its identity not from its *distinctiveness*—that is, in a clear demarcation between Whirlpool and NOT-Whirlpool—but from its

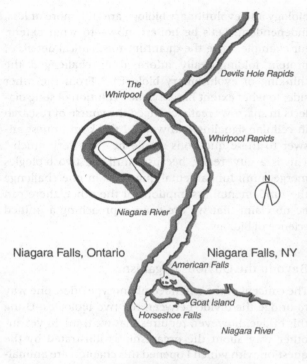

Figure 1.1 The Whirlpool is located downstream from Niagara Falls. *Inset:* Detail of The Whirlpool, showing general trajectory of flow.

persistence. Like Jupiter's Great Red Spot, The Whirlpool has persisted long enough for cartographers to put it on their maps and for landscape architects to incorporate it into their park designs. In this sense, The Whirlpool is like an organism: both are persistent in the way an eddy in the wake of a boat is not. So, perhaps we need to explore the analogy a bit further and ask: what is it about organisms and permanent eddies like The Whirlpool that confers on them persistence?

The Whirlpool is a persistent feature of the Niagara River for two reasons. First, the flow of the Niagara River provides a steady source of energy and matter to keep it swirling. Second, the flowing water interacts with specific structural features of the riverbed to channel the flow in a particular way: the dogleg turn of the river forces the water to change direction as it flows past—water, having mass and inertia, will resist this change—and the old plunge pool provides a venue for the dissipation of the water's inertia before the water is forced into the sharp turn. Both act together to modify the field of potential energy driving water down the river. The result is The Whirlpool.

For as long as it persists, an organism also modifies energy flowing through it, albeit in a very different way. An organism's persistence comes from the tangible boundary separating it from its environment. Even though it seems quite solid, an organism's outermost boundary is actually very permeable, allowing a steady stream of matter and energy to pass continually through it. But, the boundary is not passively permeable, as a sieve would be. Rather, it exerts adaptive control[2] over the flows of matter and energy across it. Here is the real breakdown in the analogy between a permanent eddy like The Whirlpool and an organism. Turn down the source of potential energy driving The Whirlpool (which the engineers of the New York Power Authority can do by diverting water away from the Falls), and The Whirlpool disappears. Turn down the potential energy driving matter and energy through an organism, and the organism will alter the nature of the boundary separating it from its environment so that it can maintain that flow. It is not the boundary *itself* that makes an organism distinctive, but what that boundary *does*. In other words, the boundary is not a thing, it is a *process*, conferring upon the organism a persistence that endures as long as its boundary can adaptively modify the flows of energy and matter through it.

A curious and paradoxical consequence arises from following the analogy between eddies and organisms

as far as we have: the obvious and seemingly clearly demarcated boundary separating the organism from its environment disappears. To see why, take the analogy just one step further. An eddy is a consumer of energy, taking in kinetic energy in flowing water and dissipating it as heat. An eddy like The Whirlpool has an indistinct boundary because this inward flow of energy and matter also influences flows elsewhere in the Niagara River. The strength of this influence diminishes with distance: its effects are easy to see close to The Whirlpool's center but become less distinct the further upstream or downstream you look. Nevertheless, the presence of The Whirlpool leaves an imprint on the flows of matter and energy that extends rather far from the obvious center of its activity. In the jargon of thermodynamics, The Whirlpool is at the center of a field of potential energy that both drives energy or matter through it and that, in turn, is influenced by The Whirlpool's presence.

Now consider a bizarre question. What would have to happen to make The Whirlpool behave more like a living thing? We have already ruled out adaptive control of the flow of matter and energy across a tangible boundary, like that which occurs in an organism, because turbulent eddies have no tangible boundaries. Suppose, however, that one night, when the New York Power Authority engineers divert water away from the Falls, The Whirlpool effects a change in the shape of the riverbed surrounding it, perhaps by forcing the riverbed downstream to sink in response to the diminished potential energy upstream. In this fanciful scenario, The Whirlpool might persist even in the face of the changing field of potential energy. In other words, if The Whirlpool could persist by adaptively modifying structural features of the environment surrounding it, the distinction between The Whirlpool and an organism—the adaptive control of the flows of energy and mass—would disappear. Could The Whirlpool then fairly be said to be "alive"? Well, that would be stretching the analogy further than even I am comfortable with, but I hope you would agree that we are now in that gray area Mark Twain referred to in his famous

---

2. I am using *adaptive* in the engineering sense of maintaining a state in the face of changing conditions, like thermostatic control of room temperature. In biology, *adaptation* is a loaded word, in part because it has been used carelessly and recklessly (what animal is not "wonderfully adapted to its environment"?) and in part because it implies a purposefulness that is anathema to many evolutionary biologists.

wisecrack about the identity of the author of *The Iliad* and *The Odyssey:* it was either Homer or another blind Greek poet with the same name.

However, it is precisely this "fuzzy" boundary between living and nonliving that is at the crux of the physiology of the extended phenotype. If The Whirlpool can be nudged closer to the realm of the living by conferring upon it the ability to adaptively modify its environment, then what should we think about organisms that do the same? If an organism modifies its environment for adaptive purposes, is it fair to say that in so doing it confers a degree of livingness to its apparently inanimate surroundings? If we agree, just for the sake of argument, that it does, then the boundary between organism and NOT-organism, the boundary that seems so tangible—so obvious—to our senses of vision and touch, dissipates into an indistinct blur, much as a turbulent eddy merges imperceptibly into the water surrounding it.

The Physiology of the Environment

The idea that organisms are integral with the world outside them, like the notion of the extended phenotype, is not new but neither is it an idea that sits comfortably with modern biology, especially neo-Darwinian biology. Consider the simple example of adaptation to some aspect of the environment, say temperature. Generally, organisms seem to have evolved to function well at the prevailing temperatures they normally experience. So, for example, a desert pupfish and an Antarctic ice fish live in very different temperature regimes, yet they each seem to function well in their own environments. Take a pupfish and an icefish and move them to each other's environments, however, and you will soon have two dead fish. In short, the two species have adapted to function well in their respective, albeit very different environments.

We have a pretty good idea how this process of adaptation works. The conventional story goes something like this: A cohort of individuals exists in an en-

vironment with a certain temperature. Because there will be variation in how well the individuals function at the prevailing temperature, there will be variations in the ability of the members of the cohort to reproduce. To the extent that these functional variations are genetic, the genetic attributes that confer "good" function will translate into high fitness and will be passed on to the next generation. Those that confer "poor" function will not be. The result over many generations will be adaptation, in the evolutionary sense.

What happens to this pretty picture, though, if you suggest that there is no real division between an organism and its environment? The notion of adaptation to the environment thus becomes problematic, because how can an organism adapt to itself? Even more strange, in this view the environment and not just the organism, never mind the genes *in* the organism, can be subject to selection and adaptation. In other words, the environment, and not just the organism, can have fitness. This kind of thinking gives many biologists fits, as is clear in dogmatic statements like the following: "Adaptation *is always* asymmetrical; organisms adapt *to* their environment, *never* vice versa"(emphasis added).[3] Nevertheless, the problem of just what the organism is and its proper relationship with the environment is too big (dare I say too obvious?) to be confined by dogma, and biology, fortunately, is returning to this problem in a serious way.

This book is undertaken very much in that spirit, and it is built around the simple idea that structures built by animals are akin to The Whirlpool's "adaptive modification" of the bed of the Niagara River. By structurally modifying the environment, I will suggest, organisms manipulate and adaptively modify the ways energy and matter flow through the environment. In so doing, they modify the ways energy and matter flow through them. Thus, an animal's physio-

3. G. C. Williams, "Gaia, nature worship, and biocentric fallacies," *Quarterly Review of Biology* 67 (1992): 479–486.

logical function is comprised really of two physiologies: the conventionally defined "internal physiology," governed by structures and devices inside the integumentary boundary of the organism, and an "external physiology," which results from adaptive modification of the environment.

I have organized my argument for this view into roughly three parts. The first section, comprising Chapters 2 through 4, will build the notion of a physiology that extends outside the conventionally defined boundaries of the organism. Chapter 2 delves into a basic discussion of what physiology is and of the thermodynamic principles governing all physiological function, whether it be internal or external. My overt agenda in that chapter is to convince you that external physiology can exist, that the environment can have physiology. Chapter 3 continues the line of thinking begun in Chapter 2 but focuses more specifically on how external physiology can work. I conclude Chapter 3 with a brief and very general discussion of how, practically, animal-built structures can modify the flows of energy and matter in the environment. Chapter 4, the end of the beginning, explores the apparently spontaneous emergence of orderliness in living systems and outlines a specific example. The interesting feature of this example will be the emergence of physiological function operating in the environment at a scale many times larger than the organisms that generate it. This imposition of orderliness at a large scale, I shall argue, is at the heart of the ability of organisms to be architects and engineers of their environment.

Chapters 5 through 11 represent the biological heart of the book. Each chapter explores how particular animal-built structures function as external organs of physiology. Chapter 5, for example, examines the indivisible link between permanent structures like coral reefs, or the "bodies" of sponges, and the flows of energy and matter in the environment. Chapter 6, on the tunnels dug by invertebrates in marine muds, argues that these structures are devices for tapping one of the largest potential energy gradients on the planet, the oxidation-reduction potential between our oxygen-rich atmosphere and the reducing muds that are a remnant of the early anoxic Earth. Carrying this discussion to the terrestrial sphere, Chapter 7 considers how earthworms manipulate the physical properties of the soil environment: in so doing, they make the soil an "accessory kidney" that enables them to survive an otherwise forbidding environment. Chapter 8 looks at woven structures, like silken webs of diving spiders and certain types of aquatic cocoons, that serve as accessory lungs and gills. Again, the theme in this chapter is the functioning of an external physiology to create an environment in which the organism's internal physiology may be maintained. Chapter 9 takes an unusual turn, proposing that leaf galls are animal-built structures that serve to modify leaf microclimates. As part of a rather speculative discussion, I suggest that galls change for the energy budgets of leaves in favor of the parasites infesting them. Chapter 10 presents animal-built structures as communications tools, focusing on the "singing burrows" of mole crickets. Finally, Chapter 11 explores the interaction between structure and physiological function in the nests of social insects, culminating in a discussion of what I regard as the most spectacular animal-built structures on the planet, the large mound nests constructed by certain species of African termites. These mounds, I assert, are not simply houses for the colony but are accessory gas-exchange systems that confer on the termites the power of adaptation to a wide range of environmental conditions. The interesting twist here is that the mounds function at a scale many times larger than the creatures building them. How they create such a system is a fascinating problem in biology, one that cannot be fully understood, I think, without understanding the external physiology that underlies the phenomenon.

For the final section, Chapter 12 returns to the theme of the extended phenotype and the many ways in which animal-built structures illustrate it. I frame the chapter around a discussion of the *Gaia hypothesis*, which asserts that the Earth is a singular living thing, an entity whose biota are engaged in a massive global physiology. Gaia is, I shall argue, simply the extended phenotype taken to its logical conclusion. I must note, however, that a physiological approach practiced on a global scale leads one to conclusions about evolution, natural selection, and adaptation that, I think it fair to say, sit uneasily with mainstream evolutionary biology.

CHAPTER TWO

Physiology Beyond the Organism

Remember the following sentence, for it is at the heart of this book. Physiology is the science of how living things *work*.

Physiology these days is a science biased toward the small, concerned with how individual cells and particular molecules in cells work. There is an obvious reason for this—we are the beneficiaries of a scientific revolution, one that heated up through most of the early part of the twentieth century and finally came to a boil in 1953. That was the year James Watson, Francis Crick, Maurice Wilkins, and Rosalind Franklin figured out (or more properly, told the rest of us about) the structure of DNA. This was a watershed event because embedded within the structure of DNA was the answer to many of biology's fundamental questions—what made us what we were, what controlled our development, what made the complicated chemistry inside cells work right virtually every time? Biologists had been asking these questions for centuries, of course, but understanding the structure of DNA enabled them, for the first time, to ask the questions in a scientifically meaningful way. Since 1953, the progress of molecular biology has been nothing short of triumphant. Like the Copernican revolution in its time, molecular biology is *the* crowning intellectual achievement of our time.

So, it is easy to forgive my molecularly oriented colleagues when they occasionally slip into arrogance, exemplified by the quotation at the head of this chapter. They've earned it, really. Nevertheless, it is grating: there are, after all, many other interesting ways to think about the world and how it works. For example, consider all the ways one could ponder the workings of automobiles. At the very small scale, we could focus on details of how individual molecules of hydrocarbon fuel are oxidized in the complex environment of the engine's cylinders. Now, this is unquestionably fascinating. It is easy to imagine how it could be crucial to understanding how automobiles work. But small-scale questions such as these are embedded in other important questions, like: how does an automobile interact with a road surface or with the air mass it travels through? Personally, I would be hard pressed to say that the *only* interesting questions left in automobile design concerned hydrocarbon molecules. I would be similarly hard pressed to agree that the only interesting questions left in biology are to be found at the molecular scale.

So, what are the other interesting questions?

Arthur Koestler has written of the "Janus-faced" nature of life,[1] meaning that life always presents two faces to the world. At no matter what scale one looks at it, life may always be described in terms of a nested hierarchy of what biologists call organizational levels. Organisms represent one level of organization, but these comprise numerous organ systems (cardio-respiratory, digestive, nervous), which themselves are made up of organs (say, hearts, lungs, blood vessels, blood plasma, and cells), which in turn are built up from assemblages of tissues (muscle tissue, connective tissue). One can carry this outlook all the way down to the level of individual molecules that, assembled in just the right way, make a living thing. Turning our

gaze around, however, we see also that organisms are themselves embedded in larger assemblages. Individual organisms interact with other organisms, whether of the same species or others, and these assemblages of interacting organisms make up the populations, communities, ecosystems, and biomes that collectively constitute the living world.

For the last century or so, biology has mostly relied upon only one of its Janus faces. Biologists today look at life with a determined inward gaze, starting from the organism and working down through the various levels of organization, ultimately ending at molecular biology. Suppose we unmask Janus's other face, however, and take a deliberate outward look? What then do we see? Well, lots of things, really, but one prominent feature is bound to be the macroscopic world in which organisms live—the environment—and how organisms interact with it. One of the premises of this book, in fact, is that it is here, beyond the organism, where the really interesting questions in biology lie.

Action and Interaction

The science of how organisms interact with their environments goes by various names, including physiological ecology, ecophysiology, and environmental physiology. These outward-looking fields of biology have been largely eclipsed by the triumphant march of molecular biology over the last fifty years, but they have been respectably active. Unfortunately, despite their outward perspective, the work of environmental physiologists and ecophysiologists has been largely one-sided. Speaking of an *inter*action between an organism and its environment clearly implies two things. One, of course, is the effect of the environment upon the organism, and the second is the effect of the organism upon the environment. Most of modern environmental physiology is focused on the first—that is, the *effect of* environmental temperature (salinity, wind speed, radiation, pH) *on* such-and-such an animal, plant, organic process. Rarely are effects acting the other way considered. Nevertheless, the action of the

---

1. In Roman mythology, Janus was the god of beginnings and ends. The Romans placed him at the creation of the world and regarded him as the supreme deity, more highly even than Jupiter. His insigne typically showed him with two faces *(Janus bifrons)*, looking in opposite directions, a figure that supposedly symbolizes the confused state of the primordial world as it emerged from chaos. This design was commonly placed over thresholds of gates and doors, symbolizing the transition between the inner and outer spaces of a home or city. In ancient Rome, the cult of Janus was politically powerful, as is reflected in the name of the month (Januarius) that inaugurates the new year.

organism on its environment has to be important, and, furthermore, it has to be physiological. In other words, a true *inter*action of an organism and its environment must result in an extension of physiology outside the conventionally defined boundaries of the organism.

I want to convince you this kind of physiology does exist and that it is, in principle, no different from the physiology that goes on inside an organism. Fortunately, my task will be an easy one, because it relies on a well-established principle about how the insides of organisms work: organisms, like all other things in the universe, are governed in their operations by the laws of that most fundamental of the physical sciences, thermodynamics. Physiology inside an organism, therefore, is fundamentally a problem in thermodynamics. This is an uncontroversial and, as far as I know, a universally accepted assertion about life. To demonstrate that physiology extends outside the organism, all one needs to do is broaden the perspective a bit and show that the principles of thermodynamics don't stop at the organism's skin.

Thermodynamics and the Creation of Orderliness

Here we return to the sentence I asked you to remember at the start of this chapter. In thermodynamics, the word work has a very precise definition. We will be getting into the concept of work in more detail momentarily, but suffice it to say that work in the thermodynamic sense is only done when energy is made to flow. How energy flows, and how it can be made to do work in the process, is the subject of thermodynamics. Thermodynamics is sometimes presented as a very abstruse and difficult subject, but its basic principles are fairly simple (see Box 2A). These principles govern one of the two fundamental things that organisms do—they channel energy through their bodies and create orderliness in the process. (The other fundamental thing organisms do is to encode and transmit information about how to make copies of themselves—the molecular stuff.)

Order is something most of us intuitively understand. I am one of those people that are politely referred to as "organizationally challenged"—in the bad old days, I would have been called a slob. My desk is littered with piles of paper, my laboratory has remnants of half-completed projects laying around everywhere. I occasionally try to put my life in order, but I soon stop, because it is just too much work! And this illustrates a very important point about order—it takes *work* to create it and to maintain it. So while I am lying down until the urge to tidy my office passes, I console myself by reflecting that I am in harmony with one of nature's fundamental laws.

This is more than a glib joke—if left alone, the tendency of the universe seems to be toward increasing *dis*order, not its opposite. The creation of orderliness runs counter to what, in all our experience, seems to be a fundamental feature of the universe. Nevertheless, order does appear, it appears in many forms, and when it pops up, as it does in living things, it demands explanation. Fortunately, the creation of order by living organisms has been explained, for the most part, and a relationship between energy and orderliness is at the crux.

The Work of Creating Order

To illustrate how energy and order are related, let us look at a chemical reaction that is arguably the most important one on Earth: the photosynthetic fixation of carbon dioxide and water into glucose (a sugar) by green plants.

The reaction as it occurs in plants is very complex, but we can state it in a simple form that relates its reactants (on the left) to its products (on the right):

$$\text{light} + 6CO_2 + 6H_2O \rightarrow C_6H_{12}O_6 + 6O_2$$

$$\textit{disorderly} \quad \rightarrow \quad \textit{more orderly}$$

Below the chemical formula for the reaction, I have made an assertion: the reaction not only produces glu-

2A

The Laws of Thermodynamics

Thermodynamics is concerned with the relationships between energy, heat, and work. With its origins in the empirical problem of how to make engines work efficiently, classical thermodynamics was fleshed out during the Age of Steam. This is why many of its units and concepts (the joule, the watt, the kelvin, the Carnot cycle) are named after the engineers and physicists that made steam power a practical tool.

It turns out, though, that the rules governing the energy relations of steam engines also govern any system—including living things—that does work of any sort. For example, one can express the work done by a steam engine as a function of the fuel energy put into it. One can do the same for an "organism engine," such as a harnessed horse, fed with hay, that works to drive a millstone. Interestingly, the work done by both engines will be constrained in some rather fundamental ways, suggesting that the two are governed by the same laws. Indeed, one of the earliest indications that this might be so came from a study comparing the energy costs of boring cannon barrels by steam power and by horse power. Whether the fuel was coal or oats, the energy cost of boring a cannon barrel was very similar. This similarity means that thermodynamics is central to our understanding of how organisms work, and even how they are able to exist at all.

Thermodynamics has as its foundation three laws, numbered First, Second, and Third. All concern the behavior of a universe, which is composed of a system and its surroundings. These terms can be rather slippery, and being careless about their meanings can make some very simple ideas seem very difficult. For example, the thermodynamic universe can encompass something as small as a molecule, a cell, or an organism; especially in biology, we rarely refer to the thermodynamics of the *entire* universe, as the common meaning of the term implies.

The First Law of Thermodynamics, which constrains the quantity of energy in the universe, is sometimes designated as the law of conservation of energy. It simply states that the total amount of energy in the universe is a constant. It does not limit the energy either to the system or to the surroundings, but the sum of all energy in the universe must be constant. It does not constrain the form the energy can take (that is, it can be potential or kinetic energy, heat energy, electrical energy), nor does it constrain the flow of energy between the system and its surroundings.

The Second Law of Thermodynamics is also known as the law of increasing entropy. This simple law has some marvelously subtle implications for life. The Second Law states that whenever energy does work—whether the system does work on the surroundings, or vice versa—some fraction of the energy is lost to random molecular motion, or entropy (sometimes referred to as "disorder"). Thus, in any universe where there is work being done, there will be a relentless increase in the universe's entropy. It is important to remember that the Second Law does not force increase of entropy on either the system or the surroundings. Likewise, there is nothing in the Second Law that prevents a decrease in entropy (or an increase in order) in some *part* of the universe. Any decrease in the entropy of one part must be accompanied by a greater increase in the entropy of another part, however. The Second Law demands only that the universe experience a *net* increase of entropy. Organisms, which are highly ordered systems that can be thought of as transient "pools" of low entropy, exist only by disordering the universe in which they exist.

The Third Law of Thermodynamics is a bit more esoteric, but it is important in that it gives us a thermodynamic definition of temperature. Put simply, it states that there is a lower limit on the temperature of any universe, the point at which random molecular motion, or heat, falls to zero. This is the basis of the absolute, or thermodynamic, scale of temperature, designated by units of kelvins, or K. The zero on the absolute temperature scale is often referred to as absolute zero, because a temperature lower than 0K is impossible (there is no such thing as negative motion). In terms of the more frequently used Celsius or Fahrenheit scales, which are zeroed at "convenient" temperatures, absolute zero is −273.15°C or −459.67°F.

cose and oxygen, it also produces orderliness from disorder.

Orderliness increases by this reaction because a large number of simple molecules is reduced to a smaller number of more complex molecules:

light + 6CO$_2$ + 6H$_2$O →   C$_6$H$_{12}$O$_6$ + 6O$_2$
12 molecules   →   7 molecules
(6 containing carbon) → (1 containing carbon)

The increased orderliness means it is easier to keep track of the atoms on the equation's right (orderly) side than on its left (disorderly) side. Look at the carbons. On the left side of the equation, each carbon atom is locked up in one of six carbon dioxide molecules. Nothing about one particular carbon atom tells us anything about what the other five carbon atoms are doing—where they are, how fast each one is going in what direction, and so forth. For a complete description of the system on the equation's left side, we need information about each carbon atom. But when six carbons are brought together into a single glucose molecule, knowledge of one carbon atom tells us a lot about what the other five are doing. Less information is required to describe the equation's right side, and it is therefore more ordered.

Energy comes into the picture as light. "Light" is a bit of a misnomer—it is simply our word for a particular type of electromagnetic energy that comes bundled as particles called photons. Having light come in photons is handy because it allows us to treat energy in chemical reactions as we treat atoms. We know, for example, that the production of one molecule of glucose by a green plant (that is, one with chlorophyll) requires about 48 photons of red light. This lets us rewrite our photosynthetic reaction:

48 photons + 6CO$_2$ + 6H$_2$O → C$_6$H$_{12}$O$_6$ + 6O$_2$

We also know a lot about how photons carry energy. For example, the energy in a photon, expressed in joules (J), can be calculated from Planck's law:

$$E = h\nu \hspace{3cm} [2.1]$$

where h is Planck's constant (6.63×10^{-34} J s) and ν is the photon's vibration frequency (s^{-1}).[2] For red light, for example, the vibration frequency is about 4.3×10^{14} s^{-1}, and the energy carried by a photon of red light is about 2.8×10^{-19} joule. We can now write the photosynthetic reaction with the energy term made explicit:

(48 photons × 2.8×10^{-19} J photon^{-1}) + 6CO$_2$ + 6H$_2$O → C$_6$H$_{12}$O$_6$ + 6O$_2$

or:

1.4×10^{-17} J + 6CO$_2$ + 6H$_2$O → C$_6$H$_{12}$O$_6$ + 6O$_2$

So, not only can we say that it takes energy to create order in this reaction, we are able to say with some confidence just how much energy is required. And this enables us to illustrate another important feature of how organisms work.

If we turn the photosynthetic reaction around, we get a chemical reaction representing the breakdown of glucose into carbon dioxide and water. This happens when, for example, we burn wood (which is basically glucose), or when we burn sugar as fuel in our bodies in the process of metabolism:

C$_6$H$_{12}$O$_6$ + 6O$_2$ → 6CO$_2$ + 6H$_2$O + energy

more orderly   →   *disorderly*

In this reaction the carbons in glucose are returned to their initially disordered state, and in so doing energy is released. This energy had to come from somewhere. Its source, of course, was the energy initially supplied in the form of photons from the sun. Production of order is a means of storing energy the organism can use later on.

---

2. If you are not familiar with scientific notation or with the conventions for expressing units of measure, you may wish to take a few moments now and peruse Box 2B.

Scientific Notation and Units of Measure

The numbers in this book (in scientific writing of all sorts, really) may look odd to readers unfamiliar with mathematics, but in fact the notation is straightforward and the abbreviations fairly easy to learn. Scientific notation is simply a convenient way of writing very large or very small numbers. It expresses numbers as a multiple of a "conveniently written" number and a power of ten. For example, we could write 230 as the product of 2.3 × 100. However, since 100 is the same as ten squared, or 10^2, we could also write 230 as 2.3 × 10^2. Expressing this particular number in scientific notation does not really offer any advantage to us, but imagine having to write 2,300,000,000 (two billion, three hundred million). We could express this as the product of 2.3 and one billion. Because one billion is the same as ten multiplied by itself nine times (try it on a hand calculator), or 10^9, we could write the number much more compactly as 2.3 × 10^9. Expressing very small numbers is also easily done. For example, 0.023 is the same as 2.3 × 0.01. However, 0.01 is the same as ten *divided* by itself twice, and mathematically, this can be written as 10^{-2}. Thus, we can conveniently write the very small number 0.0000000023 as 2.3 × 10^{-9}. Computers have an even more compact way of writing large or small numbers: they simply express the power of ten as a number preceded by an *e* (for *exponent*). Therefore, 0.0000000023, or 2.3 × 10^{-9}, may appear on a computer printout or on the display of a hand calculator as 2.3e-9.

Whenever one deals with real quantities, rather than pure numbers, one must be able to describe what those numbers quantify. For example, a number like π is a pure number. It can be calculated from the ratio of two lengths (the circumference and radius of a circle), but the number itself does not signify a length. A number that does indicate a length, however, must include a descriptor of some sort, in other words, a unit of measure. Units usually are written after the number. So, we would describe the radius of a circle as *x* meters, or feet, or whatever.

Units that describe fundamental quantities, like mass, length, time, or temperature, have simple units. One often must deal with compound units, however, combining two or more simple units. A speed, for example, is a ratio of a length or distance traveled and a time required to travel it. One unit of speed could therefore be meters per second, which could be written in abbreviated form as m/s. Scientific convention dictates another form, though, as the multiple of distance and the inverse of time. So, the compound unit m/s could also be written as m × (1/s). Since an inverse of a quantity is equivalent to the quantity raised to the power of −1, we can write the unit of velocity as m s^{-1}.

This might seem the ultimate in obfuscation: what's wrong with a simple ratio, like m/s? There is a good reason for using this style, though. Sometimes units become very complicated. Weight, for example, is actually the product of a mass and an acceleration imparted to the mass by the gravitational attraction of the Earth. The unit for mass is straightforward, the kilogram (kg). An acceleration, on the other hand, is a change of speed with time, or (meters per second) per second (no, I did not inadvertently repeat "per second"). The unit in simple ratio form is m/s/s. Or should it be (m/s)/s? Or should it be m/(s^2)? And when we multiply acceleration by mass to get a weight (properly designated with the newton, N), the potential for confusion abounds. Is it kg × m/s/s? kg × (m/s/s)? kg × ((m/s)/s)?

Or we could simply write it as kg m s^{-2}. By writing the units with the inverses as negative powers, the unit can be specified unambiguously and without the need to decode a multitude of parentheses.

You will also see many of the unit designators with prefixes, such as millimeters (mm) and kilojoules (kJ). These represent multipliers to the basic units. For example, the standard unit of length is the meter (m). Sometimes it is inconvenient to express a length in meters, though. The distance between, say, New York City and San Francisco is 4,713,600 meters, but it is more common to express as at 4,714 kilometers (km), where *kilo* stands for "1,000 times."

Certain prefixes are used quite frequently, especially those that specify multiples of 10^3 or 10^{-3}, as listed in the accompanying table. Sometimes a unit of measure is used even though the number could be "simplified" by using another. So, for example, we express the distance between San Francisco and New York as 4,714 km, even though we could more easily (and properly) write it as about 4.7 Mm—kilometers are simply more conventional than megameters. Common numbers like distances would never be expressed in megameters or nanomiles—it would bring us too close to the Dilbertian world of pocket protectors and calculator holsters.

Prefix	How it is said	Multiplier
T	tera-	10^{12}
G	giga-	10^9
M	mega-	10^6
k	kilo-	10^3
–	–	$10^0 (= 1)$
m	milli-	10^{-3}
mc or μ	micro-	10^{-6}
n	nano-	10^{-9}
p	pico-	10^{-12}
f	femto-	10^{-15}
a	atto-	10^{-18}

Whether this way of storing energy is efficient is an interesting question, because it illuminates a constraint imposed by the Second Law. We can measure how much energy is released when an ordered glucose molecule is subsequently disordered. It turns out to be:

$$C_6H_{12}O_6 + 6O_2 \rightarrow 6CO_2 + 6H_2O + 4.8 \times 10^{-18}\ J$$

The energy released is in the form of heat (as when a log is burned), and it represents about 36 percent of the energy initially captured as light.[3]

We can now state two thermodynamic principles that are important to our understanding of physiol-

ogy, whether it occurs inside or outside an organism. The first concerns the forms energy can take. In the glucose examples there were two transformations of energy: light energy was transformed into a potential energy stored in the bonds of a glucose molecule, and that energy was then recovered as heat when the glucose molecule was broken up again. More transformations are possible: indeed, there is only one intrinsic limit (which we will explore in the next chapter) on the number of transformations a parcel of energy can undergo. The only constraint is that whatever transformation energy undergoes, the total *quantity* of energy cannot change. In constructing a glucose molecule, for example, 1.4×10^{-17} joules of energy were put in on the left side of the equation, and there must be 1.4×10^{-17} joules coming out on the right side. In other words, energy is conserved. This is the First Law of Thermodynamics (which shall henceforth be referred to simply as the "First Law").

Conservation of energy implies that as much energy should be recovered from destroying glucose as was put into creating it. But that obviously isn't true: burning glucose yields only about 36 percent of the energy put in. This leads us to our second important principle: *no transformation of energy that involves work is ever perfectly efficient.* In the photosynthetic reaction for glucose presented above, for example, I did not give enough detail concerning the fate of the energy. The complete equation is thus:

48 photons + $6CO_2$ + $6H_2O$ $\rightarrow$ $C_6H_{12}O_6$ + $6O_2$
1.4×10^{-17}J   = 4.8×10^{-18}J + 9.2×10^{-18}J
energy in   =   *energy in* + *heat*
          glucose

The energy from the 48 photons that was not captured as order in glucose was therefore lost as heat, warming

---

3. Those who are well versed in chemical thermodynamics will recognize that I am painting a sort of "worst-case" scenario here. The figure of 36 percent efficiency is based upon a particular set of conditions and concentrations of reactants and products known as the standard condition; specifically, all products and reactants are at a concentration of 1 mole per liter, a temperature of 25°C, and a pressure of 1 atmosphere. Cells, of course, are rarely at the standard conditions, and this allows them to be more efficient than the 36 percent figure just quoted.

Indeed, some chemical reactions in cells approach energy efficiencies of 95 percent or more. But the point made below remains the same—the reactions will never be perfectly efficient—and some of the energy in the transformation will always be lost as heat.

whatever container (universe) the reaction took place in. The role of this heat is interesting, because increasing the temperature of a substance generally disorders it. Indeed, we can state the photosynthetic equation in a more general form that describes any work-producing transformation of energy:

energy in = work energy + heat energy

Or, we can be more radical and state that:

energy in = production of order + production of disorder

Thus, whenever energy is made to do work, some portion of it ultimately ends up increasing the disorderliness of the universe. This is the Second Law of Thermodynamics (henceforth simply the "Second Law").

The ATP Cycle

When glucose is broken down to carbon dioxide and water, the energy stored in the glucose molecule is ultimately lost as heat. When glucose is burned directly, all the stored energy is converted to heat in one step. Organisms, however, take the energy stored in glucose and use only a portion of it to do chemical work; the left over energy is stored in another form, to be used later as needed. The ability to store chemical work is central to an organism's ability to use energy.

Metabolism is a process of controlled combustion of fuel, usually glucose. The energy released from glucose can be made to do physiological work only if it is coupled somehow to an energy-demanding process. In nearly all organisms, coupling is done through an intermediary chemical that carries the energy from glucose to the chemical reaction that needs the energy. This intermediary is a nucleotide, adenosine, that carries energy in bonds that link phosphate molecules to it. Most commonly, the adenosine takes part in a cyclical reaction between its diphosphate form, adenosine diphosphate (ADP), and its triphosphate form, adenosine triphosphate (ATP).

When the energy in glucose is released by metabolism, some of it is used to add a phosphate ion, usually written P_i, to ADP, to form ATP:

energy + ADP + P_i → ATP + *heat*

The energy comes from coupling this reaction, the phosphorylation of ADP, to the release of energy from glucose:

glucose + *oxygen* → *carbon dioxide* + *water* + *energy*

energy + ADP + P_i → ATP + *heat*

ATP will release energy if it subsequently loses its third phosphate group:

ATP → ADP + P_i + *energy*

Any other energy-requiring, or order-producing, reaction can be driven, then, simply by coupling it to the release of energy from ATP:

ATP → ADP + P_i + *energy*

energy → *work* + *heat*

The work thus done could be used to create other kinds of energy, like order, mechanical work, electrical potential energy, and so forth.

To recap, thermodynamics says four important things about how animals work. These are:

1. Animals use energy to produce order.
2. Order can be used as a store of energy that can be tapped at a later time to do work.
3. The amount of order produced is limited by the quantity of energy available to do work.
4. The amount of order produced is limited further by the inescapable inefficiency of any order-producing process.

Thermodynamics and Physiology: Two Examples

I would now like to describe two physiological processes and how energy drives them to create order. The first, formation of urine by the fish kidney, is an example of "blood-and-guts" physiology taking place within a well-understood organ inside a tangible organismal boundary. The second is concerned with the deposition of calcium carbonate by a reef-forming coral. This process is also reasonably well understood, but the boundaries between what physiological activity is inside and what is outside the organism are somewhat blurred. The obvious conclusion to which I want to lead you is that the common features of these two processes do not include a boundary between an organism and its environment.

Water Balance in a Freshwater Fish

The body water of fishes, like that of most vertebrates, is a weak solution of salts and other small solutes: roughly 0.9 percent of the mass of blood plasma, for example, is sodium chloride, common table salt in solution. Fresh water, of course, has far fewer solutes in solution. A fish in fresh water can be thought of as two bodies of water, each differing in composition and separated from one another by the fish's skin. This is order: the orderliness is manifest in the separation of solutes and water into two compartments, the fish and the environment. Because the universe does not like order, the Second Law will dictate that this ordered system be driven to *dis*order. Disorder in this case will arise as solutes diffuse from the fish to the surrounding water, and as water flows into the fish by osmosis.[4]

---

4. Osmosis and diffusion both refer to the movement of matter under the influence of a difference in potential energy. They differ in some crucial respects, though. In an aquatic medium, diffusion refers to the movement of solute from a region of high solute concentration to a region of low solute concentration—down the concentration gradient, it is said. Osmosis refers specifically to the movement of water from a region of low solute concentration to a region of high solute concentration. Osmosis is sometimes referred to inaccurately as water diffusion, because a solution with high solute concentration is also a solution with

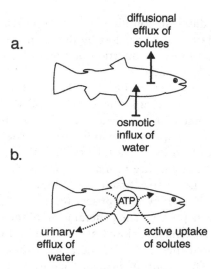

Figure 2.1 The water and solute balance of a freshwater fish. *a:* The fluxes of water and solutes are driven by gradients of potential energy between the fish and its environment. *b:* The fish's physiological response is to drive fluxes of water and solutes in opposite directions.

Both processes act to equalize the concentrations of solutes on either side of the fish's skin (Fig. 2.1). Because the volume of water in which a fish lives is many times greater than the volume of water contained in its own body, the major change of concentration will occur inside the fish.

The dilution of its internal fluids constitutes a mortal threat to the fish, and the fish's defense is to use energy from glucose to reimpose order(a more concentrated solution). First, by filtering blood through the kidneys, the fish produces large quantities of very dilute urine. This pumps water out of the fish as fast as it comes in by osmosis, keeping the total water content of the fish roughly constant. Second, the fish keeps solutes within its body by using energy to take solutes

---

low water concentration, and vice versa. Water thus moves down its concentration gradient, but it moves against the solute concentration gradient. The analogy is useful, but inaccurate. I will return to the principles of osmosis and diffusion in more detail in later chapters.

from the dilute surroundings and pump them into its body as fast as they are lost by diffusion. Again, this keeps the solute concentrations in the fish's blood high and relatively steady.

Let us now review the physiology of this process. Fishes use two organs, the kidneys and the gills, to keep internal solute and water concentrations steady in the face of thermodynamic assault. The kidneys are primarily responsible for handling the water and salts, and the gills are primarily responsible for handling salts. Let us turn first to the kidneys.

The fish kidney comprises numerous subunits known as nephrons. The general form of the nephron is a tubule, which is closed at one end and opens at the other to the outside of the body through a pore or duct.[5] The nephron is surrounded by a network of blood vessels that deliver blood to the tubule and return water and solutes to the blood. It is the job of the nephron to produce urine from blood. Production of urine begins at the closed end of the nephron, and as it is produced, it is pushed out the open end of the nephron to leave the body.

Nephrons work largely by two processes, filtration and reabsorption (Fig. 2.2). Filtration occurs at a junction between the blood and nephron involving two structures: a knot of capillaries, called the glomerulus, and a cup-shaped expansion of the end of the nephron tubule, the Bowman's capsule. The glomerulus sits enveloped in the hollow of Bowman's capsule, and together they form a porous filter between the blood and the tubule. As blood is pumped into the nephron, water, salts, and other small solutes are forced across the glomerulus into the tubule. The resulting liquid is called filtrate: its composition is essentially blood plasma minus blood cells and large proteins left behind in the blood.

Filtration has two interesting features that both help and hurt our freshwater fish in its quest to maintain its

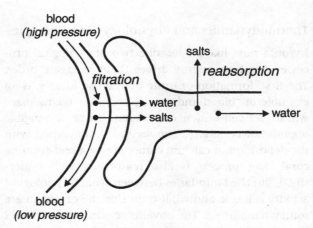

Figure 2.2 The formation of urine in the vertebrate nephron. Blood at high pressure is filtered at the junction between the capillary and the tubule. In the tubule, salts are reabsorbed back into the body, leaving only water to be collected by the duct.

internal environment. First, filtration produces a lot of filtrate. This helps the fish because it gets the excess water out of the blood. A problem arises, though, because the filtrate's concentration of small solutes, like salts, is virtually identical to the blood's. Consequently, one of the costs of producing lots of filtrate is a loss of solutes. This is undesirable, because one of the fish's problems, after all, is loss of solutes to the environment by diffusion. It does the fish no good if the problem is compounded by further loss of salts via the urine. Luckily, a solution to the problem is available: reabsorption. As the filtrate passes down the tubule, the solutes are transported out of the filtrate (reabsorbed), back into the blood. As solutes are removed from the urine, their concentrations in the filtrate decline. The urine that is produced, therefore, is voluminous—because of the high rate of filtration—and very dilute—because of the high rate of reabsorption. The urine that is excreted from the fish is essentially water, whose return to the environment offsets the osmotic flow of water into the organism.

The recovery of solutes from filtrate is only part of

---

5. Normally, the many ducts emerging from the nephrons merge into a single large tube, the ureter, that drains the assemblage of nephrons that constitute the kidney.

the fish's process of maintaining high-solute concentrations in the blood. At the fish's gills are specialized cells, known as chloride cells, that use ATP energy to transport chloride ions from the water into the blood; these ions then pull positively charged sodium ions along with them. Because salt is being transported from the dilute source of fresh water into the relatively salt-rich blood, energy is required to pump the chloride. The inflow of salts across the gills effectively offsets the diffusional loss of salts across the skin.

In the midst of all this movement of water and salts, transformations of energy and the creation of order are taking place. Filtration requires energy to make it go because filtration is an order-producing process. One form of orderliness is the separation of molecules across a membrane. In the nephron, proteins and other large solutes end up being concentrated in the blood, and water and small solutes end up in the filtrate. The energy to power this separation comes from ATP fueling the fish's heart muscle. The work done is transformed by the heart into an elevated blood pressure, another form of potential energy. This potential energy drives and separates the large and small solutes across the glomerulus (Fig. 2.3).

ATP is also used to drive reabsorption. This process, too, is productive of order and therefore requires energy to make it go. Reabsorption produces order because solutes are moved from dilute filtrate into more concentrated blood, against their natural tendency to intermix and reach an equilibrium concentration. The separation is accomplished by protein molecules, known as active transport proteins, embedded in the inner membranes of the nephron's tubules (that is, facing the filtrate). These bind solutes and then release energy from ATP. The energy so released forces the transport molecule to change shape, with the consequence that the bound solute is moved across the tubule wall, from the filtrate into the blood. The transport cells in the gills work in very much the same way. In both cases, energy is used to create order—by separating solutes across a membrane.

There is nothing in these physiological processes

that is not well understood energetically. The fish must maintain order (a high solute concentration) in the face of an environment that promotes disorder (water with a low solute concentration). Maintaining order requires work to be done. The energy to power this work comes from potential energy stored in glucose, and the energy released is transformed in several ways as it is made to do work: first into chemical potential energy in ATP, then into mechanical deformations of heart muscle proteins, then into pressure within the fish's capillaries, a form of potential energy that is used to separate proteins from smaller solutes. ATP also promotes the mechanical deformation of transport proteins, which are enabled then to pick up solutes and move them across membranes. All this is conventional physiology, and the operation of simple thermodynamic principles is clear.

Calcium Carbonate Deposition in Hermatypic Corals

Now let us turn to another interesting physiological process: the deposition of calcium carbonate by polyps of reef-forming (hermatypic) corals.

Corals are coelenterates, related to the jellyfishes and sea anemones. They come in a variety of forms, all very simple in organization, and they are often colonial, with large numbers of individuals cooperatively occupying a space together. One form this coopera-

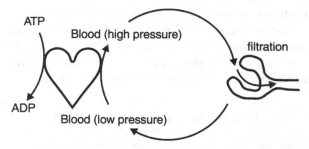

Figure 2.3 The transformations of energy in the production of filtrate. Energy in ATP is converted by the heart to blood pressure. The difference in pressure between blood vessels and the tubules of the nephron does the work of filtering the fluids of the body.

tion takes is the deposition of minerals to build structures, coral reefs being the most spectacular. I will have much more to say about corals in Chapter 5.

A coral reef consists of two parts, one living and the other not. The living part of a coral reef is a thin layer of coral polyps growing on a foundation of calcium carbonate ($CaCO_3$), or calcite, secreted and laid down by cells at the base of the polyp, which are known as the calciloblastic ectoderm—the calciloblast, for short. As the generations of polyps come and go, the reef's mineralized support therefore grows. The end result of this process, multiplied across the millions of individual polyps, is the continuous accretion of layers of calcite into a massive reef.

Crucial to the ability of corals to build reefs is an interesting house guest residing in the tissues of the polyps—tiny protozoa, known generically as dinoflagellates ("terrible whip bearers," for the pair of large flagella they sport). Dinoflagellates are common members of the plankton that float about in the surface waters of oceans, lakes, and streams. When dinoflagellates occur as corals' house guests, or symbionts, they are known as zooxanthellae ("little yellowish animals"), because they contain a variety of interesting pigments, some of which impart to corals their spectacular colors.[6] The pigments also help the zooxanthellae photosynthesize, just as the green pigment chlorophyll aids green plants. There is thus a nice mutual arrangement, or symbiosis, between the coral polyps and their resident zooxanthellae. The zooxanthellae get a nice sheltered home inside the polyps' bodies, and the corals have little "food factories" living inside them, producing glucose from carbon dioxide and water and powered by sunlight.

---

6. Coral bleaching has recently become a concern for scientists, as there is an apparent epidemic of it among Caribbean corals. The bleaching results from the expulsion of the zooxanthellae from the corals' cells, which is thought to be a stress response. The potential agents for stress include disease, pollution, and high temperature. Warming of the seas may be the culprit in the current epidemic of coral bleaching, and it is thought by some to be a harbinger of global warming.

All corals are capable of producing calcite, whether or not they contain zooxanthellae. However, there is a strong correlation between the ability to build reefs—that is, to produce calcite in the enormous quantities necessary—and the presence of zooxanthellae. This implies that reef building is energetically demanding. The zooxanthellae are thought to be one explanation for why reef-building corals are common in the beautifully clear, but nutritionally impoverished, waters of the tropical oceans: they have a virtually unlimited source of energy (light captured as glucose) supplied to them by their protozoan house guests.

The odd thing about this idea is that there is nothing in the process of depositing calcite that *should* require the coral to expend energy. Calcite is a weakly soluble salt of calcium and carbonate ions:

$$Ca^{2+}(aq) \quad + \quad CO_3^{2-}(aq) \quad \rightarrow CaCO_3(s)$$

calcium ions in solution + carbonate ions in solution → calcite

"Weakly soluble" means that calcite has a strong tendency to form as a solid crystal (to go to the right side of the reaction) even if its soluble ions (on the left side of the reaction), are present in very small quantities: for calcium carbonate the required concentration is only about 95×10^{-6} molar. The weak solubility of calcite means that the process of its deposition in a reef should have the Second Law on its side: just put calcium ions and carbonate ions together and *presto!* you have calcite. Corals certainly have no problem obtaining the ingredients in sufficient quantities. In the oceans, calcium concentrations are roughly 100 times higher than they need be to support the formation of calcite. Neither is there any shortage of carbonate, which is abundant whenever there is metabolism. Wherever there is something living, there should always be plenty of carbonate.

So why should reef deposition be energetically costly? The answer lies in some peculiarities of the chemistry of carbonate when it is dissolved in water. What follows is the complete process of dissolving carbon dioxide (the product of glucose metabolism, remember) in water:

$$CO_2 + H_2O \leftrightarrow H_2CO_3$$
carbon dioxide + water ↔ carbonic acid

$$\leftrightarrow H^+ + HCO_3^- \leftrightarrow 2H^+ + CO_3^{2-}$$
↔ proton + bicarbonate ↔ protons + carbonate

In this formula, the double-headed arrows signify that the reaction is reversible. This is a fairly complicated reaction, but its lesson for our problem is simple. Carbon dioxide does not dissolve in water like oxygen does—that is, as a solute that remains uninvolved, so to speak, with its solvent. Rather, carbon dioxide reacts with its solvent, water, to form a weak acid, carbonic acid (H_2CO_3). Carbonic acid can break apart into hydrogen ions (protons) and a hydrogenated form of carbonate, known as bicarbonate (HCO_3^-). To make carbonate in the form that calcium must pair with to make calcite, that hydrogen ion must be stripped off the bicarbonate.

Energy is required because the form that solubilized carbon dioxide most likes to be in is bicarbonate. If we were to use relative type size to express the relative abundance of the different forms carbonate tends to take, the reaction above would look something like this:

$$CO_2 + H_2O \leftrightarrow H_2CO_3 \leftrightarrow \mathbf{H^+} + \mathbf{HCO_3^-}$$
$$\leftrightarrow 2H^+ + CO_3^{2-}$$

There is another problem with bicarbonate, this one for the zooxanthellae. For photosynthesis to occur, the zooxanthellae *must have carbon dioxide*. Not bicarbonate, not carbonate, not carbonic acid, but carbon dioxide. Even if the coral's metabolism produces prodigious quantities of CO_2, the zooxanthellae will be starved if this CO_2 is immediately locked up into the bicarbonate form.

So both the coral and the zooxanthellae have to work against the thermodynamically favored tendency of carbon dioxide, from one side, or carbonate, from the other, to form bicarbonate. Furthermore, they must work against this tendency in opposite directions. To deposit calcite, the coral must drive the reaction to the right, toward CO_3^{2-}. To photosynthesize glucose, the zooxanthellae must drive the reaction to the left, toward CO_2. (For a brief outline of chemical kinetics and the relevant principles behind the energetics of chemical reactions, see Box 2C).

Fortunately, there is a simple way to solve both problems, and it involves manipulating the concentration of hydrogen ions around the bicarbonate. Removing hydrogen ions from a solution of bicarbonate (making it more alkaline) forces the bicarbonate to get rid of its second hydrogen atom and form carbonate. Again, if we represent relative concentrations by type size, the reaction would look like this:

$$CO_2 + H_2O \leftrightarrow H_2CO_3 \leftrightarrow H^+ + HCO_3^-$$
$$\leftrightarrow \mathbf{2H^+} + \mathbf{CO_3^{2-}}$$
↓
remove protons

On the other hand, *adding* hydrogen ions to a solution of bicarbonate (acidifying it) will force the reaction back to carbonic acid and hence back into carbon dioxide and water:

$$\mathbf{CO_2} + \mathbf{H_2O} \leftrightarrow \mathbf{H_2CO_3} \leftrightarrow H^+ + HCO_3^-$$
$$\leftrightarrow 2H^+ + CO_3^{2-}$$

↑
add protons

Adding hydrogen ions makes CO_2 available to the zooxanthellae. So the solution for both the coral and zooxanthellae is to do work to move hydrogen ions, which they do. The details of the process are unclear,

2C

Energy and Chemical Kinetics

Chemical kinetics describes chemical reactions and the energetic forces that drive them. Central to this field of science is the idea of equilibrium between the reactants and products of a chemical reaction. Consider a hypothetical reaction in which reactants A and B react to form C and D. By convention, we write a chemical equation to represent the reaction:

$$jA + kB \Leftrightarrow lC + mD \qquad [2C.1]$$

where the coefficients j, k, l, m refer to the molar quantities of reactants A and B and products C and D, respectively.

In any chemical system, each side of the reaction will have a characteristic energy level. Usually, one side will have a higher energy level than the other. If a reaction proceeds from the high-energy side to the low-energy side, the reaction gives off energy and is said to be exergonic. The combustion of glucose is an example of an exergonic reaction. However, a reaction can also proceed from a state of low energy to high energy. In this case, energy must be fed into the reaction, which is then said to be endergonic. Any order-producing reaction is endergonic.

Usually, there is sufficient energy in a chemical system to drive both exergonic and endergonic reactions. What determines the net direction of the reaction is how fast each reaction proceeds relative to the other. The Second Law of Thermodynamics states there will always be a bias toward the exergonic reaction (producing an increase in entropy, or disorder). Consequently, reactions will always tend to run "downhill," from states of high energy to states of low energy. However, reaction speed in either direction is also determined by the molar concentrations of the reactants and products. Thus, even if a reaction were not energetically favored (if it required an input of energy, say), it could still proceed if the concentration of product was high enough to make the endergonic reactions more frequent than the exergonic reactions.

It follows that there will be some concentration of reactants and products that will drive the reaction toward products as fast as the reaction is driven toward reactants. At this point, the concentrations of reactants and products will experience no net change, and the reaction will be at equilibrium.

This condition is denoted with an equilibrium constant that relates the concentrations of the products and reactants:

$$K_{eq} = ([C]^l[D]^m)/([A]^j[B]^k) \qquad [2C.2]$$

Obviously, if $K_{eq} > 1$, the products will be favored at equilibrium over the reactants, and the reaction is energetically disposed toward the right side of the equation. Conversely, if $K_{eq} < 1$, the reactants are favored at equilibrium over products, and the reaction is energetically disposed toward the left side of the equation. The real value of the equilibrium constant, however, is that it provides a way of explicitly relating a chemical reaction's equilibrium to the energy driving it. Specifically, the change of free energy of a reaction (energy that is capable of doing work) is:

$$\Delta G = -RT \ln K_{eq} \qquad [2C.3]$$

where ΔG is the net change of free energy in the reaction, R is the universal gas constant (8.31 J mol^{-1} K^{-1}), T is the absolute temperature (K) of the reaction, and $\ln K_{eq}$ is the "natural" logarithm (logarithm to the base e) of the equilibrium constant. For purposes of comparison, the change of free energy is usually determined under an agreed-upon set of standard conditions (1 molar concentrations of product and reactant, 25°C, and pH = 7.0), which yields the net standard change of free energy, or $\Delta G^{o'}$.

Very useful to us is what energy is required to displace a reaction from equilibrium—a common problem for organisms that want to engineer or maintain a concentration inside their bodies different from that dictated by the standard change of free energy. In this case, we can simply look at the difference between the net change of free energy and the free energy of the reactants and products as they are in the animal. If we assume that the concentrations at equilibrium represent the minimum energies the system can attain, it follows that any displacement of the system from that equilibrium will require energy. If we assume that the actual concentrations of A, B, C, and D in an animal are α, β, γ and δ, respectively, the energy, ΔE, required to force the system out of equilibrium will be:

$$\Delta E = \Delta G - RT \ln[(\gamma^l \delta^m)/(\alpha^j \beta^k)] \qquad [2C.4]$$

Thus, adding or subtracting either reactants or products to a chemical system at equilibrium is the same as doing work on it and will force the reaction out of its equilibrium state. In the case of carbonate ion, adding or subtracting hydrogen ions is sufficient to drive the reaction to one or the other thermodynamically unfavorable state—high concentrations of either carbon dioxide or carbonate, rather than the favored form, bicarbonate.

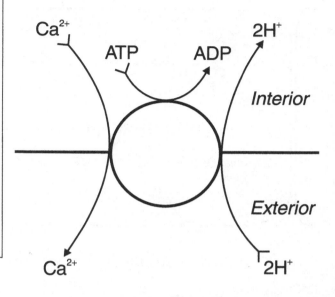

Figure 2.4 The calcium-proton antiport in the calciloblast of a hermatypic coral.

as is just which ones, the corals or the zooxanthellae, are running the show, but the elements seem to be as follows. At the outside surface of the calciloblast, where the calcite will be deposited, there is an active transport protein that uses ATP (energy from sugar) to transport calcium ions from the cell's interior to the outside of the calciloblast (Fig 2.4). This interesting type of transport proton is known as an antiport because it transports two things in opposite directions across the membrane. In the calciloblast antiport, for every calcium ion moved out of the cell, two hydrogen ions are transported into the cell. Both transport processes work against their respective concentration gradients; that is, they are creating order, and so they require energy from ATP. In general, one ATP molecule is sufficient to turn the crank once, moving one calcium ion and two hydrogen ions in opposite directions across the calciloblast's membrane. The operation of the antiport does two things for the aqueous solution between the calciloblast and the cal-

cite base: it makes it rich in calcium ion, and it moves acidifying hydrogen ions away so that the carbonate may embrace calcium without the encumbrance of that last proton. And just as marriage follows love, calcite is finally produced.

On the other side of the calciloblast membrane, the hydrogen ions that are moved *into* the cell (at energetic cost) push bicarbonate toward carbon dioxide, which the zooxanthellae can then take up and combine with water to form sugar (Fig. 2.5). The sugar can then be used by the coral to make more ATP, which is then

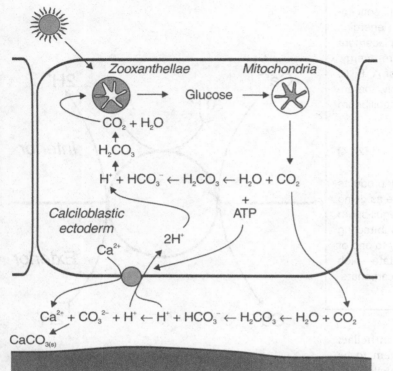

used to power the separation of more hydrogen ions across the calciloblast membrane, which keeps calcite deposition going, which keeps photosynthesis going, and so on and so on.

Physiology, the Organism, and the NOT-Organism

Let us now step back a bit and reflect. I asserted earlier that physiology is essentially how animals use energy to do order-producing work. On this very general level, there are obvious parallels between what goes on in the fish nephron and what goes on in the coral calciloblast. In both cases energy flows through the organism's body (light ultimately, then glucose, and then ATP) to create either order (partitioning ions or other molecules across a membrane in both cases) or potential energy (blood pressure in the fish nephron, con-

centration gradients of hydrogen ions in the calciloblast). The two examples differ primarily in their frames of reference. In the fish nephron, what is organism and what is NOT-organism is clear—the environment is outside the organism, and the physiological function takes place inside the animal. In a hermatypic coral, though, the boundary between organism and NOT-organism is not so clear. An important component of the physiological process takes place outside the animal, at the space between the outside surface of the calciloblast and the exposed surface of previously deposited calcite. For the coral, even, two organisms are involved, the boundary between them blurred by their symbiosis. In short, the identity of the organism is very tenuous in hermatypic corals. Yet, the whole system—reef, calcite, coral, and zooxanthellae—functions together as a physiological system.

If physiology occurs in both systems, and if the existence of physiological function is not dependent upon a clear partition of an organism from its environment, then there seems to be little reason to regard the organism as an entity discrete from its environment. This is the crux of the argument I wished to make in this chapter. If animals use energy to do work on the "external" environment, their activity is as much physiology as when they use energy to do work on the "internal" environment.

CHAPTER THREE

Living Architecture

Dante set aside a special place in the underworld for people who make the kinds of arguments I made in the last chapter. I suppose it would not be all that bad—I could be consoled by the good company, as Dante sent most of the pre-Christian (pagan) philosophers there as well. But, *mea culpa*, there I would have been sent, for my crime is sophistry.

These days, sophistry is a Bad Thing: my dictionary defines the word as "deceptively subtle reasoning or arguing," that is, glib argument intended to put something over on us. When a tobacco company lawyer claims, for example, that his company never forced anyone to light up, and so his company should not be held responsible for the social consequences of the public's use of its products, the fact that he is technically correct does not overshadow the rather transparent and cynical agenda he is trying to advance. That is sophistry, as we now know it.

Sophistry has a rather more distinguished ancestry, though. The word is derived from the name given to a group of itinerant pre-Socratic philosophers, the Sophists, or "wise ones." Their principal approach to philosophy was to use logic to explore fundamental questions about man and the cosmos. They seemed not to be popular in their day—Socrates may have been put to death for being too close to them—because they often used their formidable argumentative skills to question the morals and customs of the Athens of their day. They were tolerated by the ruling classes, though, because they were superb teachers of rhetoric, logic, and public speaking, essential skills for successful careers in public life. They were, in other words, a bit like a modern university faculty.

We know most about the Sophists from Plato's *Dialogues,* and he had little good to say about them. He disliked their fondness for argument for its own sake and what he viewed as their quibbling contentiousness. Most annoying to Plato was their great skill at using logic and rhetoric to advance falsity as easily as truth. To Plato, Sophists were most interested in empty rhetoric, specious logic, and victory in debate.

One of the Sophists' favorite rhetorical tools was a technique known as antilogic, in which a logical proposition is met with a logical counter-proposition, the implication being that both are equally valid. One can then undermine the proposition without having to explore, on its own merits, why the proposition might be correct. What frequently got the Sophists into trouble was their use of antilogic when the matter under scrutiny was a question of morality or social custom. This is a method familiar to us Baby Boomers and now (maddeningly) to our progeny:

- *Proposition (father to daughter):* No, you are not allowed to have a nose ring.

- *Counter-proposition (daughter to father):* You wore long hair when you were my age and your parents didn't stop you!

I used antilogic in Chapter 2, in fact. My argument was roughly as follows:

- *Proposition:* Physiological processes inside organisms are demonstrably governed by thermodynamics.

- *Counter-proposition:* Certain processes associated with living things, but occurring outside the body, also are demonstrably governed by thermodynamics.

- *Ergo:* There is no distinction to be made between physiological processes within the body and similar processes without.

All I have proven, of course, is that physiology, as it occurs in "the real world," is governed by the same chemical and physical laws that apply to inanimate physical systems. This is a trivial conclusion, really, but it can very easily be twisted it into absurd conclusions. For example, a thundercloud, as it forms, obeys the laws of thermodynamics as rigidly as a kidney does as it maintains the water balance of a fish. My exercise in antilogic could have as easily led me to assert that there is no fundamental distinction to be made between the origin of thundershowers and the origin of fish urine. And the assertion is as logically correct as it is nonsensical. You can see why the Sophists drove Plato mad.

All I really did in the last chapter was show that there is no logical reason why physiological processes *cannot* occur outside an organism. I am still far from showing that "external physiology" is a reasonable or even an interesting idea. What I would like to do in this chapter is to show that physiology outside the organism is not only logically possible but also reasonable.

The Inefficiency Barrier?

Conventional physiology encompasses an organism using metabolic energy to do work on its internal environment. A putative external physiology extends the reach of this work to the environment outside the animal. In the last chapter, I asserted that the laws of thermodynamics put no constraints, save one, on the extent of this outward reach. It so happens that this one constraint is a serious one, so it must be dealt with straightaway.

The supposed constraint is imposed by the Second Law, which states that any transformation of energy into useful work must be accompanied by the loss of a portion of that energy as heat:

energy in → *useful work + heat*

The problem is that physiological functions are powered only by energy being made to do useful work—with a few exceptions, heat is pretty much useless to an organism's physiology. And this leads to a problem

Table 3.1 Hypothetical efficiency of filtration work done by the fish kidney.

Process	Process efficiency	Cumulative efficiency
$CO_2 + H_2O \rightarrow$ glucose	36%	36%
glucose $\rightarrow$ ATP	38%	38% of 36% = 13.7%
ATP $\rightarrow$ increase in blood pressure	21%	21% of 13.7% = 2.87%
blood pressure $\rightarrow$ filtrate production	70%	70% of 2.87% = 2.01%

we might call the inefficiency barrier. The inefficiency barrier, if it exists, may force us to the conclusion that, although external physiology might be possible, we must conclude that it is energetically unreasonable.

Organisms, like all machines, are inherently inefficient. Furthermore, the inefficiency is cumulative, which means that the more useful things an organism tries to do with a parcel of energy, the less efficiently it will do them. This is bad news: most organisms do work in sequences of small steps between the initial input of energy to a physiological "engine" and the work ultimately done by it. If each step wastes some energy, the accumulating inefficiency may mean that little energy will be left over from internal physiology to extend its reach outward. This "inefficiency barrier" may explain why organisms have tangible boundaries even though there is no thermodynamic reason they should.

To illustrate, consider how energy was made to do the work of filtration in the fish kidney, as described in Chapter 2 (Table 3.1). The process begins with a parcel of energy initially stored in sugar and continues as that energy is made to do work in several stages before it dissipates ultimately to heat. Each step has a characteristic efficiency, always less than 100 percent, and the degradation of useful work to heat is cumulative at each step. At the end, once the filtrate is made, very little energy is left, only about 2 percent. This pyramid of inefficiency piled upon inefficiency offers a rather bleak prospect for organisms being able to power physiology outside their bodies.

I can think of two ways to confront this problem.

The first is a bit of sophistry that may earn me additional time on my infernal sentence. I will employ it anyway, because this time it is the proposition, not the counter-proposition, that is probably in error. The second is the argument that even if an inefficiency barrier precludes an organism powering its own external physiology, there are ways of circumventing this limitation.

Negation by Minimization

Another of the sophist's useful tools is negation by minimization. Equating "small" with "insignificant" is a convenient way of avoiding serious thought about something, or (even better) of diverting someone else from thinking seriously about it. We all are depressingly familiar with this technique. We see it used all the time in what we call, without irony or embarrassment, the political "debate" of our fractious age. Some examples: "The tax money spent on (my) [program X] is so small that cutting it will have no significant effect on the national debt. It would be far better to cut (someone else's) [program Y]." Or: "The taxes paid by taxpayers (like me and my friends) earning over $3 million a year are so small a portion of all taxes levied that raising them will have no measurable impact on overall tax revenues. Better to raise taxes on (those other) taxpayers (over there) that earn less than $50 thousand a year." Positing an inefficiency barrier as an obstacle to external physiology uses the technique of negation by minimization. If only a tiny fraction of the energy passing through an animal is left to do useful work, then anything else an organism could do with it, like powering external physiology, is not worth noticing.

In politics, this kind of argument is smoke and mirrors. So it is in biology. The obvious antidote to negation by minimization is to show that what we are being asked to ignore is not so negligible after all. So, let's take a closer look at the propositions that underlie the inefficiency barrier. Consider that chain of inefficiencies I just outlined. Are organisms really *that* inefficient?

Well, yes, and no. Consider the claim that the conversion of glucose to ATP is about 38 percent efficient. Where does this number come from? It is a fairly straightforward calculation based upon a quantity known as the standard free energy (See Box 2C), abbreviated $\Delta G^{\circ\prime}$ (pronounced "delta gee nought prime"). The standard free energy quantifies the energy in a chemical reaction that is available to do work. For the oxidation of glucose to carbon dioxide and water, for example, the $\Delta G^{\circ\prime}$ is 2.82 MJ mol^{-1}. As energy is transferred from glucose to ATP, its release from glucose must be coupled to the addition of a phosphate to ADP. This reaction has a $\Delta G^{\circ\prime}$ of about -30.5 kJ mol^{-1}, that is, roughly 30.5 kJ of energy is required to phosphorylate a mole of ADP. If a cell completely oxidizes a mole of glucose, it will produce thirty-six moles of ATP. Phosphorylating 36 moles of ATP therefore requires 30.5 kJ mol^{-1} × 36 mol = 1.09 MJ of energy. The efficiency of conversion of glucose to ATP, therefore, is simply 100 × (1.09 MJ/2.82 MJ) = 38 percent.

Standard free energies are rigorously measured quantities, so it would seem our calculation of inefficiencies is built upon a rock-solid foundation. It is not as solid as it might appear, though. The problem with standard free energies is not that they are inaccurate, it is that they are incomplete. A standard free energy of a reaction is so called because it is quantified under certain standard conditions. Measurements are made under standard conditions so that different reactions may be compared and predictions may be made about whether a particular reaction will go. As useful

as this practice is to biochemists, it unfortunately says little about the actual energy yields of particular reactions when they occur inside cells. Indeed, by manipulating the conditions in which reactions take place, it is possible to make significant improvements in yields of useful work, even from reactions that supposedly are very wasteful of energy. For example, the synthesis of a single peptide bond operates at a standard efficiency of about 23 percent.[1] Changes in concentrations of products and reactants, local pH, and temperature that may occur in an organism's cell may raise efficiencies to a theoretical maximum of around 92–96 percent (depending upon how the costs of infrastructure, like DNA and mitochondria, are factored in).

Reactions in most cells do not approach this theoretical limit, but they nevertheless do much better than their standard efficiencies would suggest. Bacteria, for example, operate at total conversion efficiencies[2] of around 60 percent. Protozoan cells do a little poorer on average, but still pretty well, about 50 percent. Most animal cells are similar, operating at 50–60 percent efficiencies. Some predatory protozoa seem to push the envelope, with efficiencies of around 85 percent. So, the Second Law seems a bit less formidable a barrier to an external physiology, at least when we take into account the conditions that may occur in a living cell.

This is all very nice theoretically, but does it mean that organisms will have energy to spare once they have taken care of their physiological housekeeping? On the face of it we might say no: would it not make more sense for an organism to keep the energy inside

---

1. A peptide bond links two amino acids into a dipeptide. A protein, consisting of a linear chain of n amino acids, contains within it $n - 1$ peptide bonds. Synthesis of a peptide bond requires about 20.9 kJ mol^{-1}. Two ATPs and 1 GTP are required, providing energy of 3 × 30.5 kJ mol^{-1} = 91.5 kJ mol^{-1}. The efficiency is therefore 100 × (20.9/91.5) = 22.8 percent.

2. The total conversion efficiency is the efficiency of converting energy in food to new organism. These figures therefore reflect the net efficiencies of a sequence of steps, including at minimum the conversion of food energy to ATP and the use of ATP to power the synthetic reactions required for growth.

to be used for other useful chores, like growth or reproduction? It turns out there is a convenient way to examine this matter independently, by looking at how efficient the energy transfers are *between* organisms. Such transfers of energy usually involve one organism being eaten by another, what biologists delicately call trophic interaction.

The efficiency of a trophic interaction can be roughly gauged by measuring how much of one organism (the eaten) is needed to support another (the eater). For example, green plants convert light energy to glucose, use it to power growth, and in the process accumulate stored energy, the so-called biomass. Herbivores (predators of plants, really, although it hardly seems sporting to call them that) take this energy for their own support. If the use of energy by an herbivore is completely efficient, a certain biomass of plant tissue should be able to support the same biomass of herbivore. If, on the other hand, the herbivore is only, say, 10 percent efficient, a certain plant biomass will only an herbivore biomass one-tenth its size.

If we take this argument to its logical conclusion, we could arrange all organisms in nature in a kind of trophic pyramid (Fig 3.1), where (assuming an arbitrary efficiency figure of 10 percent) the total biomass of herbivores is 10 percent of the plant biomass, and the biomass of primary predators (things that eat herbivores) should be 10 percent of herbivore biomass (or 1 percent of plant biomass), and so forth. The pyramidal shape of these trophic interactions is enforced by the supposedly inevitable wastage of energy as heat as it passes through living organisms. The greater the internal wastage, the less energy will be available for powering the physiology of other organisms, and the more sharply constrained the trophic pyramid should be.

What, then, do trophic interactions tell us about just how efficient organisms are? This issue has long been a staple of ecological research, and it turns out there is quite a lot to tell—so much, in fact, that there is no way to summarize it without also trivializing the topic. Nevertheless, one can say, as a crude generalization,

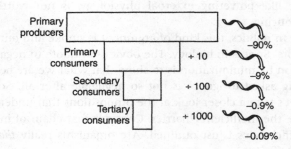

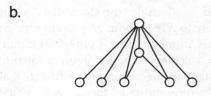

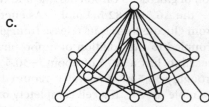

Figure 3.1 The supposedly inefficient flows of energy through ecosystems. *a:* The passage of energy from one type of an organism to another inevitably involves the loss of some energy as heat. The efficiency is arbitrarily set to 10 percent in this example. *b:* A simple pyramidal food web in an intertidal community on the Washington coast. *c:* A food web from an intertidal community in the Gulf of California. [*b and c after Pimm (1982)*]

that there is both good news and bad news for the idea that organisms are inherently inefficient. For example, some types of organisms, like mammals and birds, which divert considerable quantities of energy to the deliberate production of heat, are decidedly inefficient, operating at "ecological efficiencies" of 2–3 percent. Other types of vertebrates seem to operate at efficiencies around 10 percent. Most invertebrates, excluding the insects, do better, with efficiencies ranging from about 20 percent to 35 percent. Insects operate at the greatest ecological efficiencies of them all: with 39 percent efficiencies for herbivore species and up to a whopping 56 percent for insect carnivores (another reason why insects will ultimately conquer the world).

In general, highly efficient organisms not only power more work inside their own bodies, but they also produce more work energetically "downstream." Ecologically, this relation is sometimes reflected in the length of a chain of trophic interactions, essentially how many times bigger fish can eat littler fish. Ecosystems with poor productivity, for example, support simple trophic pyramids that can power only a few trophic exchanges before the energy is all dissipated as heat (Fig. 3.1). Highly productive ecosystems, on the other hand, "frontload" sufficient energy into the system, powering parcels of energy through many more trophic levels, as many as 10–15 or so. A productive ecosystem is better described as a complex food web than as a pyramid.

Clearly, inefficiencies exist in the transfers of energy both through and between living things, and these inefficiencies occur at every level—cellular, organismal, and ecological. Also, the inefficiency is cumulative and multiplicative. As a consequence, there is a limit on the extent that physiological work can reach beyond the organism's conventionally defined outer boundary. Despite this limitation, there seems to be little reason to minimize the possibility of an external physiology to the point of negligibility. Efficiencies of physiological work are subject to improvement through engineering, and there is every reason to expect—both *a priori* on the basis of natural selection, and *a posteriori* on the basis of empirical observation—that species will generally improve the efficiency of their internal physiology. Certainly, energy flowing through organisms demonstrably supports the physiology of other organisms energetically downstream from them. Why, then, should we suppose this energy could not be used by an organism to support its own physiology outside (and also energetically downstream from) itself?

Circumventing the Inefficiency Barrier

Another interesting way an organism can work around its internal inefficiency is to use energy that does not pass through its own chemically based physiology. By this means an organism can perform some interesting energetic jujitsu.

Organisms channel energy through themselves, using it to do work in the process. Consequently, organisms are positioned in the middle of a stream of flowing energy, much in the manner of a hydroelectric plant in a river. Conventional (internal) physiology relies on a particular form of energy flow—specifically, that from chemical potential energy stored in the orderliness of complex chemicals like glucose and ATP. Aside from the initial capture of light energy by plants, nearly all the energy flowing through organisms is in the form of various chemical transformations: glucose to ATP, and ATP to various tasks, like synthesis, transport, and mechanical work. Ultimately, it is all dissipated as heat. This energy stream, because it primarily involves chemical transformations under the direct control of organisms, we might call the metabolic energy stream (Fig. 3.2). The inefficiency barrier, whether it is big or little, is a problem peculiar to the metabolic energy stream.

Energy flows through the environment in other forms, however. The potential rate of energy transfer from the Sun to the Earth is prodigious—about 600 W m^{-2}, averaged through the year.[3] Of this, only a relatively small fraction, on the order of 1–2 percent, is

---

3. To put this number in perspective, compare it with the energy consumption rate of a small four-person house located in the northeastern United States (mine). My family's average yearly energy consumption is about 19,500 MJ for electricity, and about 84,300 MJ for natural gas, a total of about 104,000 MJ. This worked out to a total energy consumption rate of about 3.3 kW. Our house occupies a small plot about 350 m^2. Thus, our typical house consumes energy at a rate of about 9 W m^{-2},

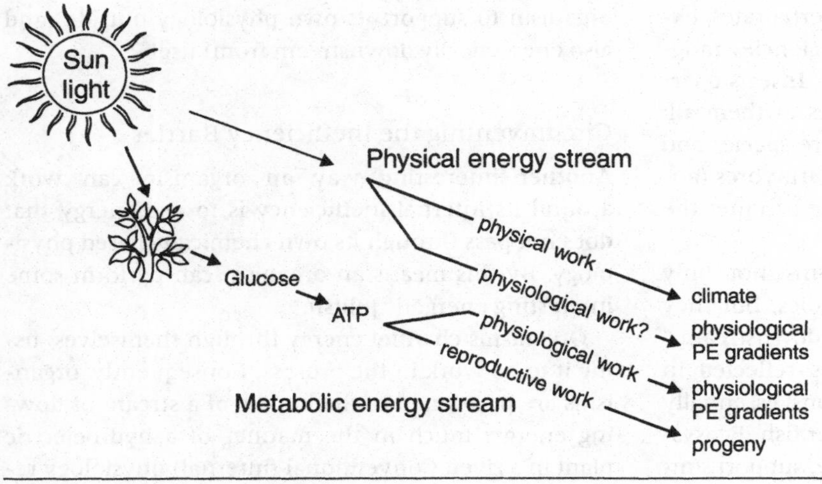

Figure 3.2 The multiple energy streams in the environment.

captured by green plants. The rest, if it is not reflected back into space, is available to do other things. The excess can be considerable: although some natural surfaces reflect as much as 95 percent of the incoming solar beam, many natural surfaces reflect much less (Table 3.2), on average about 15–20 percent. The remaining absorbed energy is then capable of doing work, like heating up surfaces, moving water and air masses around to drive weather and climate, evaporating water, and so forth. Eventually, it all dissipates ultimately as heat, but it does work along the way, forming a physical energy stream that runs roughly parallel to the metabolic energy stream (Fig. 3.2).

If organisms could tap into this parallel stream of physical energy, they could power external physiology while circumventing the efficiency limits of the metabolic energy stream. Capturing even a small part of the Sun's energy to do physiological work would confer enormous benefits to an organism. For example, temperature regulation is one of the major metabolic costs for warm-blooded animals like birds and mammals.

---

roughly 1.5 percent of the solar energy delivered. This abundance of energy reaching the Earth's surface is one of the things that keeps advocates of solar energy hopeful.

Maintaining a steady body temperature means producing heat as rapidly as it is lost to the environment, and the colder it gets, the greater this cost will be. The energy for production of heat is supplied by ATP, that is, it comes right out of the metabolic energy stream. This is a waste—it is akin to heating your house by burning dollar bills. Roadrunners (*Geococcyx californianus*) are an interesting counter-example to this rule. On cold mornings, when metabolic costs for heat production would ordinarily be high, roadrunners will sometimes be seen sunning themselves. Obviously, heat the roadrunner absorbs from the sun is heat that will not have to be generated from ATP, and savings in metabolic energy will result. Roadrunners' bodies are even modified to increase the benefit they derive from sunbathing. Their back skin has several melanized patches that can be exposed directly to the sun when the back feathers are lifted up, like slats in a venetian blind. The additional heat from the sun can lower a roadrunner's metabolic costs for thermoregulation by about 40 percent. It takes little imagination to see that animals might build structures external to their bodies to take advantage of such benefits, and we will be exploring many of them in the later chapters of this book.

Table 3.2 Surface reflectivities of some representative natural surfaces (after Rosenberg 1974).

Surface	Short wave reflectivity (%)
Fresh snow	80–95
Old snow	42–70
Dry sandy soil	25–45
Dry clay soil	20–35
Peat soil	5–15
Most field crops	20–30
Deciduous forest	15–20
Coniferous forest	10–15

Energy and Evolution

The concept of the extended phenotype is incomplete without an understanding of the underlying physiology. Genes can act outside the bodies of the organisms containing them only if they can manipulate flows of matter and energy between organisms and their environment. Let us now build upon this fountain a more formalized framework of energy, evolution, and physiology that will support the rest of this book. The framework will include two elements: the energy streams that power physiology, whether inside or outside the organism; and the types of "external organs" and "external organ systems" that can manipulate these energy streams to power physiological work.

I begin by making a baldly teleological assertion:

> Anything an organism does is "for" one and only one purpose: to ensure that its genes are copied and passed on to progeny.

If organism X does this marginally better than organism Y, organism X will leave more progeny than Y, passing along whatever it did right to its progeny so that they can do it right also. This is simply natural selection restated.

Evolution by natural selection is predominantly a matter of genes and their transmission, but adapta-

tion and natural selection have an indissoluble link to physiology, and hence to energetics. We know that reproduction takes energy—energy must be used to make eggs and sperm, to get eggs and sperm together, and to get the resulting zygote to grow to the point where it can begin capturing its own energy to power its own reproduction. So part of the energy stream flowing through an organism must power reproduction, or more properly reproductive work (Fig. 3.2). Organisms that do more reproductive work will be selected in preference to those that do less.

Reproductive work requires infrastructure to support it, and building, maintaining, and operating this infrastructure also costs energy. Much of an organism's operating energy budget is devoted to providing the right physical environment for reproductive work to occur. Usually, this involves maintaining an environment inside the body that differs in some way from the organism's external environment. Accomplishing this task has some important energetic consequences.

The internal environment of an animal is described by various physical properties, like its temperature, salt and solute concentration, acidity, pressure. Conditions outside the animal are also reflected in these properties. Usually, conditions vary only slightly in the internal environment, while outside they vary more widely. For example, the daily excursion of environmental temperatures is about 10–20°C through the day, but the range can be much larger, depending upon the season and locale. Whether through active regulation or simple inertia, the range of body temperatures experienced by animals during the day will be somewhat smaller. For mammals and birds, at most about 5°C separates the highest body temperature commonly experienced from the lowest.[4] Furthermore, environmental temperatures often exceed the maximum or minimum tolerated temperatures of ani-

---

4. Obviously, I am leaving out hibernators and birds that go into torpor. Birds and small mammals in this state will allow body temperatures to drop to within a few degrees of freezing.

mals. Animals as a group tolerate body temperatures up to about 45°C, though the limit is higher in a few and may be as low as about −2°C, lower if ice formation can be suppressed or managed somehow. In contrast, daytime surface temperatures in the tropics can exceed 70°C, and winter temperatures even in the temperate zones can fall well below −10°C.

Such disparities mean that conditions in an animals' internal environment will frequently differ from conditions outside. These disparities, in turn, establish potential energy (*PE*) differences that can drive a flux of matter or energy across the boundary separating the organism and its environment (Fig. 3.3). For example, an animal that is warmer than its surroundings will lose heat at a rate proportional to the difference in temperature between the body and the surroundings.[5] Similarly, an animal whose body fluids contain a concentration of a solute X higher than that in the environment will lose X from the body as it diffuses down the concentration difference.[6] Differences in solute concentration or pressure can drive water across an animal's boundary with its environment, and so on. Matter and energy flow down such potential energy differences spontaneously, in conformity with the dictates of the Second Law. Consequently, we will call these kinds of fluxes thermodynamically favored fluxes *(TFFs)*.

If an animal's internal environment differs from the external environment, it will experience a *TFF* (Fig. 3.3), and its internal environment will change—a loss of heat from the body will result in a drop of body temperature, for example. Maintaining the internal

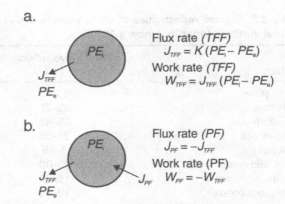

Figure 3.3 The energetics of maintaining the internal environment. *a:* A potential energy difference (*PE*ᵢ−*PE*ₑ) drives a thermodynamically favored flux, J_{TFF}, across the integument. The environment does work on the organism, at a rate W_{TFF}, the product of the flux and the potential energy difference driving it. *b:* To maintain the internal environment, the organism must do work on the environment, which drives a physiological flux, J_{PF}, of matter across its outer boundary, at a rate W_{PF}.

environment will therefore require work to drive an equal flux of matter or energy opposite to the *TFF* (Fig. 3.3), which we shall call the physiological flux *(PF)*. Maintenance of body temperature in the cold requires a physiological flux of heat into the body, which is released by the metabolism of glucose or fats, for example.

Conventionally, physiological fluxes, like the performance of any physiological work, are powered by the metabolic energy stream (Fig 3.2). In Chapter 2's example of the fish in fresh water, the potential energy gradient driving the *TFF*s of water and solutes arose from the different concentrations of solutes and from the resulting differences in osmotic concentration inside and outside the animal. The fish then had to use metabolic energy to drive *PF*s of water out and solutes in. In this case, the metabolic energy powers two kinds of work, and the metabolic energy stream divides into two roughly parallel substreams (Fig. 3.2). Reproductive work makes progeny. Physiological work powers

---

5. More properly, the animal experiences a potential energy difference in the form of heat content, which is equivalent to the temperatures, the specific heats, and the masses of the respective bodies exchanging heat; that is, $\Delta PE = \Delta Tc_pM$, where ΔT = a difference of temperature (K), c_p is a specific heat (J K⁻¹ kg⁻¹), and M is the mass (kg).

6. The potential energy difference is related to concentration difference in the following way: $\Delta PE = RT\Delta C$, where R = the gas constant (8.314 J mol⁻¹ K⁻¹), T = the temperature of the solution (K), and ΔC = the difference in concentrations (mol l⁻¹) inside and outside the body.

the fluxes of matter and energy that are necessary to maintain the internal environment.

Adaptation involving conventional physiology can be fitted comfortably into mainstream Darwinian thinking. As far as we know, the only type of energy that can power reproductive work is the chemical potential energy in food. That is, reproductive work must come from the metabolic energy stream. Presumably, this is because the only means by which organisms transmit information about themselves is chemical (through DNA), and the maintenance, synthesis, and replication of DNA are chemical processes that require chemical energy to drive them. If reproductive work requires maintenance of a particular internal environment, then infrastructure that provides this environment efficiently will be favored by natural selection, because proportionally more energy will be available for reproductive work if less energy is needed for "overhead." To the extent the infrastructure is coded for genetically, conventional Darwinism is adequate for explaining physiological adaptation.

Keep in mind, however, that much of the physiological work is fairly prosaic—maintaining gradients of temperature, pressure, solute concentration. Thermodynamically, there seems to be no fundamental reason why the metabolic energy stream must do all this work. Indeed, if some other source of energy could be used to power physiological work, even more of the metabolic energy stream might be diverted to reproductive work. Even if this diversion is small, its evolutionary consequences could be large.

How Structures Can Live

Making energy do work means capturing and channeling it so that it flows in a controlled way down a potential energy gradient. Usually a structure of some sort—whether it be an imperfection in a silicon crystal, an enzyme molecule, or a crankshaft—directs the flow of energy. If physiological work is to be powered by the physical energy stream, there must be a structure of some sort that can capture and channel the energy.

Most engineering done by humans is devoted to figuring out clever ways of manipulating energy. Surprisingly, there turn out to be just four: energy flows can be resisted, they can be rectified, they can be switched and they can be stored for use later. These common patterns will help us organize our thoughts about how animals can build structures to do physiological work.

One of the great achievements of nineteenth-century physics was to show that energy can exist in many interchangeable forms. In itself, this is a remarkable feature of the universe: we just have to let that claim stand for the time being. I offer it for practical reasons, for it provides a very powerful tool for thinking about how energy moves and works. The tool is this: if all forms of energy are potentially interchangeable, it follows that a description of how one form of energy does work will also describe how other forms of energy do work. Understand one and (on at least one level) you understand them all.

Let us illustrate by relating two superficially unrelated processes: the flow of electrical current across a resistor and the flow of water in a pipe. The flow of current can be described by Ohm's law, which relates how current (energy) flows under a voltage (potential energy) difference between two points:

$$I = \Delta V / R_e \qquad [3.1]$$

where I is electrical current in amperes, ΔV is the voltage difference driving the current, and R_e is an electrical resistance, in ohms, that impedes the current. The flow of water down a pipe, on the other hand, is described by Poiseuille's law:

$$V = \pi r^4 \Delta P / 8L\eta \qquad [3.2]$$

where V is the volume of water flowing through the pipe per second, L and r are the pipe's length and radius, respectively, η is the water's viscosity, and ΔP is the difference in pressure between the two ends of the pipe.

Poiseuille's law is a complicated equation that seems to bear little relation to Ohm's law. However, we can simplify Poiseuille's law by lumping several of the terms together into one: a hydraulic resistance, R_h, that describes those things that impede the flow of water through the pipe. These include the dimensions of the pipe (it is harder to get water to flow through a long or narrow pipe than it is through a short or wide pipe) and the fluid flowing through it (it is harder to get a viscous fluid like syrup to flow through a pipe than it is to get water to flow). If we do a little algebra on equation 3.2, we can easily formulate the hydraulic resistance so:

$$R_h = 8L\eta/\pi r^4 \qquad [3.3]$$

We can now rewrite Poiseuille's law as follows:

$$V = \Delta P/R_h \qquad [3.4]$$

With the equation in this form, Ohm's law and Poiseuille's law look very similar. Indeed, we could write both using common terms:

flux of matter or energy
= potential energy difference / resistance to flux   [3.5a]

Some prefer to express a resistance as its inverse, or conductance, in which case we can rewrite as follows:

flux of matter or energy
= conductance × potential energy difference   [3.5b]

The generality implied by these equations makes possible a kind of intellectual piggybacking: if we know a lot about one kind of energy flow, then we know a lot about all kinds of energy flow. If, for example, electrical engineers have been more clever than hydraulic engineers at figuring out how to make electrical energy do work, hydraulic engineers can take these clever ideas and adapt them to their own problem of making water do work.

I shall do the same here, using concepts, terminology, and symbols familiar in electrical engineering to discuss how structures built by animals could alter or manipulate flows of physical energy and make them do physiological work. This is, in fact, a very widely used method known as the electrical analogy, and it is one I use often through this book. Furthermore, just as the number of possible electrical circuits is vast, so too will be the analogues that animals could build for powering their external physiology. Let me close the chapter now by illustrating, in a very general way, how externally built structures might be employed by animals to make external physiology work.

Assume first that there is a potential energy *(PE)* difference between the environment and the Earth—it could be sunlight, wind, gradients in temperature or water vapor. Formalize this difference in words:

PE difference = environmental PE − Earth's PE

or as an equation:

$$\Delta PE = PE_{env} - PE_{Earth} \qquad [3.6]$$

If it drove a flux of matter or energy in just the right way, this potential energy difference could be made to do work. The rate at which the work is done, or the power, will be the product of the flux rate and the potential energy difference driving it. In words:

power = work rate = flux rate × PE difference

As indicated in equation 3.5a, the flux rate *(J)* is determined by the *PE* difference and a resistance to the flux:

$$J = \Delta PE/R_e \qquad [3.7]$$

Thus, any structure that imposes a resistance to the flow of energy between the environment and Earth is potentially capable of powering work.

In the absence of our hypothetical structure, the flow of energy down the *PE* difference between the environment and Earth could be represented as an analogue electrical circuit, or equivalent circuit (Fig. 3.4). The equivalent circuit is simplicity itself: two points, each of which has a particular potential energy, bridged by a resistor that limits the rate that energy can flow.

Suppose now that an animal builds a structure that is interposed between PE_{env} and PE_{Earth} (Fig. 3.4). There are several ways we could represent the effect of this structure on the flow of energy. One thing the structure might do is simply to impede the flow of energy between PE_{env} and PE_{Earth}. In an equivalent circuit, this can be represented by a new resistor, which we shall call R_{str}, that is connected in series with R_e (Fig. 3.4). The point of connection between R_{str} and R_e represents a new potential energy, PE_{str}, which, of course, introduces a new potential energy difference that can drive a flux between the structure and environment or between the structure and the Earth. When PE_{env} is greater than PE_{Earth}, the energy in the environment will do work on the structure. Conversely, when PE_{env} is less than PE_{Earth}, the energy in the Earth will do work on the structure.

A resistor is a passive component because it is, by definition, insensitive to control: a resistor will as easily (or as poorly) impede the flow of energy in one direction as it will in another. In many circumstances, this feature of resistance is irrelevant, but often it undermines the possibilities for adaptive control. Suppose, for example, that our hypothetical structure is useful to the animal that built it only when the environment does work on it, not when the Earth does work on it. In other words, the structure is useful only when energy flows in one direction (*environment → structure → Earth*), not in the other (*Earth → structure → environment*).

This is a common problem in making electrical circuits work properly, and one of the many solutions electrical engineers employ is to interpose a device that allows current to flow only in one direction and not the other. These devices are known as rectifiers, and the process whereby a current is limited to a cer-

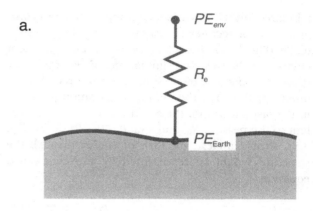

a.

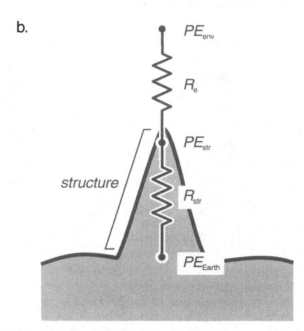

b.

Figure 3.4 *a:* Energy flows down a potential energy difference between the environment (PE_{env}) and the Earth (PE_{Earth}). The direction of energy flow will be determined by which potential energy is greater, and it will be restrained by some resistance, R_e. *b:* A hypothetical structure can be made to do work by imposing an additional resistance, R_{str}, between the potential energies of the Earth and environment.

tain direction is known as rectification. The most common type of rectifier is a simple device known as a diode (Fig. 3.5a). Just as a resistor in an equivalent circuit symbolizes the limitation of energy flow down a gradient, a diode symbolizes the rectification of energy flow. For the kind of rectification proposed in the last paragraph, the structure would be equivalent to the resistor, R_{str} and a diode, D_{str}, in series (Fig. 3.5). At its simplest, the flow of energy through the structure would be represented by the conditional equation:

$$J = (PE_{str} - PE_{Earth})/R_{str} \quad [\text{IF } (PE_{str} - PE_{Earth}) > 0] \quad [3.8]$$
$$J = 0 \quad [\text{IF } (PE_{str} - PE_{Earth}) < 0]$$

A diode is a simple logic device that "decides," on the basis of its direction, whether to let a current flow across it. There may be circumstances, however, in which the decision is best based upon other criteria. In our hypothetical structure, capturing environmental energy may be useful only under some conditions and not others. This would imply there is an alternative pathway for the energy to flow when the organism does not want it to flow through the structure. Again, this is a common problem in electrical circuits, and the device used to solve it is the switch. Switches come in a variety of forms—for our purposes, we will use a common type, the transistor. In a transistor, a small current fed into a terminal called the base (or gate) switches a

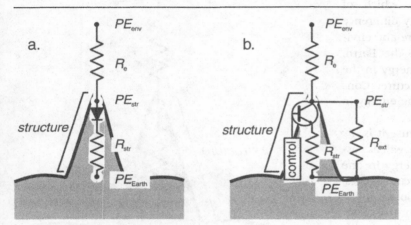

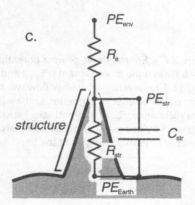

Figure 3.5 *a:* Structures can be made to do adaptive work by rectifying the flow of energy across the structure, symbolized by the diode. *b:* More sophisticated adaptive work can result if the structure can adaptively control energy flow as if through a transistor. *c:* Structures can store environmental energy in a capacitor, C_{str}.

larger current across two other terminals, the collector and the emitter. Current flows only one way in a transistor—from the collector to the emitter. Consequently, a transistor can be thought of as a switchable diode controlled by the current at the gate. Adaptively regulate the gate current and you adaptively regulate the flow of current through the transistor.

Consequently, we could analogize energy flow through the structure as a transistor, Q_{str}, controlled by some kind of input, whether from the animal that built the structure or from the environment acting on the structure (Fig. 3.5b). At its simplest, energy flow through the structure would be represented by another conditional equation:

$$J = (PE_{str} - PE_{Earth})/R_{str} \quad \text{[IF (Condition = TRUE)]} \quad [3.9]$$
$$J = 0 \quad \text{[IF (Condition = FALSE)]}$$

Finally, in all these equivalent circuits, circumstances may arise in which the energy that comes into the structure, from whatever source, is stored for use at a later time. If the energy coming in acts to heat the structure, for example, the increased temperature is actually stored energy that can be used to do other work when the *PE* gradients between the structure and environment are unfavorable. Suppose, for example, a structure uses heat energy to evaporate water and move it around in some way for the animal's benefit. If the structure can store this energy (as a high temperature) during the daylight hours, when the sun is contributing lots of light energy, it could be used to power work after the sun goes down.

Again, storing electrical energy is something that electrical engineers must do all the time. They store electrical energy in capacitors, which can be placed into an electrical circuit in a variety of configurations, depending upon what might need to be done. For our purposes, we will illustrate in an equivalent circuit the simplest configuration: a resistor in parallel with a capacitor, C_{str} (Fig. 3.5c). Here, when the *PE* gradient is favorable for the environment doing work on the structure, energy flows in. Some of the energy does work by flowing down R_{str}, and some is stored in the capacitor. When the *PE* gradient for the environment doing work on the structure is not favorable, the stored energy will then flow out of the capacitor and down R_{str}, continuing to do work.

By now, I hope that the path to a notion of physiology outside the organism, if not clear, is at least reasonably visible. In the next chapter we turn to "real biology," which will occupy the rest of this book.

CHAPTER FOUR

Broth and Taxis

In Chapter 2, I committed the sin of sophistry, and I have just spent Chapter 3 trying to exculpate myself. I confess that I committed another intellectual sin in the effort—I relied on an "occult force" as an explanatory tool.

Now, this sounds very bad: these days, the word *occult* evokes images of magic tricks taken rather too seriously, or things even more sinister. But being occult is not so bad, really. The literal meaning of the word is simply "unseen" or "obscured from view." An occult force, therefore, is simply an unseen cause of something. In fact, occult forces are used by scientists all the time because they are actually very handy. For example, gravity is an occult force, in that it is unseen and it remains mysterious in certain ways. Yet we know pretty well how it behaves, well enough to calculate positions of planets or predict trajectories of spacecraft with uncanny accuracy. And no one (at least no one *I* know) is prepared to dismiss gravity as some crackpot idea or to lump Isaac Newton together with occultists like Uri Geller.

I invoked an occult force as a means of linking together orderliness and energy. Although I tried to show with my fairly simple examples that energy and order are related, I implied a causal relationship between the two that, like the workings of gravity, seems almost magical:

Whenever energy flows through a living thing, order is created.

One cannot contemplate questions biological for very long without being struck by this apparently mystical ability to create order.

In this chapter, I'd like to lift the veil a bit on the relationship between energy and orderliness. I will do so by exploring, in some detail, a remarkable example of how order arises from disorder, seemingly spontaneously, simply because energy flows through a living entity. This is something organisms commonly, albeit wondrously, do, as when a cell assembles a complex and highly ordered protein from simpler amino acids. In the example to follow, though, the order appears not within the organisms themselves but in their environment, on a scale many orders of magnitude larger than the organisms producing it. In describing this large-scale orderliness, I hope also to enlarge on my claim that structural modification of the environment can power a physiological function outside the body.

Large-Scale Orderliness in Cultures of Microorganisms

First let me describe the phenomenon I have in mind. The organism is a swimming protozoan, something like the flagellated *Chlamydomonas* (Fig. 4.1). These organisms are fairly easy to culture in the laboratory, and if you have studied zoology or protozoology you probably have heard of them. The phenomenon occurs in a liquid culture of *Chlamydomonas*, at a density of about a million cells per cubic centimeter. You'll want to view the petri dish with transillumination, that is, lit from below so light passes through it to your eyes above it.

If you stir the culture up, it will appear uniformly cloudy, or turbid, like a jar of apple juice that has begun to ferment in the refrigerator. The reason the culture looks cloudy is that the billions of cells floating in the culture each reflect and refract light passing through them. When the cells are randomly distributed through the culture, the scattering of light beams through it is also random, and turbidity is the result.

Turbidity of the culture, arising as it does from a random, disordered distribution of the protozoans, should be favored by the Second Law. While we might have helped the culture become random by mixing it up, once it was randomized, the Second Law says it should have stayed that way. But in fact, the culture will stay

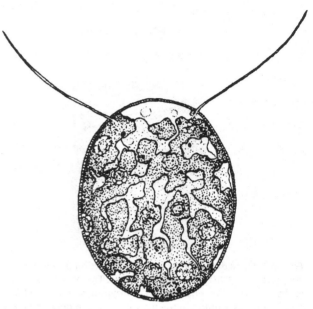

Figure 4.1 The single-celled microorganism *Chlamydomonas*. [From Kudo (1966)]

turbid only for as long as we keep mixing it. If we stop mixing, its appearance begins to change dramatically. Within a few minutes, the uniformly cloudy suspension begins to develop spots, focal areas of darkness, easily distinguishable visually from brighter areas surrounding them (Fig. 4.2). Eventually, the dark spots begin to merge and ramify, forming beautiful patterns of interleaved light and dark bands. The dark areas, of course, are regions where microorganisms have concentrated at densities high enough not just to scatter light coming through the culture but to block it. The light areas, just as clearly, are regions where organismal densities are low and light beams have a "straight shot" through the culture to our eyes.

The Difference between Thunderstorms and Organisms
So, what's going on? A closer look at the structure of these spots or bands gives us clues about how they are generated. When viewed from the side, the dark spots are revealed to be plumes, flowing columns of

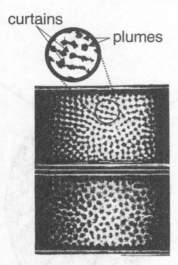

curtains

plumes

Figure 4.2 The development of bioconvection patterns in a suspension of *Chlamydomonas nivalis*. The photograph looks down on suspensions of *Chlamydomonas* in two flat culture dishes. Culture in the top dish is 7 mm deep, 4 mm deep in the bottom dish. The inset shows plumes, indicated by black spots, and curtains, indicated by thick black lines. [*After Kessler (1985a)*]

This kind of flow pattern, familiar to anyone who has baseboard heaters in the home, is known technically as a convection cell.[1] When a baseboard heater warms cool air near the floor, the air becomes less dense, and therefore buoyant, so it rises to the ceiling. As the heated air rises, it loses its heat to the cooler walls, windows, and surrounding air until it again becomes dense enough to sink to the floor and eventually make its way back to the heater. The result is a doughnut-shaped pattern of air flow in the room, where air rises at the walls, collects at the ceiling, and is forced down in a concentrated plume at the center. A similar process, albeit in a much more com-

microorganisms, like the downdrafts from a storm cloud. Similarly, the dark bands are downward-flowing aggregations, but in this case they are organized into curtains, walls of downward flow rather than columns (Fig. 4.3). Once these concentrated plumes or curtains reach the bottom of the container, the microorganisms are dispersed sideways. Eventually, they make their way back to the top, sometimes by concentrating in upwardly driven plumes, sometimes simply by swimming to the top in the spaces between the plumes, eventually to be collected again into a downward-moving plume. The net effect is a circulatory movement of microorganisms between the upper surface of the culture liquid and the bottom of the culture dish (Fig. 4.4).

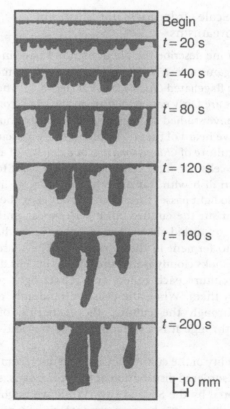

Begin
t = 20 s
t = 40 s
t = 80 s
t = 120 s
t = 180 s
t = 200 s

10 mm

Figure 4.3 The development of bioconvection plumes as viewed from the side, in a culture of the planktonic protozoan *Stenosemella nucula*. [*After Kils (1993)*]

---

1. *Convection* literally means "to carry" and it refers to the transport of something (for example, heat, mass, or momentum) by fluid flow.

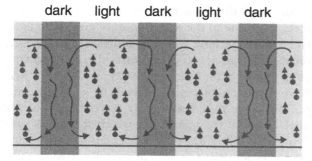

dark   light   dark   light   dark

Figure 4.4 Schematic side view of three bioconvection plumes. The dark zones contain concentrated streams of microorganisms flowing downward in the culture medium.

plicated form, drives the pattern of winds in thunderclouds.

We ought to pause, now, and consider the facts carefully. Convection cells in rooms and storm clouds are well-understood phenomena of fluid mechanics. Their similarity with the convection cells in our protozoan culture might tempt us to look for similarity in causes. That would be a mistake, for the following reason. To generate the convection cell in a room, you must do work on the room air, which you accomplish by heating it with the baseboard heater. In the generation of a storm cloud, the work is done by sunlight heating air close to the ground. All this is perfectly in line with the Second Law: the ordered patterns of air flow appear only when work is done on the system by some external agency. Stop doing the work (switch the heater off, let the sun go down), and the orderly convection cell disappears. Let me emphasize that what we observe in the petri dish is exactly opposite: the order appears only when we *stop* doing work on it. So there seems to be something to explain.

Because this phenomenon differs from the ordinary convection we see in, say, a room heated by a baseboard, it has a special name to distinguish it: bioconvection. We know a lot about bioconvection, in part because it is interesting in and of itself, and in part because it has some commercial applications, of which

more momentarily. For bioconvection to occur, certain requirements must be met:

1. *The liquid culture must contain swimming microorganisms*—cells incapable of moving under their own power will not produce bioconvection in culture.

2. *The organism must swim with some kind of taxis*—that is, it must have a tendency to swim toward or away from something.

3. *The organism must be slightly more dense than the culture it is in.* This is easy: most cells are 5–10 percent denser than the water they live in.

4. *The organism should have a center of mass that is offset from its center of buoyancy.* This last requirement sounds rather obscure, but it is crucial to understanding how bioconvection works. We will explore it in detail momentarily.

Each of these requirements forms a piece of a puzzle. Before putting the pieces together, we must first understand them individually.

"Smart" versus "Dumb" Gravitaxis in Chlamydomonas *Chlamydomonas* move about their culture by doing a sort of breaststroke with their locomotory organs, a pair of flagella mounted at the "head" of the cell (Fig. 4.1). The speeds they can attain are, in their own way, impressive: a *Chlamydomonas* can swim at a top speed of about 200 micrometers (μm) per second, or 1 millimeter every 5 seconds.[2] The swimming is not random, however. The cells are able to sense conditions in the environment and to use this information to direct their swimming. As a general rule, the tendency is for the cell to stay where it is if conditions are "good" or to swim away from "bad" conditions, toward areas of the culture where conditions are "better."

---

2. Scaled up to the size of a human being (with an average body length of 1.5 m), this works out to a swimming speed of 135 km h^{-1}.

This tendency to swim preferentially toward or away from a particular type of environment is common enough to deserve a name, *taxis* (from the Greek *tassein*, "to arrange"). Taxes (plural of taxis, not what the government takes from you every year) come in a variety of forms, and they are named according to the way the arranging is done. For example, phototaxis is swimming toward or away from light (literally, "arranged by light"). Gravitaxis ("arranged by gravity," sometimes erroneously called geotaxis) is swimming up or down with respect to the Earth's gravitational field. By convention, taxes can be either positive or negative. For example, positive phototaxis is swimming toward light, while negative phototaxis obviously is swimming away from light. Similarly, positive gravitaxis is swimming down, or toward the Earth, while negative gravitaxis is swimming up, or away from the Earth.

Taxes may be managed in a number of ways. In some cases, the taxis is "smart": the cell has a sensory apparatus that takes in information about the environment and directs the swimming organs to drive the cell where conditions are good or away from conditions that are bad. For example, light falling on one side of an algal cell can influence the pattern of beating of its flagella, making it swim toward the light. In other cases, the taxis is "dumb," as in the magnetotactic bacteria *Aquaspirillum magnetotacticum*. These mud-dwelling spirochaetes contain within them a small crystal of magnetite. The Earth's magnetic field exerts a torque on these small crystals, just as a magnetic field imposes a torque on a compass needle. Thus, the bacterium's body is pointed in a particular direction by the magnetic field, and the bacterium simply swims in whatever direction it is pointed. In *Chlamydomonas*, the taxis is partially dumb and partially smart. The protozoan does make a sensory assessment of the environment—is there sufficient oxygen? or is there sufficient light?—and swimming is initiated if the answer is no But *Chlamydomonas* swimming is dumb in that the distribution of mass within the cell biases its direction of swimming. Let us now see how.

All bodies that have mass will experience a downward force due to the acceleration of gravity. In analyzing how this force acts, it is convenient to treat an organism's mass as being concentrated in an imaginary point known as the center of gravity, or CG (Fig. 4.5a). If we suspend a *Chlamydomonas* cell by its flagella, the cell will come to rest so that its center of gravity sits directly below the point from which the cell is suspended. In *Chlamydomonas*, the center of gravity happens to be located at the end of the cell opposite to the flagella.

When an object of a certain density, say an air-filled inner tube, sits in a fluid that is more dense than air, say water, the downward gravitational force acting on the inner tube is offset by the upwardly directed force of buoyancy.[3] Just as gravitational forces act through a single center of gravity, so too do the buoyant forces act through a single point, the center of buoyancy (CB; Fig. 4.5a). Generally, objects suspended in a fluid will orient so that the respective centers of buoyancy and gravity align vertically with respect to one another, with the center of buoyancy located directly above the center of gravity. If they do not align in this way, the object will experience a net rotational force, or torque, that will return the object to the orientation that aligns these two points properly.

Like all cells, *Chlamydomonas* have a density that is slightly higher than water's. Therefore, there will be a net gravitational force that will make the cell slowly sink. Because of the disparate centers of gravity and buoyancy, an undisturbed *Chlamydomonas* will also be oriented so that its flagella point upward (Fig. 4.5a). In the absence of any other force, a *Chlamydomonas* that is

---

3. Anything embedded in a fluid medium will experience buoyancy proportional to the mass of the fluid it displaces. Thus, a person standing in air will experience a buoyant force lifting him up, but air being much lighter than flesh, the force will be small. The weight of an object does not simply result from gravity, but from the difference between gravitational force and buoyant force. The difference constitutes the body's specific gravity. Thus, the specific gravity of any substance is less in water than it is in air.

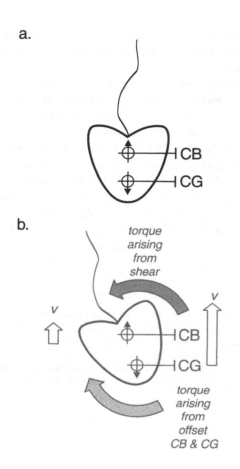

a.

b.

torque
arising
from
shear

torque
arising
from
offset
CB & CG

Figure 4.5 Orientation of a microorganism with noncongruent centers of buoyancy (CB) and gravity (CG). *a:* In still or uniformly moving fluid, the center of gravity will always come to rest directly below the center of buoyancy. *b:* When the microorganism is subjected to shear (indicated by the different lengths of the velocity vectors, *v*), it will be rotated by a torque proportional to the magnitude of the shear (darkly shaded arrow). The rotation displaces the centers of buoyancy and mass from their resting positions, which imparts an oppositely directed torque to the microorganism (lightly shaded arrow). The organism will come to rest at the angle where the two torques are balanced.

not swimming will always exhibit a positive gravitaxis. In other words, it will sink. When it swims, however, because it is pointed up, it will always swim up, in a negative, albeit a "dumb," gravitaxis.

Hydrodynamic Focusing in Cultures of Chlamydomonas
We now turn to another piece of the puzzle, a phenomenon known as hydrodynamic focusing. Unlike bioconvection cells, which arise spontaneously in shallow cultures, hydrodynamic focusing requires work to be done on the culture, and it is work of a special kind.

As before, you start with a culture of swimming protozoans like *Chlamydomonas*. This time, however, the culture is placed in a device consisting of two vertical glass tubes that are connected so that the culture can be made to circulate between them, flowing downward in one tube and upward in the other (Fig. 4.6). We will be focusing our attention on the tube containing the downward flow. If a well-mixed culture is circulated between these tubes, it starts, as it did in the flat culture dish, as a uniformly turbid suspension. After a few moments, however, the cells in the downwardly flowing tube focus in the center into a densely packed column of cells. This is hydrodynamic focusing.

Hydrodynamic focusing is of great commercial interest because it offers a cheap and convenient method for harvesting microorganisms in culture. When microorganisms are grown commercially, they must at some point be harvested, usually at great expense. For example, many commercial cultures must be processed as a batch, in which a culture is started, allowed to grow for a time, and then stopped for harvesting. Hydrodynamic focusing offers a way around this problem: if you could place a small siphon at the center of one of the downward-flowing columns of culture (Fig. 4.6), you could drain off a highly concentrated suspension of cells without having to stop the whole batch. However, the practical applications of hydrodynamic focusing concern us less than how the plume originates.

The focusing arises from an interaction between the organism's swimming and the characteristic patterns of fluid flow through tubes. When fluid flows through a tube, its velocity is not everywhere the same (Fig. 4.7). Fluid in the center moves faster than does fluid near the tube's walls. If velocity is plotted against the tube's radius, we see a characteristic velocity profile. In ideal circumstances, velocity is null at the inside walls and reaches a maximum at the tube's center. How does this distribution of flow affect the *Chlamydomonas* suspended in a culture broth?

If a *Chlamydomonas* sits anywhere in that tube but at the exact center, the flow on one side of the cell (that facing the center) will be slightly faster than the flow on the other side (that facing the wall). This disparity of velocities, called shear, will impose a torque on the cell that ordinarily would cause it to rotate (Fig. 4.5b).

You will recall, however, that disparate centers of mass and buoyancy make the protozoan resist this torque. Consequently, the cell will tilt, coming to rest at an angle where the torque imposed on the cell by the shear in the velocity profile is exactly offset by the torque arising from the misalignment of the centers of gravity and buoyancy. Because the shear is highest at the tube's walls, the tilt will be greatest near the walls. Because gravity always biases *Chlamydomonas* to point upward, the shear in the downward-flowing tube will point all the cells toward the center. Because *Chlamydomonas* swim wherever they are pointed, they will concentrate at the center of the tube, forming the plume (Fig. 4.6). The opposite happens in the upward-flowing tube. Here, the upward-pointing veloc-

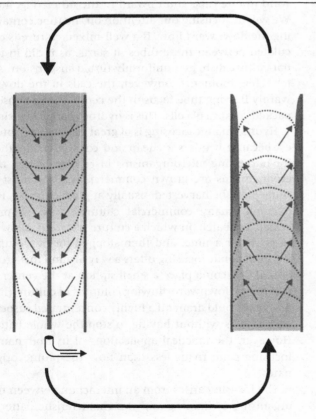

Figure 4.6 Hydrodynamic focusing by microorganisms like *Chlamydomonas*. In the culture flowing downward in the pipe, cells concentrate in a dense plume at the center. When the culture flows upward, cells migrate to the walls of the tube. Development of the velocity profile is indicated by the series of dotted lines. Vectors (small arrows) indicate the forces driving the cells.

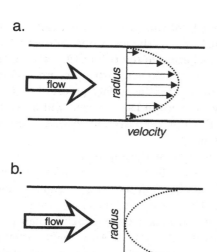

a.

flow

radius

velocity

b.

flow

radius

shear

Figure 4.7 The profile of velocity of a fluid flowing inside a tube. *a:* Velocity in a tube assumes a parabolic velocity profile, ranging from $v = 0$ at the tube walls and the maximum velocity (indicated by the longest arrow) at the center. *b:* Distribution of shear in a tube with a flowing fluid. Shear is greatest where the velocity gradient, dv/dr, is steepest (at the walls) and falls to null at the center.

ity profile tilts all the cells in the column toward the walls. The cells now congregate at the walls of the tube rather than at the center. This phenomenon is known as gyrotaxis—literally, "arranged by a vortex."

How Do Bioconvection Plumes Arise?

We can now return to the remarkable appearance of bioconvection plumes in a petri dish of *Chlamydomonas*. Like all interesting phenomena, this one has two explanations: a short simple one that masks a more complicated, but more interesting, explanation. The short simple explanation is that the order is imposed by the energy flowing through the organisms. In this case, the work being done comes not from an external agency, like a mixing spoon or a baseboard heater, but from the chemical potential energy in the culture broth being channeled through the cells:

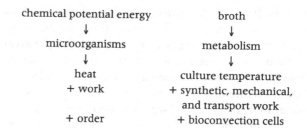

chemical potential energy → microorganisms → heat + work + order

broth → metabolism → culture temperature + synthetic, mechanical, and transport work + bioconvection cells

Of course, this explanation relies on the occult process that is the cause of the trouble in the first place. Unmasking the real answer requires us to break the problem down into three steps:

1. Large-scale gradients in oxygen concentration arise in the culture as a consequence of the organisms' collective metabolism.

2. In response to these gradients, the organisms redistribute themselves in the culture according to the oxygen concentration gradients. This imposes a new large-scale gradient in potential energy in the form of a spatial redistribution of mass within the culture.

3. Gravity acts on this new *PE* gradient to oppose the redistribution of mass, generating in the process large-scale flows within the culture.

Let us now turn to just how these steps work together to generate bioconvection.

STEP 1: ESTABLISHING THE GRADIENTS IN OXYGEN CONCENTRATION

First let us imagine a culture of randomly distributed microorganisms, contained in a petri dish that we can either open to, or close off from, the air as we wish. Let us suppose initially that the dish is closed to the air.

We will first focus on a single microorganism and the small parcel of fluid (culture broth) that surrounds it.[4] For various reasons, I will treat each microorgan-

---

4. The radius of a microorganism like *Chlamydomonas* is about $4 \mu m$, and since it is roughly spherical it will occupy a volume of

ism and its associated parcel of fluid as if they were an inseparable unit.[5] As the microorganism consumes oxygen, it necessarily takes it from its associated parcel of fluid. The concentration of oxygen in this parcel, which we designate as the oxygen partial pressure, or pO_2, therefore declines by some small amount proportional to its depletion. If we measure the oxygen concentration in our parcel at several times thereafter, we will see the pO_2 declining with respect to time, the rate of decline being exactly equal to the rate at which the microorganism is consuming oxygen.

The same thing will happen everywhere in the culture: wherever there is a microorganism, the pO_2 in its surrounding parcel of liquid will be declining. If we assume, just for convenience, that every microorganism is consuming oxygen at the same rate, the pO_2 is pretty much the same no matter where in the culture it is measured. In other words, oxygen is uniformly distributed and uniformly depleted through the culture. To put it more formally, there will be no gradient in oxygen partial pressure in any spatial dimension: no matter which way a microorganism faces, whether it be up, down, north, south, east, or west, it will confront the same oxygen partial pressure everywhere. Consequently, the movements of oxygen in the culture are dominated by small-scale variations of oxygen partial pressure between an individual microorganism and its associated parcel of fluid.

When we open the top of the culture to the air, however, there is a remarkable change. No longer do all the organisms in the culture confront the same oxygen concentration no matter which direction they face. The unfortunate ones deep in the fluid still are

surrounded, initially at least, by parcels of liquid with the same low pO_2. The lucky ones at the surface, however, are adjacent to the rich oxygen source of the air. There is now an oxygen partial pressure difference between the air and the surface parcel of liquid. As dictated by the Second Law, oxygen will move from the air into the topmost parcel, increasing its oxygen concentration.

This increase at the top level will set in motion a sort of "bucket brigade" for the transport of oxygen deeper into the culture. The bucket brigade works like this. The movement of oxygen from the air into a surface parcel makes its pO_2 higher than that of the fluid parcel just below it. An oxygen partial pressure difference between the parcels is established that will drive oxygen downward at a rate proportional to the partial pressure difference between the parcels. The same thing happens for each pair of parcels we encounter from the top to the bottom of the culture, that is between parcels just below the surface and parcels below *them*, and so on. The end result is a large-scale gradient in oxygen partial pressure that drives a large-scale flow of oxygen downward into the culture.

This downward movement of oxygen is inefficient, for two reasons. First, the oxygen moves downward by diffusion, which is very fast over short distances, such as across the radius of a cell, but is very slow over "long" distances, such as the few millimeters from the top to the bottom of the culture. Second, each parcel of liquid contains an oxygen-consuming microorganism. Because the microorganism in a surface parcel will consume some of the oxygen passing through that level, less oxygen will be available to move to the next parcel below. It is as if the microorganisms were each exacting an "oxygen tax" against the movement of oxygen molecules past them. The end result will be a vertical gradient of oxygen partial pressure in the culture, with high partial pressures at the top and lower and lower partial pressures toward the bottom, as each microorganism extracts its "oxygen tax." Unlike the closed culture, where oxygen in all the parcels was de-

---

roughly 2.7×10^{-10} cm^3. If the culture density is about 10^6 per cm^3, each *Chlamydomonas* will have about 10^{-6} cm^3 of culture fluid to occupy, or roughly 3,700 times the volume of the cell itself, or a cube roughly 100 μm on a side, 25 times longer than the length of the cell itself.

5. At very small scales, objects and the fluids they are embedded in are held together to a great extent by the fluid's viscosity, which contributes a sort of "added mass" to the object. Consequently, very small objects in fluids always tend to travel with the fluids surrounding them.

pleted uniformly, each parcel's pO_2 in the open culture is now steady with respect to time, reflecting the replacement of oxygen in the parcel as it is consumed by the contained microorganism. Also, the cells at the top levels enjoy higher, and presumably more equable, partial pressures, because their oxygen consumption rates can be met without the intervention of lots of "middlemen" each taking their piece of the flux as oxygen flows down to the bottom. The poor cells at the bottom, on the other hand, must rely on the trickle-down from the top, and so must operate at lower partial pressures.

Let us now put this scenario into a more general context of energy. Oxygen in the culture can move from one place to another only if there is a source of potential energy to move it: oxygen molecules have mass, and it therefore requires work to move them around. The potential energy gradient moving the oxygen comes from the difference in oxygen concentrations between two points in the culture. Indeed, the way we express oxygen concentration as its partial pressure is a direct measure of the potential energy that does the work of moving oxygen.

In the closed culture, the distribution of potential energy is diffuse: gradients in oxygen partial pressure extend no further than that between a microorganism and its associated parcel of fluid. If the microorganisms are distributed diffusely through the culture, so too will be the small-scale gradients in potential energy in the culture. When we open the culture to the air, however, we have now imposed a large-scale potential energy field, the top-to-bottom gradient in oxygen partial pressure. Part of this new potential energy field comes from having one surface of the culture exposed to a rich source of oxygen (to use our jargon, the environmental potential energy), and part comes from energy being channeled through living things (manifest in the "oxygen tax" extracted as oxygen moves past the microorganisms). As a result, a gradient of potential energy is imposed on the culture on a much larger scale than the gradients affecting the living things in the culture.

STEP 2: REDISTRIBUTING THE MICROORGANISMS

The next step depends on the locomotory activities of the microorganisms. Remember that the negative gravitaxis in *Chlamydomonas* is partially "dumb"—that is, it relies on the cell's distribution of mass always pointing it up. However, swimming in these organisms is also activated when the cell finds itself in poor conditions. *Chlamydomonas* in the top layers will power locomotion only to the extent they need to stay there—life is good, so why move? Cells in the bottom layers, however, will divert more energy through their locomotory organs, powering the higher levels of locomotion needed to climb to the top. Thus, the large-scale gradient of oxygen partial pressures also elicits a large-scale gradient in the magnitude of the metabolic energy stream flowing through the culture. Work is done on the culture as a whole, but its distribution reflects the distribution of oxygen partial pressure. The farther the cell must travel, the more work is done. As a result of this work, cells will accumulate in a thin layer at the surface levels of the culture.

Because these cells have a net density that is slightly greater than the water they live in, their accumulation at the surface of the culture makes the top layers of culture fluid heavier than the lower layers. This concentration of higher density is obviously unstable, in the same way that a brick balanced on the top of a broomstick would be. Unlike the brick, which would fall immediately if it were not held up somehow, the mass at the top of the culture can be stabilized to a degree. When unstable distributions of mass in fluids, called inversions, do collapse, they do so in a fairly controlled way that generates the orderly bioconvection cells.

STEP 3: GENERATING THE BIOCONVECTION CELLS

The factors that stabilize and destabilize inversions can be easily appreciated with an analogy. Imagine a pan of water being heated on a hot plate. The heat imposes an unstable inversion on the water: the water at the bottom, being hotter than the water at the top, is more buoyant. The warmer water at the bottom will therefore "want" to rise to the top, but its rise will be blocked by the layers of cooler, denser water above it, which sit on it like a pot lid.

Inversions are stabilized by the fluid's viscosity, which quantifies how resistant a fluid is to flow. In the inversion that develops in a pan of heated water, the buoyant water at the bottom pushes upward against the heavier water on top with some force, let us call it F_b. Ordinarily, that buoyant force would drive an upward flow. For this to happen, however, the buoyant water at the bottom must flow through the heavier layer of water on top, and this will be resisted by the water's viscosity; let us call it F_v. As long as the buoyant forces pushing the water up are smaller than the viscous forces resisting it (that is, $F_b << F_v$), the inversion will be stable. If the buoyant forces match or exceed the viscous forces resisting it, however, the inversion will break up. The events just as the inversion begins to break up are of particular interest to us, so let us consider them in some detail.

It is very hard to heat a surface like the bottom of a pan absolutely uniformly. There will inevitably be some parts of the pan bottom that are heated more than others. There, the water will be warmer, and a bit more buoyant, than it will be in other, less well heated areas. The horizontal boundary between the layers of the inversion will therefore not be perfectly flat but will exhibit transient humps, where "bubbles" of locally more buoyant water rise slightly higher than adjacent parcels of cooler and slightly less buoyant water.

If one continues to heat the pan, eventually enough energy will be imparted to one of these bubbles so that its buoyant force overcomes the viscous forces resisting its upward flow. It will then "punch through" the layer of heavier water keeping it down. The hole punched through the dense upper layer now opens up a pathway for the bottom layer of "suppressed" warmer fluid everywhere else in the pan. The more buoyant water on the bottom of the pan can now rise through the hole opened for it, and the denser top water helps the flow along by pressing down on it with its greater weight. The result is the sudden appearance of a vigorous upward-flowing plume of warm and buoy-

ant water, leading ultimately to the turnover of the inversion.

When *Chlamydomonas* migrate to the uppermost layers of the culture, an inversion of sorts also occurs, even though the culture's temperature is uniform. The accumulation of organisms at the surface forms a dense layer of fluid, which gravity should make sink were it not spread out on top of a less dense layer below. The stability of this inversion is governed by the same physical rules that govern the temperature inversion in a heated pan. In this case, however, it is random fluctuations in density of microorganisms that lead to the inversion's breakdown. Locally dense collections of cells form "bubbles" (anti-bubbles, really) of dense culture that will tend to sink into the less densely populated and more buoyant culture medium below it. Usually, these anti-bubbles are buoyed up until they are dispersed by Brownian motion.[6] At some point, however, one of the anti-bubbles will become dense enough to overcome the buoyant force holding it up, and it will begin to sink.

The mere presence of a sufficiently dense anti-bubble is not enough to generate a bioconvection plume, however. The maximum increase of density of the fluid is limited by the density of the microorganisms themselves. Even if the organisms were packed as tightly as possible, the top layers will be only about 5 percent denser than the water: in reality, the density differences will be less. What is needed in addition is that the anti-bubble be composed of living microorganisms: an anti-bubble of dead cells is never dense enough to generate a plume. Bioconvection plumes, however, appear reliably and rapidly (within minutes) in living cultures.

To understand what happens next, remember the mechanism of hydrodynamic focusing: a downward-

---

6. Brownian motion is named after the Scottish botanist Robert Brown. In 1858, Brown reported a peculiar random motion of microscopic particles suspended in fluids. The motion itself is caused by random variations in the forces exerted on these particles by the atoms of the fluids in which the particles are suspended. This force is significant only on the small scale of cells and molecules, for which it is a randomizing force.

pointing velocity profile will tilt cells toward its center, with the degree of tilt being directly proportional to the shear. So armed, let us now follow the development of a bioconvection plume in detail.

Suppose first an anti-bubble of densely packed microorganisms is sufficiently dense to start it sinking slowly in the culture fluid (Fig. 4.8). If these cells were dead, the increased density would be short-lived, because Brownian motion would quickly disperse them. In an anti-bubble of living cells, however, a very different dynamic occurs. When the anti-bubble begins to sink, it drags adjacent parcels of fluid along with it, and these drag adjacent parcels along with them, and so on and so on. The result is a velocity profile that extends some distance away from the sinking anti-bubble: the velocity is highest at the center of the sinking anti-bubble and falls off gradually away from the center (Fig. 4.8). Any living microorganism caught in this flow field will be rotated toward its center, just as it is when the culture experiences hydrodynamic focusing. The surrounding *Chlamydomonas,* because they swim in whatever direction they are pointed, will therefore congregate at the center of the anti-bubble. This will, in turn, increase the number and density of cells in the anti-bubble, which will increase its sinking speed, which will increase the steepness of the velocity gradients, which will tilt more cells toward the center of the field, which will increase the anti-bubble's density still further and . . .

You get the idea, I hope. The initial sinking of the bubble sets up the conditions for promoting its own sinking rate, a condition we call positive feedback. The positive feedback only works with living cells. The macroscopic result is the generation of robust convection plumes through the culture, within minutes, each plume centered on a transiently formed anti-bubble of organisms dense enough to start sinking. Eventually, these plumes will compete with one another for microorganisms to feed it, and it will come as no surprise that the initially stronger plumes will eventually incorporate the initially weaker plumes into them, forming the curtains of bioconvection that develop (Fig. 4.3).

Where Physiology Comes From

I have just described the apparently spontaneous production of large-scale patterns of fluid flow in a culture of swimming microorganisms. The process is not spontaneous, however: work is still being done on the system, just as it is in a room with baseboard heaters. It is

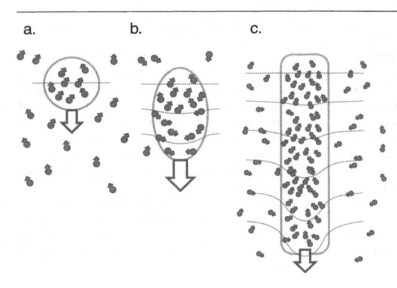

a. b. c.

Figure 4.8 The progressive development of an "anti-bubble" of dense microorganisms in a culture medium. *a:* A local grouping of cells, delimited by the dotted circle, may develop through random movements of the microorganisms. *b:* The incipient anti-bubble will begin to sink, because of the slightly higher density of the group. *c:* As the anti-bubble sinks, it sets up a shear field around it, which orients organisms outside the anti-bubble toward its center.

just that the work is coming from a different source—namely, stored energy in the culture broth—and the energy is traveling through the system in a different way, through the metabolism of the microorganisms (Fig. 4.9). At first, the work is done by the microorganisms altering the distribution of oxygen in the culture, which is followed by work done on the fluid by the microorganisms themselves. All steps in the process are perfectly rational, explainable, and consistent with the Second Law. Yet order, not disorder, is the result.

To the microorganisms themselves, bioconvection has a more important consequence than the generation of pretty patterns of light and dark bands in a petri dish. The parcels of surface fluid that are dragged down with the plume carry with them the rich load of oxygen picked up while the parcel was at the surface. Their sinking transports oxygen deep into the culture at a rate much faster than diffusion: the sinking rate of a plume can be as fast as two or three millimeters per second, roughly ten times as fast as the microorganisms can swim. At this speed, oxygen and carbon dioxide are transported in a way similar to the transport of these gases by blood circulating through our bodies. *Voila*, we have a circulatory-respiratory system!

The phenomenon of bioconvection also raises some interesting questions about the evolutionary origins of physiology. As I have said earlier, we usually associate physiology with a high degree of coordination and control. In an organ like the intestine, for example, the muscle cells that drive material through the tube must contract in a highly coordinated manner, so much so that the intestine has embedded in it an accessory nervous system that controls the activity of the organ's muscles and multitudinous secretions. Yet, the *Chlamydomonas* culture is a collection of relatively independent and autonomous cells that, as far as we know, do not coordinate their activities with other inhabitants of the culture. Nevertheless, they become organized into structures many times larger than themselves: the typical diameter of a bioconvection cell is a few millimeters, the cells themselves about a thousand times smaller. Furthermore, these structures do physi-

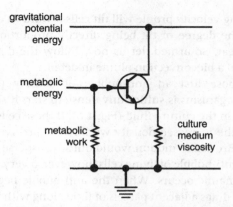

Figure 4.9 The flows of energy in the development of large-scale order in cultures of swimming microorganisms. Metabolic energy, through the agency of powering locomotion, controls work done by the large-scale potential energy in gravity.

ological work at a large scale, transporting oxygen and carbon dioxide throughout the culture.

Could this kind of self-organization represent the evolutionary origin of physiology, the ephemeral superstructure upon which the more sophisticated physiology of multicellular organisms has been built? Certainly, it is not much of a stretch to see how such "superorganismal" physiology would benefit the microorganisms that are associated with it. But then how do we reconcile this diffuse physiological function with the conventional Darwinian model of adaptation described earlier? Certainly it is easy to posit that a particular gene governs the shape or distribution of mass within a *Chlamydomonas* or the efficiency and responsiveness of the proteins that make up its flagella. It is also easy to imagine that natural selection would favor those genes that give a *Chlamydomonas* just the right distribution of mass or flagellar responsiveness that promotes bioconvection. Genes could result in the "superorganismal physiology" phenotype, however, only if (a) they act in concert with numerous copies of the same gene scattered among different individuals and (b) if their action results in the modification of the physical environment and the large-scale distributions of potential energy in it. The first condition is not hard to reconcile with neo-Darwinism—we will come back to this in Chapter 11—but the second is rather more difficult, for it forces us to ask the uncomfortable question of what is adapting to what? Is the organism adapting to its environment, as a good Darwinian would insist it must, or is the environment being adapted to it? Certainly, in the genesis of bioconvection, such environmental properties as fluid density, viscosity and the partial pressure of oxygen are as much a part of the phenotype as are the *Chlamydomonas* themselves.

A philosopher, one Bishop Berkeley,
Remarked metaphysically, darkly,
That what we don't see
Cannot possibly be,
And the rest is altogether unlarkly.
—FRANCIS AND VERA MEYNELL (1938)

CHAPTER FIVE

Then a Miracle Occurs . . .

In the last chapter, I argued that energy flowing through organisms, when it interacts with large-scale gradients of potential energy in the environment, can impose orderliness on the environment at a scale many times larger than the organisms themselves. I asked you to believe that the generation of this orderliness somehow relates to the external physiology I posited in Chapter 1.

Skeptics confronted with that argument probably will be reminded of the famous Sydney Harris cartoon showing a blackboard covered by complex equations, in the center of which is a straightforward note: "then a miracle occurs" (Fig. 5.1). The weak point of the argument I tried to develop in Chapter 4—the "miracle," if you will—was my assertion that bioconvection cells in protozoan cultures represent a primitive type of structure that does physiological work. The implication, of course, was that such orderly structures should arise any time there is an interaction of the metabolic and physical energy streams, and that the structures should always do physiological work. *Ergo*, structures built by animals are physiological organs. Right?

Well, no. As tantalizing as the phenomenon of bioconvection is, it admittedly is not a firm foundation upon which to build such an assertion. Bioconvection cells undoubtedly are orderly things, they undoubtedly do physiological work, and they undoubtedly arise from the interaction between two flows of energy, one metabolic and one physical. But are they structures? If they cannot justifiably be called that, or at least the precursors to the more substantial structures built by animals, then the whole argument is a

"I think you should be more explicit here in step two."

Figure 5.1 The problem of orderliness . . . solved! *[From Harris (1977)]*

---

dead end. What is needed is a way to bridge the gap between the tangible structures built by animals and the orderly, albeit ephemeral, "structure" of a bioconvection cell.

In this chapter I will try to bridge this gap by considering structures built by the most primitive animals, sponges and coelenterates (which include the jellyfish, corals, and hydras). Because it is crucial that we understand the difference between an animal's body (which, by convention, handles internally driven physiological work) and the structure an animal builds (which purportedly handles externally driven physiological work), we must start with a digression into just what animals are, and how they come to be. Sponges and coelenterates, particularly corals, provide an interesting venue for exploring this question, because the distinction between body and external structure among these animals is a fuzzy one. We will also

have to delve into some fairly deep questions of animal form: how it is measured, how it develops, and how it is influenced by both the environment the animal lives in and by its evolutionary history. We will then be ready to bridge the gap between the quasi-structures represented by bioconvection cells and the more substantial structures built by animals. By then, I hope to reduce the "miracle" to a reasonable possibility, or at least to a mere improbability.

What Animals Are

At present, biologists more or less agree that the living world is properly divided into six large categories, designated kingdoms. Two of them are prokaryotes (literally "before the seed"), also known as bacteria, single-celled organisms that do not enclose their genetic material (invariably DNA) within a distinct nucleus. The prokaryotic kingdoms are:

- The Archaea, primitive bacteria that probably are the most closely related organisms to the first living cells. Archaea cannot live in the presence of oxygen, and they often inhabit extreme environments, like hot springs or strongly acidic or briny environments.
- The Eubacteria, derived from the Archaea, which inhabit a diversity of environments and carry out a diversity of life styles.

I will have much more to say about the prokaryotes in Chapter 6.

Derived from the prokaryotes are four kingdoms comprising the eukaryotes (literally "true seed"), organisms that confine their genetic material (again, DNA) within a membrane-bounded nucleus. The eukaryotic kingdoms are:

- The Protista, an extraordinarily diverse assemblage of single-celled organisms, divided roughly into the "animal-like protists," which do not synthesize their own food, and the "plant-like protists," which manufacture their own food photosynthetically. Usually, protists exist as solitary cells, but some species form colonial assemblages. These probably gave rise to the three remaining kingdoms.

- The Plantae, which include all the green plants. Plants are autotrophic ("self-feeding"), that is, they produce their own food through photosynthesis.

- The Fungi, which include the mushrooms and yeasts. Fungi are heterotrophic ("other feeding"), that is, they steal food from other organisms. The Fungi, of all things, are probably the most closely related to our own kingdom.

- The Animalia, which include all the animals. Animals are invariably heterotrophic, although some enter into symbioses with autotrophs.

All the so-called higher kingdoms (the Fungi, Plantae, and Animalia) evolved from the Protista. This involved a fundamental shift in the ways cells associate with one another. Specifically, there had to be a shift from the relatively loose association of the colonial protozoa to the complex multicellular organization that characterizes the higher kingdoms.

Body Plans of the Animals
Among the animals, zoologists recognize roughly thirty-two phyla[1] (the actual number is inexact because the status of some of the phyla is in dispute).

---

1. The category of phylum is the most inclusive of the "official" system of Linnaean classification. Each category in the Linnaean system forms part of a nested hierarchy, where each category is divided into ever more exclusive categories. By agreement, these categories are, from most inclusive to most exclusive: phylum, class, order, family, genus, species. Biology students learn various mnemonic devices to keep the order of these categories straight. Unfortunately, every one I know is rude and vulgar, so I cannot share them with you here.

Phyla can be differentiated from one another by the body plans of the organisms they include. For example, most of the animal phyla, known as the triploblasts, have organs derived from three so-called germ layers: an endoderm ("inside skin"), lining the interior of the gut; an ectoderm ("outside skin"), forming the outside skin; and, sandwiched in between, a mesoderm ("middle skin"), which gives rise to most everything else. At least one of the animal phyla, the coelenterates, has only two layers, an ectoderm and an endoderm. These animals, called diploblasts, obviously differ fundamentally from the triploblasts in their body plans.

A body plan is a phenomenon of embryological development, and different body plans represent different sequences of growth from a single-celled fertilized egg, or zygote, to the adult. For example, among the triploblasts, two major body plans are recognized, each based upon a fundamental difference in the development of the digestive tract. The digestive tract of most adult animals has two openings to the outside: a mouth, which takes food in, and, at the opposite end, an anus, from which the remnants of digestion leave. During the early stages of embryonic life, though, the digestive tract starts out as a closed sac, or archenteron (literally "ancient gut"), which opens to the outside through a single opening, the blastopore. To become a proper tube, a second opening must form between the body wall and archenteron, opposite the blastopore. These two openings arise sequentially during development, and the order of appearance offers two possibilities for the digestive tract's further development. The two great lines of evolutionary descent among the animals are distinguished by which sequence was followed. In the primitive condition, known as protostomy (literally "first mouth"), the blastopore becomes the mouth and the second becomes the anus. Protostomy is characteristic of most of the so-called higher invertebrate phyla, like the annelid worms, molluscs, and the group that is arguably the pinnacle of animal evolution, the arthropods (crustaceans, spiders and their allies, and the insects). The more recent condition, derived probably from a protostomous ancestor, turns this arrangement on its head, so to speak. The blastopore becomes the anus and the second opening becomes the mouth in the condition known as deuterostomy (literally "second mouth"). The deuterostomes include the chordates, the echinoderms, and one or two other minor invertebrate phyla.

Body plans arise from a sequence of developmental events controlled by a genetic developmental program. These programs ensure a body plan's integrity as it is passed to future generations, and it is through modification of these programs that body plans evolve. Let us briefly explore how.

First, a genetic developmental program modifies lines of descent of the many generations of cell division that are required to build a multicellular adult from a zygote. A line of descent, in this case, is similar in principle to the line of descent that gave rise to you: each cell in a line of descent has a parent cell, and if the cell itself reproduces, it will produce one or more daughter cells. A particular line of descent is traced sequentially through the parent and daughter cells of several generations, and all lines of descent form a sort of "family tree" of development. During embryonic growth, lines of descent are modified in three basic ways. These are: growth, or rates of cell division; differentiation, essentially the activation of particular genes so that cells in a particular line of descent specialize to do one or a few things well; and apoptosis, a programmed degeneration and death of particular cells and, by implication, whatever lines of descent that might have arisen from them.

Second, development operates by a sequence of contingent events. To become an adult, for example, each of us must successfully negotiate our way through childbirth, childhood, puberty, and adolescence. If our passage through any stage is unsuccessful or abnormal, the aberration will affect all subsequent stages of our lives. Suppose we designate, rather arbitrarily, the sequence of developmental events as follows:

$$Zygote \rightarrow A \rightarrow B \rightarrow C \rightarrow D \rightarrow E \rightarrow F \rightarrow \ldots \rightarrow Adult \qquad [5.1]$$

The successful completion of each step is contingent upon the successful completion of all the previous steps. The development of the vertebrate spinal cord, for example, requires the embryo to fold in on itself in a particular way. If this folding process does not occur, or does not occur successfully, normal spinal cord development will not occur, nor will all the subsequent events that depend upon normal spinal cord development. The folding of the embryo that precedes spinal cord development in turn relies on a prior step involving migration of particular cells to particular locales in the embryo.

The evolution of the various animal phyla is largely a matter of how these developmental programs have been modified through lines of evolutionary descent. Generally, modifications arise from mutations either in the genes controlling the fates of particular lines of descent of cells within the embryo or in the genes that control the contingent events.[2] For example, suppose a mutation arises that controls the transition $C \rightarrow D$ in the developmental sequence in equation 5.1. The new developmental program will now look like this:

$$Zygote \rightarrow A \rightarrow B \rightarrow C \rightarrow D' \rightarrow E' \rightarrow F' \rightarrow \ldots \rightarrow Adult' \qquad [5.2]$$

with the mutation at one step affecting the development of all subsequent steps. The mutation has led to a branching of the evolutionary line of descent: the lin-

2. The genes that control major developmental events are known as homeotic genes. Mutations in homeotic genes can sometimes result in radically different body plans. Animal body plans were thought to arise from the appearance of these homeotic genes, and indeed, homeotic genes of distantly related organisms are very similar, indicating common origins. For example, fruit flies and humans share five of these homeotic genes, indicating that our common ancestor, something a lot like a rotifer or a bryozoan, had them too.

eage of the common ancestor, *Adult*, continues through those individuals that lack the mutation, while the descendants of individuals with the mutation form the new lineage *Adult'*. Generally, the earlier in embryogenesis the developmental mutation occurs, the more dramatic the change of body plan will be (Fig. 5.2). The radically different body plans of the protostomes and deuterostomes, for example, arise from differences that appear very early in embryonic development. Modifications that occur later in the program result in less dramatic differences. Differences between some breeds of dogs, for example, result from modifications in the growth rates of the bones of the skull, when bone growth ceases, and so forth, and these events occur fairly late in development.

Body Plans of Sponges

The concept of the body plan becomes a little shifty, though, when it is applied to the organisms that presumably are at the roots of the animal kingdom. The sponges, for example, members of the phylum

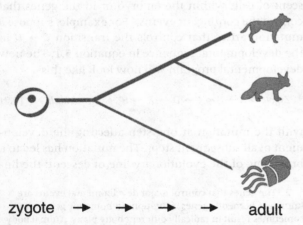

zygote → → → → adult

Figure 5.2 Major differences in body plan, like those between the nautilus and dog, imply a divergence early during the development of the adult from the zygote. Animals with more similar body plans, like the fox and the dog, develop along the same pathway until late in development. The divergence of body plan frequently reflects an organism's evolutionary history.

Porifera, are considered the most primitive of the animals, just one step up from colonial Protozoa, perhaps even super-colonial Protozoa. Some zoologists actually put them into a unique grouping called the Parazoa (literally, "beside the animals") rather than with the animals proper, the Metazoa ("higher animals"). It is only through mutual agreement and long-standing convention rather than good evidence that we currently lump the sponges in with the animals.

One of the problems in deciding just where the sponges' affinities lie is identifying just what a sponge's body plan is. Compared with the more tightly constrained development of the Metazoa, the rules of development for sponges are, shall we say, flexible. Development in a sponge starts conventionally enough, with a single-celled zygote giving rise to multiple types of cells (Fig. 5.3). Unlike in more complex animals, where differentiation results in dozens or hundreds of specialized cell types, in sponges cells differentiate into just three or four: flat pinacocytes that form a protective sheet on the outside; archaeocytes that migrate around the body of the sponge like amoebas, laying down tendrils of fibrous spongin or secreting mineralized bodies known as spicules; and choanocytes, cells that use flagella to drive water currents through the sponge. These cells organize themselves, according to certain rules of association, into a hollow tube pierced by numerous pores that connect the outside of the animal with a hollow interior cavity, the spongocoel. Water flows into the spongocoel through the many openings in the side, called ostia, and out through a single opening, the osculum, at the top. As the water flows through the sponge, suspended particles of food are captured by the choanocytes.

It is questionable whether this organization strictly qualifies as a body plan. On the one hand, certain families of sponges have characteristic body forms, implying that developmental programs do shape their development. The asconoid sponges, for example, are organized into simple columnar tubes. In contrast, the leuconoid sponges typically are spherical, with the spongocoel reduced and the ostia arranged in ramify-

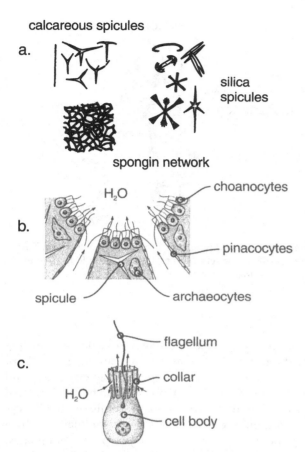

calcareous spicules

a.

silica spicules

spongin network

H₂O — choanocytes

b.

pinacocytes

spicule — archaeocytes

flagellum

c.

H₂O — collar

cell body

Figure 5.3 The components of sponges. *a:* Connective elements. Spicules are mineralized connective elements made either of calcite or silica. Certain sponges lack spicules but are held together by networks of a fibrous protein, spongin. [*From Storer et al. (1979)*] *b:* Cell types. *c:* The choanocyte filters microorganisms and other food items from the stream of water driven through its collar by the flagellum. Solid arrows (on collar of choanocyte) indicate direction of movement of captured items of food [*b and c from Hickman and Hickman (1992)*]

ing networks of vessels that permeate the sponge's "body." Nevertheless, it still is hard to say that the shape of a sponge's "body" results from a developmental process typical of a metazoan animal, with its complicated series of programmed growth, differentiations, and contingent developmental events. The difference is illustrated nicely by the ways animals and sponges recover from a disruption of development. Take a sponge, scramble it into its constituent cells, leave the disordered mess in a dish of seawater, and after a few days its cells will reassociate in particular ways, as if Humpty-Dumpty put himself back together again. Animals can do something similar: cut the limb off a developing chick, for example, and soon a new limb will regenerate. The similarity is only superficial, however: re-association of sponge cells is not regeneration. In the sponge, the very cells that were separated from their neighbors soon find the same or similar neighbors to reassociate with. In limb regeneration, the developmental process of limb formation is done all over again, but with a whole new set of cells going through the same process of growth and differentiation as the original limb cells underwent.

Among animals, development is tightly constrained by genetic programs, but not completely so: the embryo's environment during development also has effects. In the jargon of developmental biology, body form is the result of both genetic (inherent) and epigenetic (environmental) factors. Among the Metazoa, it is clearly the genetic control that dominates, as is reflected in the high degree of fidelity of the body plan: a dog's offspring will always be puppies, recognizable as puppies irrespective of what environment they develop in. Environmental influences surely play a role: a period of starvation during development may make the body smaller or more slender, or it may alter the way the brain develops, but even dog embryos starved during development still are born as puppies and grow into dogs.

Among sponges, however, it seems to be the epigenetic factors that are dominant, while the genetic control is relatively weak. For example, the basic growth form of sponges of the genus *Haliclona* is a ramifying series of the sieve-like tubular elements typical of the asconoid sponges. However, the growth forms of *Haliclona* vary widely with the types of water flow they experience as they grow. When *Haliclona* grow in still waters, the branches are long, thin, and few in number. When *Haliclona* grow in flowing water, the branches are short, squat, and numerous, at the extreme merging into the stout growth form typical of a leuconoid sponge. One can even change an individual sponge's body form simply by moving it from one habitat to the other. Indeed, what form a sponge takes seems to be less strongly determined by the kind of sponge it is (by whatever genetic legacy directs its development) than by the environment in which it lives.

Body Plans of Coelenterates and Corals

In contrast to the sponges, coelenterates are unambiguously animals. As such, genetic factors assert themselves more strongly during development than they do in sponges. Like sponges, coelenterates start as a simple cell and grow into a hollow tube closed at one end (Fig. 5.4). Development of the digestive tract stops at the archenteron stage (although it is called the coelenteron in the adult). The coelenteron opens to the outside through a single opening, which serves as both mouth and anus, and is surrounded by a two-layered body wall. The inside layer of cells, lining the coelenteron, is the endoderm, while the one facing outward is the ectoderm.

Unlike sponges, in which the "body plan" of one type seems to merge almost imperceptibly into the "body plans" of other types, the coelenterates can be differentiated pretty reliably by their body plans. For example, coelenterates exhibit a life history known as alternation of generations. The basic diploblastic body plan of coelenterates exists in two forms: an asexual polyp, typically sedentary, frequently nonreproductive or reproducing by budding of the parent animal; and a sexual medusa, a swimming form with sexual organs. The life history of most coelenterates alter-

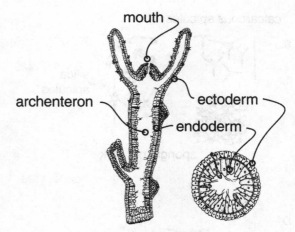

Figure 5.4 *a:* The diploblastic organization of a *Hydra*. Two layers—the outer ectoderm and the inner endoderm—surround the closed primitive gut, the archenteron. [*From Storer et al. (1979)*]

nates between these two stages: sexually reproducing medusae give rise to embryos that develop into polyps, which grow and from which medusae eventually bud again. The classes of coelenterates can be differentiated by which of these two life stages dominates the life cycle. Among the Anthozoa ("flower animals"), which include the sea anemones, the polyp stage dominates, and the medusa stage is relegated to fairly insignificant animals, in some species being lost altogether. Among the Scyphozoa ("cup animals"), which include the free-swimming jellyfish, the dominant stage is the medusa. These forms exist even in the face of wide variation in environmental conditions, suggesting the genetic factors in development are relatively robust in comparison with the epigenetic factors.

Coelenterates have not been entirely tamed by the genetic bridle, though. Some coelenterates, most notably the corals, exist as colonial organisms whose "body" is made up of a more or less obligatory association of individual polyps. Development of the individual polyps might be under fairly tight genetic control, but the development of their association seems to be less tightly constrained. Thus, one sees a very strong environmental component to the growth forms of corals, similar to the pattern seen in sponges, even if the individual polyps retain characteristic patterns of growth. Corals like *Millepora*, for example, ramify as they grow, resulting in arborescent forms reminiscent of the sponge *Haliclona*. Just as in *Haliclona*, the corals residing in still waters develop long, thin branches, while those in rougher waters are stouter, squatter, and more highly branched. This is a common feature of many types of corals, and again, it seems that epigenetic factors dominate the morphology of the association, if not the individuals that constitute it.

Both sponges and corals, then, build structures that seem to arise without undue complicating influence of genetic constraints on development. They seem to fit the bill nicely for an examination of how metabolic energy interacting with energy in the environment can result in permanent, tangible structures—edifices, furthermore, that do physiological work. Admittedly, we are stretching the definition of *structure* a bit far for the sponges: when you look at a sponge, are you seeing an animal's body, or are you seeing a structure built by a colony of cells? I'm simply going to adopt the latter view and ask your indulgence: I really have no good reason for doing so other than it helps me make my argument.

Modular Growth and Fractal Geometry

Sponges and corals experience a type of growth, modular growth, that is more similar to the growth pattern of trees than that of animals. Modular growth, as the name implies, is the successive addition of identical or similar modular units to an existing organism. The structure these organisms take on as they grow is determined mainly by how fast and where new modules can be added. The process is easiest to illustrate with corals (Fig. 5.5). Each polyp secretes beneath it a layer of calcite, which accumulates as a roughly columnar

structure known as a corallite. A polyp together with its associated corallite form a modular unit known as a zooid. A coral "animal" consists of all the polyps and their associated corallites; in other words, it exists as a collection of the modular zooids. Growth of the coral occurs either by growth of a zooid, usually by the extension of the corallite, or by multiplication of the zooids.

Among sponges, modular growth arises because the spicules of sponges often are organized into polygonal

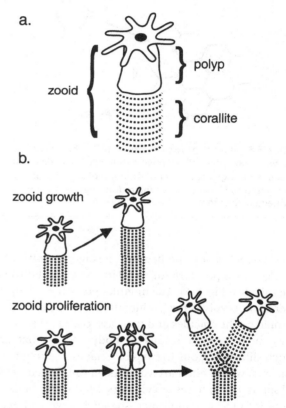

Figure 5.5 The organization and basic pattern of growth of a coral. *a:* The zooid consists of the polyp and the crystalline corallite. *b:* Zooids grow by simple accretion of new layers of calcite onto the corallite. Should the polyp divide, two new zooids are formed, resulting in a branch.

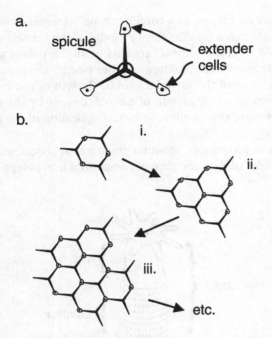

a.

spicule

extender cells

b.

i.

ii.

iii.

etc.

Figure 5.6 Modular growth in a sponge. *a:* The spicule grows as extender cells deposit either calcite or silica at its tips. *b:* A sponge grows by proliferation of spicules and their associated cells, which combine to form networks of interconnected spicules.

arrays—rectangles, pentagons, hexagons—with the sponge's cells proliferating to fill in the spaces between the spicules (Fig. 5.6). The modular growth of sponges, therefore, involves the proliferation of the unit arrays formed by the spicules and their associated cells.

Modular growth results in body shapes that are much different from the typical shapes of higher organisms, which often are variants on simple geometrical solids, like spheres or cylinders. Modular organisms grow into more complicated shapes known as fractal objects, so called because the structures of their bodies exhibit fractal geometry. Fractal geometry is a very powerful way of describing the complex shapes frequently found in nature. Understanding the growth forms of sponges and corals requires that we come to

grips with it. Like most powerful tools, fractal geometry is superficially intimidating, but it is in principle very simple.

The basics of fractal geometry are most easily appreciated by considering a common task: measuring the length of a curve. One might estimate a curve's length by "walking" a pair of dividers (or some other measuring device) along it and counting the number of steps required to span the curve (Fig. 5.7). The curve's estimated length is simply the product of the distance separating the divider's points and the number of steps. This procedure yields only a rough approximation of the curve's length, though, because dividing the curve into a series of straight-line segments leaves out the "curvy bits" that contribute some portion of its true length. To improve the approximation, one moves the points of the divider closer and repeats the measure-

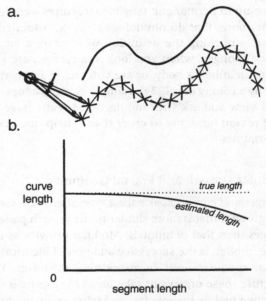

a.

b.

curve length

true length

estimated length

0

segment length

Figure 5.7 Estimating the length of a smooth curve. *a:* Dividing the curve into a series of straight-line segments. *b:* With shorter and shorter segment lengths, the estimated length of the curve asymptotically approaches the curve's true length.

ment (Fig. 5.7). If one keeps repeating the estimate, moving the divider's points closer together each time, the estimated length will converge onto a particular value, which is the curve's true length. Indeed, the exact length of the curve, and not merely a good approximation, can be had essentially by setting the points of the divider infinitesimally close. Those who remember their calculus will recognize this procedure as taking the limit of the estimate: in the jargon of the calculus, the curve is differentiable.

In traditional geometry, every differentiable object has a dimension, which we designate with a number D. A line or a curve, for example, has only a dimension of length, so $D = 1$. Planes or surfaces, on the other hand, because they have an area, have dimension of length squared (l^2), which we can express as a dimension of $D = 2$. Solid objects, because they have dimensions of length cubed (l^3), have a dimension of $D = 3$. Obviously, D is the power of the length dimension needed to describe the object.

The dimensions of simple curves, planes, and shapes are easy to grasp. Fractal objects require us to stretch the concept of a dimension a bit. You will have noticed that the dimensions of lines, surfaces, or volumes are, respectively, 1, 2, and 3—that is, they are whole numbers, or integers. Get ready to stretch, now: fractal objects, in contrast, are objects whose dimensions are not integers but fractions. This is a tricky concept to grasp, but again it is easily illustrated by the problem of how one measures the length of something.

This time, consider a curve found in nature—the usual example in most introductions to fractal geometry is a coastline, say of Papua New Guinea (Fig. 5.8). You can measure the coastline's length as we measured the differentiable curve's, by dividing it up into a series of linear segments and multiplying the segments' lengths (corrected, of course, for the scale of the map) by the number of lengths needed to span the coastline. You can take these measurements with increasing degrees of precision, either by altering the distance between the divider's points or by employing several maps drawn at smaller and smaller scales, get-

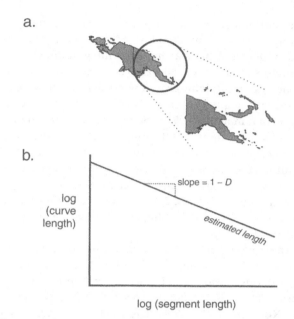

Figure 5.8 Estimating the length of a fractal curve. *a:* A map of a coastline reveals ever more finely detailed curves as its magnification is increased. *b:* Unlike a smooth curve, a fractal curve's estimated length increases continuously and without bound as segment length decreases.

ting, as we did before, increasingly fine estimates of the length. Once you run out of maps of sufficiently fine scale, you can go down to the shore with a surveyor's chain and keep repeating the measurements on still finer scales, all the way down to the level of individual grains of sand if we so chose. Plotting the estimate of coastline length against step length, you will see a remarkable result—no matter how fine the scale used for our measurement, no matter how infinitesimally small the lengths, your estimate will never approach a "true" value, as our previous example did for the length of the simple curve. Rather, the estimated length just keeps increasing *ad infinitum*, never converging on a "true" value (Fig. 5.8).

This inability to measure the true length of a natural shape like a coastline (or a leaf margin, or the lining of the lung, or . . . I could go on forever) leads us to the central idea in fractal geometry. The fact that you cannot arrive at the coastline's "true" length means that it does not have a simple or integral dimension. Structures that have well-behaved dimensions, like 1, 2, or 3, are, by definition, differentiable, and so it is possible to apply the methods of the calculus to estimate their magnitude. The inability to define a length's "true" value means that the dimension of the structure or object is not an integral number, that is, it is neither a curve ($D = 1$) nor a surface ($D = 2$) but something in between. It has, in fact, a fractional dimension (or fractal), where $1 < D < 2$. In fact, most coastlines have fractional dimensions D that fall somewhere between 1.2 and 1.3.

The connection between measuring coastlines and the growth of sponges and corals is that modular growth often results in a fractal object. The similarity is easily illustrated by one of the simplest fractal objects, the Koch curve (Fig. 5.9). The Koch curve starts as a simple triangle, obviously with three corners, or apices. The curve is generated by repeated addition of a triangle in the middle of each of the original triangle's legs. With each step, the curve grows into an elaborately branched structure. The analogy with modular organisms is obvious. Anything that grows by repeated addition of some sort of modular unit will end up generating a fractal object of some sort. Indeed, the growth of sponges and corals has been simulated by computer models that use fractal geometry: the results are startlingly lifelike.

Accretion and Modular Growth

Modular growth of sponges and corals is a process akin to accretion, which is simply growth by addition of new material to an existing surface. The most familiar example of accretive growth is crystallization. A crystal of table salt (NaCl), for example, grows by the addition of sodium and chloride ions—the "modules" in this case—to an existing surface of a sodium chloride crystal. This surface serves as a faithful template for future growth: the arrangement of atoms in the crystal provides "niches" for the addition of new ions. Through the repetition of this process—niches open up and then are filled—the crystal grows. Because the arrangement of the niches arises from the arrangement of sodium and chloride atoms in the existing crystal, the shape of the crystal is maintained as it grows: NaCl crystals, no matter what their size, are always cubic.

Sponges and corals must have a different type of accretive growth, though: as we have seen, their shapes are variable and are strongly influenced by the conditions in which growth occurs. Why should the process of accretive growth in one case, crystallization, preserve shape, while in another, the growth of a sponge or coral, it does not?

Accretive growth is a two-step process. First, material must be delivered to the surface from the solution in which it is immersed. In the case of a sodium chloride crystal, sodium and chloride ions must travel from the solution to the surface of the growing crystal. Second, once the material is at the surface, it must find its appropriate "niche" there and settle in. Each step affects the shape of a growing surface differently; depending upon which step is more easily accomplished, accreting surfaces will either maintain their shapes as they grow or not. When the settling-in time of new material is slower than the rate at which new material can be delivered, accretion growth preserves shape. When delivery rate is slower than the settling-in time,

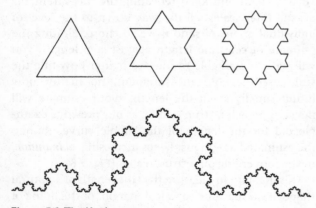

Figure 5.9 The Koch curve.

however, the shape of an accreting surface will be influenced by conditions that affect the delivery rate. Areas that experience more rapid delivery will grow faster than areas that do not, resulting in a change of shape of the growing surface. The variable morphology of sponges and corals suggests that their growth is dominated by limitations on delivery of new material to the growing modules. To understand their growth, we must then understand what these limitations are.

What's Different about Gradients

The rate of delivery of material to a growing surface is ultimately determined by the process of diffusion, already encountered in an informal way in Chapter 4. Let us examine the concept formally now, for understanding diffusion is the key to understanding the growth forms of accretive growers like sponges and corals.

Diffusion is the spontaneous movement of matter down a concentration gradient. In textbooks, diffusion usually is illustrated by a box separated into two compartments by a permeable barrier (Fig. 5.10): let us designate them with the Roman numerals I and II. Each compartment contains some substance that is capable of moving by diffusion, either molecules in solution or components in a mixture of gases. Each compartment will therefore contain some concentration of this substance, C. If the concentration in compartment I differs from that in compartment II, there will be a concentration difference, $C_I - C_{II}$.

The concentration difference is essentially a potential energy difference, and as such it is capable of doing work. Work is done when the substance is moved across the barrier separating the compartments. For example, if compartment I is richer in oxygen than compartment II, oxygen molecules will move spontaneously from I to II. Because work is proportional to the potential energy driving it, we can express a proportionality:

$$J \propto -(C_I - C_{II}) \qquad [5.3]$$

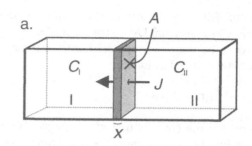

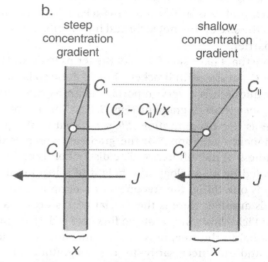

Figure 5.10 The distinction between concentration difference and concentration gradient. *a:* The standard model of diffusion involves a flux, *J*, of a substance across a barrier of thickness *x* and cross-sectional area *A* between two compartments, I and II. *b:* Concentration gradient, $(C_I - C_{II})/x$, can be altered independently of the concentration difference simply by changing the boundary's thickness.

Obviously, the flux, *J*, varies directly with the concentration difference. Note that *J* can be either positive or negative: the sign of the number simply indicates the direction of the flux, whether it is moving from compartment I to II (if C_I is greater than C_{II}) or *vice versa* (if C_I is less than C_{II}). Note as well that there will be no flux if the concentration difference is *nil* (which happens to be the state of maximum disorder).

The relationship between concentration difference

and flux driven by diffusion can be expressed in one of nature's fundamental laws, Fick's law of diffusion. For the case of the box separated into compartments, Fick's law is stated as follows:

$$J = -DA[(C_I - C_{II})/x] \qquad [5.4]$$

where A is the surface area of the permeable barrier separating the compartments, x is the thickness of the barrier, and D is a diffusion coefficient that depends upon the size of the molecule and how it interacts with the barrier.[3]

Note that I have set off two of the terms in Fick's law, $(C_I - C_{II})$ and x, with brackets. I have done so because their quotient is a very important quantity known as the concentration gradient (Fig. 5.10). This is not the same as the concentration *difference* used in the proportionality equation 5.3: the gradient expresses the steepness of the difference. The distinction is easily illustrated by an analogy: the height of a loading platform is one thing, the steepness of a ramp leading up to it is another. Here is the crucial part, so read carefully: Fick's law states that the flux rate is directly proportional to the steepness of the concentration gradient, and not necessarily to the magnitude of the concentration difference. Indeed, one can alter the flux independently of the concentration difference simply by altering the thickness of the barrier separating the two compartments. Make the barrier thin (steepen the concentration gradient) and flux increases, even if the concentration difference is unchanged.

The distinction between a gradient and a difference offers us a clue about how the shape of an accreting surface can change as it grows. At the heart of this mechanism is a simple principle: in a system where diffusion limits accretive growth, local rates of growth will be proportional to the local concentration *gradient* driving flux of the diffusing material to the surface. This principle can be developed into a general model for accretive growth, known as diffusion-limited accretion (DLA), which is particularly suited to modeling the growth of modular organisms. Let us now flesh out this model with an example.

Assume that a surface, say a layer of coral polyps, grows by accretion of some substance, say calcite, and that the delivery of new material is slower than the accretion rate (the "settling-in time"). In other words, diffusion rate limits the growth by accretion. To understand growth by diffusion-limited accretion, we need to understand in some detail the movements of calcium through the liquid in contact with the accreting surface. Imagine a thin layer of solution above the surface (Fig. 5.11). As calcite accretes to the surface, calcium moves from solution to the surface at a rate $J_{sol \rightarrow surf}$. As calcium leaves the thin layer of solution, its concentration there declines. This sets up a concentration difference between the thin layer of solution adjacent to the surface and the layer above it, and this difference drives a flux, J_{diff}, downward into the layer of liquid adjacent to the surface. At equilibrium, the concentration of calcium in the surface layer will settle down to some low value, forming an unstirred layer of calcium-depleted liquid at the surface. This unstirred layer forms a diffusion barrier for the movement of calcium from solution to the surface.

The picture changes if we now allow water to flow very smoothly over the surface. When water flows over a stationary surface, it slows down and forms a characteristic velocity profile known as a boundary layer. The fluxes of calcium across a boundary layer are substantially different from those across an unstirred layer. When water flows over the surface, it is as if a tile of water were allowed to reside over the surface for a while, and while "in residence" it deposits some quantity of calcium. Once the calcium in this tile is depleted, it is pushed out of the way and replaced by another tile, which brings in a fresh load of calcium.

---

3. Fick's law is more generally (and properly) expressed as a differential equation: $dJ = -DdC/dx$. This be solved for various geometric configurations—for example, across a flat barrier with thickness x and surface area A. The two-compartment model outlined in the text is the simplest, where flux is uniform and in one direction. Diffusion in different configurations, such as diffusion through tubes or in open space, can be solved by integrating the general form of Fick's law.

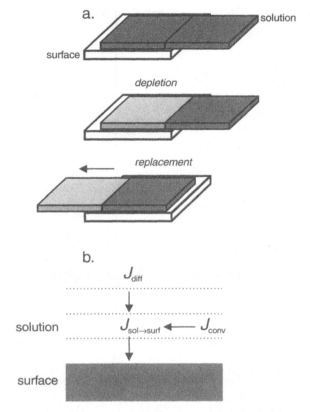

Figure 5.11 Diffusion-limited fluxes at an accreting surface. *a:* Material in a thin surface layer (dark shading) is depleted the longer it sits adjacent to the surface (light shading). Convection is analogous to sliding the depleted layer away and replacing it with a fresh layer. *b:* The concentration of material at an accreting surface is therefore determined by three fluxes: $J_{sol \to surf}$, the flux of calcium from solution to the surface; J_{diff}, the flux driven by difference in concentration in different layers of the liquid; and J_{conv}, the flux driven by convection.

This process is represented by an additional flux coming in from the side, a convection flux, J_{conv} (Fig. 5.11). Concentrations of calcium at the surface will, on average, be higher in flowing solution than they would be in an unstirred layer, and the greater concentration will increase the rate of deposition of calcium to the surface. In the end, concentrations of calcium will

form a steady-state boundary layer at the surface. Because J_{conv} is proportional to the flow rate, the concentration boundary layer will parallel the boundary layer's velocity profile (Fig. 5.12). Speeding up the flow, and therefore increasing the velocity boundary layer, will in turn make the concentration gradient in calcium through the boundary layer steeper. In accordance with Fick's law, the higher concentration gradient will sustain a higher accretion flux of calcium to the surface, and the surface will grow more quickly. The opposite sequence of events applies to slowing the velocity.

So far, this model is fairly uninteresting, because the ideal conditions I have posited (perfectly smooth surface, well-behaved flows and concentration differences) will promote only uniform growth at the surface. Nothing has yet been presented that would promote a *change* of shape of the surface, which is what needs explaining. Things become interesting, however, when we look at what must happen at a surface imperfection that elevates some parts of the surface above their surroundings.

Describing what happens is easiest if flows over the surface are depicted with streamlines. A streamline simply graphs the trajectory followed by a small parcel of water as it flows. Streamlines themselves do not indicate the parcel's velocity. For example, streamlines above a perfectly smooth surface will form an array of parallel, evenly spaced lines (Fig. 5.13), even though

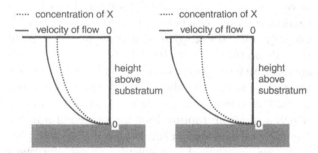

Figure 5.12 Velocity and concentration boundary layers at an accreting surface. With slow flow *(left)*, the velocity and concentration profiles are similar. If flow increases *(right)*, the concentration profile becomes steeper as well.

the velocity in the boundary layer changes with height (Fig 5.12). Streamlines sometimes indicate *changes* of velocity, however. If we introduce an imperfection into the surface, let us say a hemispheric bump, the streamlines will be crowded together above it (Fig 5.13). This crowding together indicates that the velocity of water above the bump has increased, in just the same way velocity of water in a hose is increased when it passes a partial blockage, like a sprinkler nozzle.

Streamlines packed together in this way indicates that the velocity boundary layer above the bump has steepened, and if accretion is happening, the concentration boundary layer will have steepened as well. You can now see the really interesting part: accretion growth will be fastest at the bump, because the concentration gradient driving accretion flux will be steepest there. This has the interesting consequence of magnifying small imperfections in an accreting surface (Fig. 5.13). In short, diffusion-limited accretion changes the shape of an accreting surface by amplifying existing imperfections of shape.

Diffusion-Limited Accretion in Sponges and Corals

DLA growth models have been applied to all sorts of fractal growth systems, ranging from the growth of crystals to development of lightning bolts. They have also been used to explain the patterns of growth and form among the sponges and corals, but with an interesting twist. Obviously, accretive growth in, say, a coral is not simply a matter of crystalline enlargement: as we saw in Chapter 2, the deposition of calcite in a corallite requires metabolic energy in the form of ATP. Consequently, the growth of a corallite will depend not only on the rate at which calcium ions and carbonate ions can be delivered to the calcioblast, but also on the rate at which metabolic energy can be delivered there. This rate, in turn, is determined by the rate at which a coral can capture food from its environment. The situation is similar in a growing sponge, because the addition of new spicules or spongin fibers and the cells to fill the spaces between them also requires energy. Although Fick's law of diffusion does

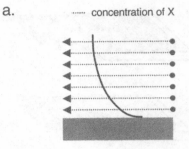

a. ---- concentration of X

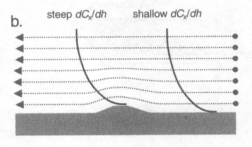

b. steep dC_x/dh shallow dC_x/dh

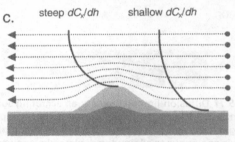

c. steep dC_x/dh shallow dC_x/dh

Figure 5.13 Magnification of surface imperfections at an accreting surface. *a:* In a flow field with uniformly spaced streamlines, the concentration boundary layer is uniform over the entire surface. *b:* A surface imperfection, like a small hump, compacts the streamlines over the surface, and the concentration gradients (represented by the differential expression dC_x/dh, where h = height) at the hump are steepened locally. *c:* The locally-steep boundary layers accelerate accretion at the imperfection.

not strictly apply, because capture of food particles by a filter feeder like a coral polyp is not really diffusion, the two processes are similar enough that a DLA model should still work. Remember the justification for the electrical analogy in Chapter 3—most rates of energy transfer operate through fundamentally similar processes. Fick's law is, at root, an equation that describes how potential energy does work.

Let us see how one of the common morphological features of sponges and corals, a branch, can be explained by a DLA growth model. Suppose that on a surface comprising several modules there is a slight imperfection. That is, the surface will consist of areas (bumps) that are elevated above other areas (valleys). Because of the steeper "concentration gradients" in nutrients that develop above the bumps, the modules on the bumps will capture more energy and power more growth than will the modules located lower down, in the valleys. The bumps will grow, magnifying the difference between their position and that of the valleys. If a growing bump gets large enough, imperfections in *it* may become large enough to initiate growth of new branches. The repetition of this process results in the development of the arborescent form common among sponges and corals.

The variation of form seen between sponges and corals inhabiting still water and those inhabiting flowing water also can be explained by a DLA growth model. Cylindrical bodies grow in two dimensions. Longitudinal growth at the ends elongates the cylinder. Radial growth at the outer surfaces makes the cylinder wider. What form the cylinder takes as it grows, either long and thin or short and squat, depends upon the relative rates of growth of these two dimensions. Obviously, if the rate of longitudinal growth is greater than the radial growth rate, a cylinder will grow to be long and thin.

A cylinder standing upright in a boundary layer will distort the streamlines of flow in two dimensions, vertical and horizontal. Distortion in the vertical dimension, as has already been illustrated, will promote elongation of the cylinder. Distortion of flow in the horizontal dimension will promote radial growth of the cylinder. In an environment with no flow, or slow flow, the greatest disturbance of the streamlines will occur at the highest points on the sponge, and elongation growth will be favored. The result will be a long and spindly sponge. At higher flow velocities, the boundary layer is thinner and the concentration gradients are steeper. Nutritive water will be brought to points lower on the cylinder, promoting their growth as well as growth at the tip. The result will be a shorter, squatter morphology.

Architecture and Physiology in Sponges and Corals

The time has now come to see whether I have successfully bridged the gap between bioconvection and the solid structures of corals and sponges. To reprise the problem, I asserted in Chapter 4 that bioconvection cells represent a primitive type of structure, one that arises spontaneously from the interaction of a metabolic energy stream with a large-scale environmental gradient in potential energy. I suggested this model might be an organizing principle that would carry over to other, more tangible structures that animals build. In other words, external physiology of the type exemplified by bioconvection should be a common feature of life, animal life included, and perhaps it should be the apparent absence of an external physiology among certain animals that is the special case, not its existence.

The weak points of this argument are three: that there is a fundamental similarity between bioconvection cells and a putative external physiology that might occur at the larger scales characteristic of animals; that this fundamental similarity can be made incarnate in the structures that animals build; and that the structures so built are capable of doing physiological work. Up to now, I have been concerned mainly with laying the foundation I need to shore up these weak points. Let us proceed, then, point by point.

Obviously, bioconvection cells and the growth

forms of sponges and corals are very different things, but they share a fundamental similarity: they are both the result of a modulated positive feedback process. I touched upon positive feedback briefly in the last chapter: I shall expand the topic now, and then turn to what a modulated positive feedback is.

As the name implies, any kind of feedback involves a feedback loop, a pathway that carries a flow of matter, information, or energy back on itself. Consider a technique popular in rock music, where a microphone (mounted, say, on a guitar) feeds into an amplifier that in turn drives a speaker (Fig. 5.14). If the microphone is placed near the speaker, sound energy will travel in a loop, from the speaker to the microphone, then to the amplifier, and back again to the speaker. You can see why it is called positive feedback—sound energy is amplified with every pass it makes through the loop.

Positive feedback also operates in the genesis of bioconvection cells and in the growth of corals and sponges. In the case of bioconvection cells, the positive feedback operates in the context of hydrodynamic focusing: the sinking of an anti-bubble draws microorganisms into it, which causes the anti-bubble's density to increase, which further increases the sinking rate, and so on. In the growth of corals and sponges, the positive feedback operates in the context of diffusion-limited aggregation: a slight elevation on a growing surface steepens the local boundary layer gradients driving accretion, which promotes local upward growth, which steepens the boundary layer gradients further, and so on. In both cases, a process (sinking of an anti-bubble, growth of a coral zooid) sets up the conditions for increasing its own rate of change.

Positive feedback has come to have a bad rap, and not just because it is frequently employed as a cheap trick to cover up a lack of musical talent. Anyone who has ever sat in an auditorium with poor acoustics has been subjected to the ear-splitting screech that results when someone on stage carries a microphone too close to the speakers. However, positive feedback can, under the right conditions, be a very creative and order-producing force. A skilled musician can use positive feedback to make an instrument emit sounds of

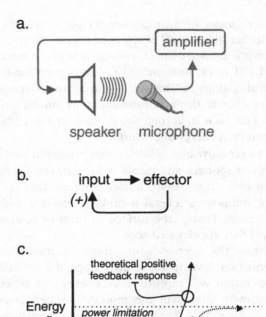

Figure 5.14 Positive feedback. *a:* A common positive feedback loop involves a microphone, an amplifier, and a speaker emitting sound back to the microphone. *b:* A schematic of a generic positive feedback loop: an effector increases the magnitude of the input to the effector. *c:* A positive feedback drives the energy flux rate through a system to increase exponentially. Theoretically, the energy flux rate increases without bound. In actual positive feedback systems, the response is limited by the power available to drive the system.

remarkable beauty. He does so by modulating the feedback loop so that only certain types of sounds travel through the loop and not others. As we shall see momentarily, a bioconvection cell arises from a similar modulation process. It is important, therefore, that we take a few moments to understand how a modulated feedback loop works.

The sound energy emitted by a guitar comes initially from a plucked string, which has a characteristic frequency of vibration. The string's vibration in turn sets the guitar's sound box vibrating at the same frequency, which in turn enhances the vibration of the string: the string and sound box together form a resonant system. The consequence of this mutual reinforcement is a louder sound than would be emitted by the string alone.

Resonance is not positive feedback, though, it is only a means of strengthening a particular frequency of vibration: a resonant vibration will die out as the energy fed initially into the string dissipates. Positive feedback occurs when an additional source of energy, in the form of the amplifier and speaker in our example of an electric guitar, is added to the resonant system. The presence of the resonant system (string and sound box) in the feedback loop now modulates what the positive feedback can do: sound can emerge from the speaker at any frequency, but only sounds at the guitar's resonant frequency will pass through the loop. Sounds at other frequencies will fail to set the string and sound box into resonant vibration, and the sound energy carried by those frequencies will dissipate away as heat: to introduce some jargon, these other frequencies have been "choked." Sound emerging from the speaker *at* the resonant frequency, however, will amplify the vibration of the sound box, which will reinforce the vibration of the string, which will feed more energy to the speakers, and so on. The result is an amplified pure tone, sustained by and limited in volume by the amplifier's ability to deliver power to the loop. We now see the high-pitched screech emerging from a simple microphone and public address system for what it really is: *un*modulated feedback resulting from the absence of any resonant system to choke the passage of most frequencies through the loop. The result is the indiscriminate amplification of multiple, and often dissonant, frequencies of sound.[4]

Modulated positive feedback also operates in the

genesis of bioconvection cells and in the growth of corals and sponges. Modulation is evident in the regular spacing of bioconvection cells in the culture. There is a resonance of sorts operating here, but this time it acts through the physical properties of the liquid culture—its viscosity, its density—and through the density, size, and shape of the microorganisms themselves. These properties promote the emergence of convection cells of a certain size and chokes the development of cells that are either larger or smaller than this favored size.

A similar process, albeit at a much larger scale, operates in the accretive growth of sponges and corals. In this case, the modulation is evident in the appearance of imperfections on the growing surface and not others. Corals and sponges commonly branch as they grow, and although different corals may branch in characteristic ways—at the tips or along the stems (Box 5A), say—the branches themselves frequently emerge at characteristic distances from one another. This implies that the growth of certain surface imperfections is favored over others, and that distance from a branch is somehow important in determining whether or not growth is favored. In all likelihood, the modulation arises from an interaction between a growing branch and the physical properties of the fluid in which it is immersed. A coral branch in flowing water will commonly generate turbulence in its wake, hiving off turbulent eddies at a frequency determined by its size and by the velocity, viscosity, and density of the water flowing past it (Fig. 5.15). Just as the peaks and troughs of a sound wave are separated by a particular distance,[5] so too are these eddies separated from one another by a characteristic distance. The eddies will interact most strongly with the surface behind the branch at this characteristic distance. This is why, for

---

4. In the case of a microphone and public address system, the properties of the electronic components tend to pass high frequencies preferentially, with a *high-pitched* screech being the result.

---

5. Specifically, this distance is the wavelength, λ, which can be calculated from the sound wave's frequency, f (cycles per second) and the speed of sound, c (in meters per second): $\lambda = c/f$. The wavelength of a middle C ($f = 880 \text{ s}^{-1}$) in air ($c = 330 \text{ m s}^{-1}$) is therefore 0.375 m; that is, every peak pressure in a sound wave is separated from the peak pressure in an adjacent wave by 37.5 cm.

How Corals Grow

Corals grow as calcite, secreted at the base of a polyp, which accumulates, usually in the form of a column. Growth proceeds according to a few simply defined rules, and these basic processes themselves may modulate the growth by diffusion-limited accretion (DLA), outlined in the text.

Many corals are autotrophic to varying degrees: they use photosynthesis by resident zooxanthellae as a source of energy to drive calcite deposition. Their growth is influenced not by one but by two potential energy gradients—gradients in nutrients brought in by flowing water and in light—having independent effects on growth. *Montastera* corals in shallow water, where light levels are high and the light is without a strong directional component, grow into a hemispherical shape. In deeper waters, where the direc-tion of the incoming light is mostly from above, *Montastera* will form stout columnar branches, because the upper surfaces receive the most light and, hence, nutrients from zooxanthellae. Upward growth is amplified by the boundary layer effects described in the text. In shaded areas, or in deep water where light does not penetrate, the greatest input of nutrients will be at the edge of the coral, where the polyps face into the flow. Lateral growth of the coral into a horizontal plate will therefore be favored.

A few rudimentary growth models can explain a lot of the diversity of growth types among corals. One of the more fruitful mimics the growth of trees, which also grow from the tips like corals. The analogy between trees and corals is strengthened by the fact that some corals move nutrients between polyps through connections called coenosarcs, in just the same way that trees move nutrients between its parts through the phloem and xylem. Tree biologists have formulated four basic models for the growth

Figure 5A.1 Schoute's model for arborescent growth of corals. *a:* A Schoute tree results when branching occurs at the tips of a growing branch. *b:* A representative Schoute tree: growth of the coral *Lobophyllia corymbosa.* [*After Dauget (1991)*]

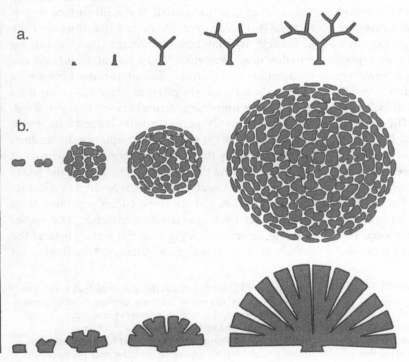

of trees, each named after the botanist who formulated it, and these seem to explain a considerable part of the variation of growth among corals.

The simplest is known as Corner's model and is characterized by simple linear growth at the tip. In a coral, the growing tip is the surface of the corallite, and the resulting structure is simply a column. Variation of shape among corals growing by Corner's model occurs when polyps detach from one corallite and set up another growth axis somewhere else. Corals growing by Corner's model typically exhibit little in the way of translocation of materials from one polyp to another.

Two of the growth models involve different rules for branching of a corallite as it grows. One, Schoute's model, relies on simple bifurcation of a growing tip (Fig. 5A.1). If a single polyp, after growing a corallite for some time, buds to produce two polyps, each of which starts to grow its own corallite, a bifurcated branch is formed. The result of repeated cycles of growth and budding is a ramifying structure. Schoute's model also applies to groups of polyps that form a sheet of growth. Fission of these groups into multiple zones of growth also results in ramifying structures, whose shape depends upon the ratio of the rates of longitudinal growth and budding (Fig. 5A.1).

The other branching model, known as Attim's model, initiates branching not at the tips but at the sides (Fig. 5A.2). This common mode of growth is made possible by the fact that coral polyps frequently form a sheet of living tissue that covers a columnar corallite. At some point, a lateral polyp might initiate budding, at which point a branch of growth will begin to extend laterally from the column. Repeated cycles of longitudinal growth and lateral budding will also produce, as in Schoute's model, an arborescent structure, but the interesting difference is in the kind of control needed to achieve the same end. Attim's model, because it relies on manipulating the growth and reproduction of the lateral polyps, involves some sort of mechanism that selectively represses or stimulates the growth of certain polyps. The same may occur in Schoute's model, but control of this type is not necessary, since a Schoute tree can result from simple growth and budding at the tip. As noted in the text, coordi-

a.

b.

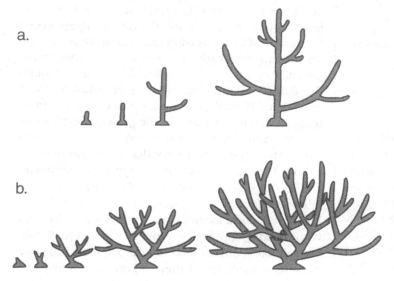

Figure 5A.2 Attim's model for arborescent growth of corals. *a:* An Attim tree results when branching occurs along the length of a growing branch. *b:* A representative Attim tree: growth of the coral *Acropora formosa*. [*After Dauget (1991)*]

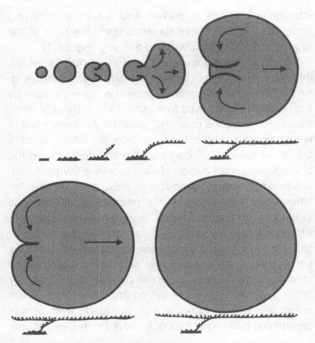

Figure 5A.3 Wood-Jones model for laminar growth of corals. The coral *Acropora hyacinthus* forms a flat, circular plate in the mature form. It grows by lateral addition of material to the plate, which eventually closes in on itself and forms a circular platform. [*After Dauget (1991)*]

nation of growth and reproduction between polyps is not necessary.

The last model is known to botanists as Aubreville's model, but coral biologists prefer to call it the Wood-Jones model. The three models discussed so far deal with growth in only one dimension, along the axis of a corallite. The Wood-Jones model considers growth in two axes, lateral and axial (Fig. 5A.3). In botany, Aubreville's model is frequently used to model the growth of leaves rather than stems and branches—a leaf can be thought of as a stem with a much higher rate of lateral growth than axial growth. The Wood-Jones model applies to corals that grow as massive flat plates, whose main extension is along the edges of the plate rather than along the vertical axis of a corallite column.

example, the ripples on a river bed or on a wind-blown sand dune are so regularly spaced (or for that matter why sand dunes themselves are regularly spaced). If a surface is growing by diffusion-limited accretion, the concentration gradients that feed accretion will be steepened the most at this characteristic distance, and growth at this location will be favored. Again, energy flow through the positive feedback loop driving accretion growth is modulated by the physical properties of the environment.

So I hope you are convinced that the process that generates bioconvection cells is fundamentally similar to that which drives accretive growth of sponges and corals. Both are modulated positive feedback systems. There are differences, of course, in what flows through the loops—in one it is gravitational energy driving convective flow, and in the other it is metabolic energy driving mineral transport and deposition. But at root, they are the same process.

If the processes that generate bioconvection cells and the growth forms of sponges and corals are similar, do the two structures that result similarly do physiological work? Again, I think the case is largely made. In the case of bioconvection cells, the work done driving the currents clearly does physiological work, transporting oxygen and carbon dioxide vertically through the culture at rates far faster than diffusion could carry them. Physiological work is also evident—admittedly, in a different way—in the growing branches of sponges and corals. Growth requires energy, and part of the physiological work that organisms must do to power growth is to capture energy. Most animals do this by moving their bodies, *a la* Willie Sutton, to where the food is. The process entails a host of physiological functions: locomotion, sensing where the food is, and taking appropriate actions to get there. Sessile animals like sponges and corals cannot move from one place to another, but if they cannot go to where the food is they at least can *grow* to where the food is. Thus, growth must be directed in a way that mimics locomotion. In the case of corals and sponges, growth toward food is ensured by the interaction between the

a.

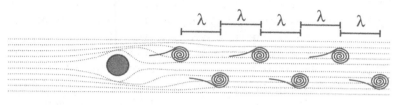

Figure 5.15 Modulation of eddy formation by viscosity. *a:* Behind an obstruction, eddies form at characteristic wavelengths (λ). *b:* Behind a growing surface, eddies will promote accretive growth at characteristic distances behind the obstruction (2λ).

b.

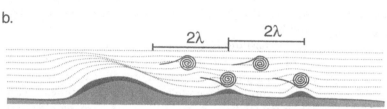

growth process and the flowing liquids that move nutrient-laden water past the organism.

How generally can these principles be extended beyond these two fairly limited examples? This question is going to occupy most of the rest of the book, so I will have to leave it largely unaddressed for now. I offer one example now, however—a teaser, if you will, that suggests strongly that external physiology is a pervasive phenomenon. Perhaps it will give you pause for thought.

The example concerns corals, which, you will recall, are fractal objects. Corals are commonly denizens of coastlines, which can be thought of as an interface between the nutrient pools of the open ocean and the organisms that live along its margins. For organisms to obtain these nutrients, the nutrients must cross this boundary. In this sense, a coastline is like the exchange barrier between, say, blood and air in the lung.

A lot of conventional physiology is concerned with exchange across boundaries, and physiologists often find themselves faced with the question: what are the structural attributes of a "good" exchange surface? If you refer back to Fick's law (equation 5.4), you can see in that equation one important "design principle" for good exchangers: make the ratio of the surface area, A, and the thickness of the diffusion barrier, x, as large

as possible. Fick's law deals with nice, differentiable exchange barriers, though. If a coastline is a fractal boundary, what can "fractal thinking" tell us about exchange of nutrients across it?

Consider a box separated into two compartments by a membrane, like the simple example used to illustrate Fick's law (Fig 5.10). In that example, I implicitly assumed that flux across the boundary was limited by diffusion. This is not the case for diffusion-limited accretion: flux is limited by a relatively slow movement of a molecule from the space above the membrane to the membrane itself. The "design principles" implied by Fick's law are of limited help here.

Here is where fractal thinking becomes useful. If movement to the boundary is limiting, the boundary's capabilities as an exchanger would be improved by putting a "kink" into it so that every molecule in both compartments is brought a little closer to the boundary (Fig 5.16). Just a single kink, though, still leaves some molecules farther from the boundary than others, so some inefficiency of design remains. This problem is easily got around, though, simply by doing it again: put an additional kink into the kinks you put in previously. In other words, you make the boundary grow as if it were a Koch curve (Fig 5.9).

You can probably see where this is leading.

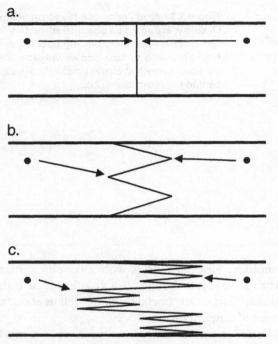

a.

b.

c.

Figure 5.16 Increasing the fractal dimension of a boundary to diffusion increases the flux rate across the boundary by decreasing the distance between the boundary and all points on either side of it.

Efficiency of the exchange between the compartments can be maximized by ensuring that every molecule in both compartments is brought as close as possible to the membrane separating them. This is done by repeatedly convoluting the membrane so it becomes a fractal boundary, forcing the dimension, D, of the cross-section to a value greater than one.[6] Now we have a different design principle: a good exchanger

6. Of course, real boundaries in the lungs are areas ($D = 2$), not curves ($D = 1$). Remember that we are dealing here with cross-sections, and that the cross-section of a surface is a curve. Convoluting an actual membrane in the way described in the text would increase its dimension from 2 (a surface) toward 3 (a volume). The cross-section would increase in dimension from 1 to 2.

is one whose fractal dimension is maximized. Still, however, dimension is limited by the fact that a curve traced by the boundary can never become an area— that is, D must be less than two. Maximum efficiency is achieved by making the fractal dimension of the boundary as close to two as possible. Indeed, in cross-sections of the highly efficient biological gas exchange membrane of the mammalian lung, the fractal dimension of the lungs' folded surfaces is between 1.9 and 2.0.

If a coastline is a boundary separating the nutrient pool of the ocean and the intertidal organisms living on its margin, how efficiently would exchange across this boundary operate? Ordinary coastlines, with their fractal dimensions of 1.2 to 1.4 or so, would seem to be not very efficient. When the coastline is fringed with a coral reef, however, the boundary becomes much rougher, and the coastline's fractal dimension increases, approaching 1.8 to 1.9 or so. This suggests that coral turns the coastlines it occupies into more efficient exchangers of nutrients and minerals than they would otherwise be. This seems to me to be external physiology on a grand scale.

What Happened with Animals?

I wanted to use this chapter to bolster the link between transient phenomena like bioconvection cells and more substantial animal-built structures, like coral reefs. Perhaps I have accomplished this, perhaps not. Nevertheless, I am now ready to launch into the "real biology" part of the book, discussions of animal-built structures that I believe are acting as external organs of physiology. Before doing so, however, I want to explore just one more question relating to the origin of animals and to what role external physiology might play in it.

I think most biologists would agree that radical differences in embryogenesis and development divide the metazoan animals from the problematic sponges and corals. Sponges' and corals' body forms are ruled by epigenetic factors. Development among animals seems much more strongly controlled by genetics, the epigenetic influences relegated to a relatively minor role. Why did this radical difference arise?

Much of the early diversification of animal body plans resulted from an increasing degree of what we might call "physiological in-sourcing." If we examine the simplest of organisms, we find that many of the major innovations in body plan exist to accommodate ever more complex internal organs of physiology. So, for example, sponges and coelenterates have no organs or organ systems whatsoever. The platyhelminthine (flat) worms possess simple organ systems for digestion and nervous function, but little else. Still higher animals are equipped with numerous and complex organ systems for water balance, digestion, circulation, gas exchange, and so forth. The development of each of these organ systems necessitated new body plans because the relatively simple bodies of coelenterates had to be folded in special ways to fit them in. Although anatomical details of folding patterns differ in the various body plans we know about, all the higher animals have fit these organs in, taking on board an increasing range of physiological function. What drove them to do this?

In all likelihood, the value of this increasingly internalized physiology was increased reliability and flexibility. External physiology has numerous advantages, among them that physiological work can be done at scales many times larger than the animals themselves. What makes this possible, of course, is the linkage between external physiology and positive feedback. If an animal positions itself in a large-scale gradient in physical potential energy, it can use positive feedback to tap enormous reservoirs of energy to do its physiological work (Fig. 5.17).

Along with the advantages of positive feedback go certain disadvantages, however. One of the biggest might have been that physiology powered in this way would have been either unreliable or inflexible or both. If, for example, you are powering external physiology with wind, and the wind dies from time to time, so does your physiology, and so, perhaps, do you. Also, if the positive feedback loop is modulated somehow,

any external physiology driven by it may be limited in the range of things it can do. Suppose, for example, protozoans could derive some hypothetical advantage from inhabiting smaller or more rapidly circulating bioconvection cells. If the size and circulation rate of the cells is limited by the water's viscosity and density, any advantages to smallness would be moot—physically, they couldn't exist.

Somehow or other, animals seem to have evolved toward a strong reliance on chemical potential energy to power their physiology. Why this happened is uncertain, but several possible reasons come to mind. Perhaps glucose is a more predictable fuel. Its availability might vary, but it is easy to store in relatively stable

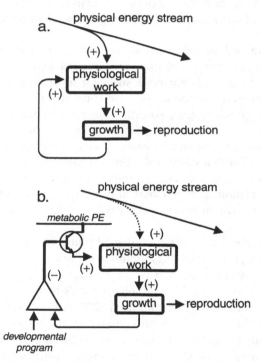

Figure 5.17 a: Modular growth of an accreting organism is governed by the positive feedback inherent in diffusion-limited accretion. b: In more conventional organisms, growth and development are brought under the control of negative feedback systems driven by the chemical energy from food.

form, and the energetics of glucose breakdown are pretty much set by its thermodynamic properties, unaffected by the vagaries of a chaotic and unpredictable environment. Perhaps the increasing degree of physiological "in-sourcing" among the higher animals reflects the conversion from a physical energy economy for powering physiology to a chemical energy economy. The greater predictability of glucose fuel may, in turn, have promoted the engineering of the internal environment to ensure the conditions for its effective utilization. And indeed, most of the major physiological functions operate to provide a degree of stability to the internal environments of animals. The phenomenon is known as homeostasis, the process of ensuring the proper conditions of temperature, pH, and solute concentration and the organized delivery and distribution of nutrients, fuel, oxidant, and wastes throughout the body.

We will be delving into homeostasis in earnest in the later chapters of this book, in particular in Chapters 11 and 12. For now, suffice it to say that many homeostatic processes operate through another type of feedback system, so-called negative feedback (Fig. 5.18). As the name implies, negative feedback is the opposite of positive feedback: it is the consequence of a process that negates the process itself. A thermostat in a room is a familiar example of negative feedback: if the room gets too cold, the thermostat activates a heater, which reverses the decline of room temperature.

Homeostasis does a lot of good things for organisms, among them enabling animals to exploit a wider range of environments than would be exploited if they did not have it. So, for example, if a rat's body temperature could not deviate far from 38°C without fatal consequence, a rat with unregulated body temperatures would be limited to a narrow selection of fairly warm environments. Give the rat a mechanism that keeps body temperature constant, though, and the rat could move into much colder or warmer environments. The benefits do not come free, of course: homeostasis car-

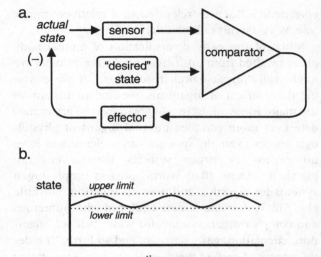

Figure 5.18 a: In negative feedback control, the effector alters the state of the system in a direction opposite to its departure from some "desired" state. b: A negative feedback system keeps the state of the system within narrowly circumscribed limits.

ries with it costs in terms of energy and infrastructure. Judging from its widespread existence among animals, though, homeostasis seems to have been a good investment. Indeed, some paleontologists assert that the appearance of homeostasis among the first animals was a key feature in their origin and the success of their descendants.

Does this mean that external physiology, mediated by structures animals build, is a phenomenon limited to animals too primitive to have developed internal physiology? I have to admit that the answer might be yes, but the fact that I am writing this book means that I think there is a substantial possibility that the answer is no, that in fact external physiology should be a widespread phenomenon. It is hard to see how it could not be, actually—even if there is a lot of physiology going on inside animals, the environmental gradients that drive external physiology will always be there, and judging from the examples I have cited so far, they should be capable of doing enormous quantities of physiological work: just contemplate the magnitude of the "engineering project" represented by a coral reef. Perhaps, then, external physiology is not a primitive condition that is swept aside by an inherently superior internal physiology. Perhaps it is always there, not always obviously visible, but operating continuously as long as the external sources of energy are there to drive it.

... Then the world seemed none too bad,
And I myself a sterling lad;
And down in lovely muck I've lain,
Happy till I woke again.
—A. E. HOUSMAN, *A SHROPSHIRE LAD* (1896), NO. 62

CHAPTER SIX

Mud Power

If you are disgusted by things that crawl around in mud, you're in good company. Karl Linné (Carolus Linnaeus), the Swedish naturalist who gave us the practice of classifying animal species, felt anything that crawled on or in the ground was a loathsome, vile creature. Aristotle also believed this and even had a theory for explaining why: animals that lived in the ground were lowly because they were furthest from the elements of air (*pneuma*) and heat (*calor*) that raise the "higher" animals (like us) to their exalted positions in nature. So our prejudice against things that crawl in mud runs long and deep, and it remains a prejudice strongly felt today. Worms are "icky." Things that crawl out of the ground are stock devices in horror movies. Our prejudice even extends to our fellow humans: many regard mining, ditch digging, grave digging, and the other, shall we say, plutonic professions as distasteful or, at the least, unrefined.

In the next two chapters, I wish to take a closer look at things that make their homes in the ground and the structures they build there. This chapter will be devoted to aquatic animals that live in so-called anoxic muds, the smelly black ooze that lurks just below the surface of mud flats and swamps. The next chapter will move onto land to examine the burrows constructed by earthworms. The functional significance of the burrows these creatures dig is best understood against the backdrop of some grand evolutionary events. In the case of animals living in marine sediments, there are two: the emergence, roughly 650 million years ago, of the Metazoa from the simpler organisms that preceded them; and the origin of photosynthesis among bacteria, roughly 2.2–2.5 billion years ago. In the case of earthworms, the major evolutionary event was the origin of terrestriality, the ability to leave water and live on land.

The Cambrian Explosion and the Emergence of Burrows

Animals probably began constructing burrows about 650 to 700 million years ago, prior to the beginning of the geological period known as the Cambrian, which dawned about 570 million years ago (Fig. 6.1). The Cambrian was a remarkable period—it was the time most of the phyla that now make up the animal kingdom first appeared in the fossil record. Even more remarkable, the emergence of these phyla occurred over a relatively short time period, in an "explosive" burst of diversification lasting just a few million years.

There is still a lot of debate over what caused the Cambrian explosion, as it has been called, but it is clear from the fossil record that life at that time was changing in a big way. One of the more significant events was the emergence of so-called macropredators. The word sounds rather lurid, but a macropredator is simply a predator that is bigger than its prey. This distinguishes macropredation from micropredation, what happens when prey are eaten by organisms smaller than themselves. In other words, if you're eaten by a crocodile, you are the victim of a macropredator. If you succumb to an infectious disease, you have been "eaten" by a micropredator.

Burrowing seems to have originated contemporaneously with the emergence of macropredation. We know this from the record of trace fossils, the remnants of burrowing activities. Trace fossils include tunnels and burrows, tracks in mud, impressions of organisms resting in muds, and so forth.[1] The trace fossil

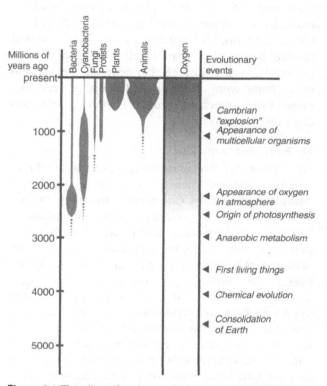

Figure 6.1 Time line of major evolutionary events in the history of the Earth. Width of figures representing the respective kingdoms indicates the relative amount of diversity. The density of shading in the "Oxygen" column indicates the concentration of oxygen in the atmosphere. [After Hickman, Roberts, and Larson (1993)]

---

1. Trace fossils that look like burrows and tracks have been found in rocks as old as 1.5 billion years, but the animal origins of these are controversial. For example, any trace fossil in rocks older than about 1.2 billion years cannot have arisen from an animal, since the first Metazoa did not appear until then. Other

record of this time indicates that burrows, and the burrowing animals that made them, were rapidly becoming more complex. For example, simple impressions were among the earliest trace fossils, left by animals resting on a soft mud. Very early in the Precambrian, these simple fossils were supplanted by tracks made by ambulatory animals slithering across soft bottom muds. After this, horizontal tunnels and short vertical tunnels began to appear. Finally, these were supplanted by vertical tunnels that penetrate deeper into muds. Some were simple vertical shafts, but complex networks of branching tunnels, chambers, and conduits also began to appear.

The Burrowing "Arms Race"

What drove this diversification and elaboration of Precambrian burrows? There are (no surprise, this) lots of possible explanations. One favored by many paleontologists suggests it was driven by the emergence of entirely new body plans, which gave organisms entirely new ways to dig in mud (Box 6A). This set in motion a self-perpetuating dynamic relationship between the burrowing animals and the new macropredators trying to eat them, a relationship that has been called an evolutionary arms race. The concept of an arms race is unsettlingly familiar to us. Having seen at least one, the nuclear arms race, play itself out, we can identify some common features of such contests. One is the evolution of increasingly complex methods of attack and defense, whereby an aggressive measure promotes a defensive counter-measure, which in turn promotes a new aggressive measure—a counter-counter-measure, if you will. The nuclear arms race, for example, began with the simple goal of

dropping really powerful bombs on enemies. From there, it evolved into a complex, almost baroque dance of intelligence and counter-intelligence, better warheads and guidance systems, more precisely targeted projectiles with explosive yields more precisely calibrated to specific military objectives, culminating in the 1980s in the remarkable Strategic Defense Initiative by the United States. Nuclear war is now so complex that we have had to take the job of killing large numbers of people and destroying their cities and crops away from human beings and turn it over to computers (although it sadly must be said that the more personalized forms of barbarism are alive and flourishing).

If the proverbial archaeologist from Mars came down in several thousand years and tried to piece together the history of the twentieth century from the artifacts we left behind, I suspect he would have no trouble doing it: the nuclear arms race has left its fingerprints all over our society, our culture, our thinking. Similarly, the paleontologist sometimes sees fossil artifacts of an ever-escalating cycle of measure and counter-measure among animals. This seems to be the case for the Precambrian macropredators and their prey. One of the first defensive measures was an increased body size—it is harder for a predator to choke down a body bigger than its own, so getting big is an obvious defense. Getting bigger still if you are a predator is an equally obvious counter-measure. Another defensive measure is evident in the appearance of the "shelly fauna," animals that secreted external armor in the form of calcite or silica shells. These were met by counter-measures from the predators, like the development of hard mouth parts that could crush or drill through their prey's defenses. These, in turn, were met by counter-counter-measures from the prey, like spines or other devices, and so on, and so on, culminating in the origin of the major animal phyla at the beginning of the Cambrian.

Increasingly diverse types of burrows probably were part of this Precambrian arms race. The best evidence for this proposition comes from a count of new types

---

supposed trace fossils have nonbiological explanations: they might be desiccation cracks or crystallization structures arising in drying sediments. The oldest trace fossils that nearly all agree were left by animals are circular impressions left in muds, designated by paleontologists as *Bergaueria*, which probably are impressions of sea anemone–like creatures resting on soft sediments. *Bergaueria* have not been found in sediments older than about 700 million years.

How to Dig a Hole

How animals dig holes has long been a matter of keen interest to paleontologists. Often, the only evidence of an animal's existence is a hole, burrow, or track it has left, a so-called trace fossil. Paleontologists are very skilled at squeezing the maximum amount of information from the slightest bits of evidence. By understanding how living animals dig holes and by correlating what they know about today's diggers with the holes that extinct animals have left behind, paleontologists are able to make remarkably detailed inferences about the lives of long lost animals.

Paleontologists specializing in trace fossils (ichnopaleontologists, as they call themselves, from the Greek, *ichnos*, for "footstep") have identified at least five common methods animals use for digging holes. These are (1) intrusion, (2) compression, (3) excavation, (4) backfilling, and (5) fluidizing.

Intrusion is probably the simplest and most primitive of all hole-building techniques, because it involves no development of new behaviors or specialized devices for digging. The animal simply intrudes itself into a substratum, usually a watery mud or clay that offers little resistance to progress of the animal's body. Once in the sediment, the animal literally swims through the substratum, using the same behaviors and devices that enable it to swim through water. Usually, intrusion does not result in a permanent burrow being left behind: the tunnel in the substratum made by a swimming animal simply collapses behind the animal as it progresses. Many nematode and polychaete worms and molluscs use intrusive digging methods to burrow into soft aquatic muds.

Compression burrowing is like driving a blunt dowel through mud. The animal forces its head (or at least the part of its body that is in front) into the substratum, simultaneously pushing sediment to the side and compressing sediment in front. As a result, the substratum at the leading edge and at the sides of the burrow is more compact and harder than the undisturbed sediment. Unlike intrusion, compression requires a specialized skill: the animal must be able to anchor at least one point of its body so that the forward compressive force will not simply push the body away in a "bounce back" effect. Consequently, compressive burrowers must be able to execute a complex series of movements. A typical compression burrower, for example, progresses through the substratum by executing a repeating sequence of push-pull movements. First, the rear of the body is formed into a penetration anchor, which holds it in place while the forward part is thrust forcefully into the sediment. Then, the forward part expands into a retraction anchor that holds the front part in place while the rear is pulled up. This is probably the most common method of burrowing among marine and aquatic invertebrates.

Excavation burrowing is still more complicated. It is common among animals that have articulated joints (like arthropods and vertebrates) and some kind of a hard digging edge, like a nail, tooth, or claw. Excavation is a method favored in soils relatively more compact than the soils that favor intrusion or compression. An excavator must detach a bolus of material from the substrate and then transport the spoil elsewhere. Excavation opens the possibility of creating burrows of complex shapes. For intrusion and compression burrowers, the burrow is typically a hollow tube. An excavation burrower can open up more complicated underground shapes, like branching burrows, chambers for turning around and other maneuvering, and galleries for storage of spoil or food or for living room for offspring or mates.

Backfill burrowing, like intrusion burrowing, typically does not leave behind an open burrow. It differs from intrusion burrowing in that it requires the development of new structures and behaviors. A backfill burrower picks up material ahead of it and transports it, as if by a conveyor belt, to its trailing edge, where the spoil is deposited. The animal's forward progress, therefore, is something like moving on a caterpillar tread. The transport mechanisms may be peristaltic waves on a soft body or cilia, but most commonly soil is transported by the tube feet characteristic of the echinoderms (starfish, urchins, sea cucumbers). Backfill burrowing is unusual in another respect. Intrusion, compression, and excavation burrowing are all most efficiently done when the burrow created is as narrow as possible. Consequently, these types of burrowing favor body shapes that are thin cylinders.

Backfill burrowing, on the other hand, is most efficiently done when the surface area of the "conveyer belt" is greatest. Thus, backfill burrowers tend to be broad and flat.

Finally, fluidizing burrowing involves elements of both intrusion and backfilling. It works by exploiting a phenomenon that is familiar (or at least should be) to most Californians, or to anyone who lives along an active earthquake fault. Soils and other substrata can, if they are shaken hard enough, become "fluidized." Put simply, a soil is fluidized when the forces that normally make soil grains stick together are overcome by some other input of energy. When this happens, the soil now flows as if it were a fluid, like air or water. In an earthquake, the source of energy is the kinetic energy released by the slippage of an earthquake fault. A burrower may fluidize soil by releasing kinetic energy by violent contractions of muscles. For example, some worms will, when disturbed, engage in violent peristaltic movements of the body. This sets water in motion, which helps fluidize the substratum, allowing the worm to sink rapidly into the mud. Similarly, many clams that live in otherwise fairly hard sediments, like mud flats, will, when disturbed, eject jets of water from their siphons. The mud surrounding it is then fluidized, and the clam literally swims away through the mud. If you have ever done any clamming, you will have seen this remarkable technique in action.

---

of trace fossils first appearing during the period (Table 6.1). There are two major bursts of diversification of burrows and burrow types, one during the Ediacaran period, roughly 650 million years ago, and one at the beginning of the Cambrian, roughly 570 million years ago. These parallel contemporaneous increases in the diversity of body types among animals (Box 6A). That these changes were probably steps in an evolutionary arms race is suggested by contemporaneous increases in the rates of extinction of new burrow types at this time: there was more testing of "novel designs" for burrows (Table 6.1).

The increasing complexity of Precambrian burrows

also shows what some of these measures and countermeasures were, and how effective they were. Obviously, if a particular type of burrow "worked," the creatures that built it survived and left descendants. Trace fossils left by them would appear over a long period of the fossil record. Conversely, a "failed" burrow type would not persist in the fossil record for very long. The difference between success and failure seemed to hinge on three factors. First, burrows began to turn vertically and deepen during the Ediacaran period, and they persisted in this trend until the beginning of the Cambrian. Vertical burrows presumably put the animals building them out of harm's way more effectively than the shallow, horizontal burrows that preceded them. The second was the development of hard body parts made from calcite (in the case of the molluscs), chitin (in the case of the arthropods), and cartilage and bone (in the case of the chordates and vertebrates). These could serve as digging tools, enabling the animals possessing them to burrow more effectively into deeper and harder sediments. This trend began just before the dawn of the Cambrian period and peaked during the early Cambrian years. Following it was the third factor in success, namely the development of more complex burrow types: simple vertical shafts evolved into branched and ramifying networks of tunnels, some with complicated systems of conduits, false entrances, and cul-de-sacs.

The Greatest Ecological Catastrophe of All Time

Burrows in muds do physiological work for the animals building them because of a legacy of events that occurred roughly two billion years ago. Let us now turn our attention back to that time, long before there were any animals.

I sometimes ask my students to identify the greatest ecological catastrophe of all time. Usually, they point to some recent newsworthy event, like the wreck of the *Exxon Valdez* in Prince William Sound or the nuclear disaster at Chernobyl in Ukraine. Some point to the extraordinarily high current rates of species extinction, presumably driven by human encroachment on

Table 6.1 Patterns of appearance and extinction of trace fossils during the transition between the Precambrian and Cambrian periods (Crimes 1994). Time frame is in units of millions of years before present (Haq and van Eysinga 1998). Percent extinctions is calculated with respect to the total number of genera present. The two bursts of diversification that are supposedly the result of an arms race are indicated by **bold** type.

Period	Time frame (million years before present)	First appearance	Extinctions	Total genera	Percent extinctions
Upper Cambrian	505–495	8	3	80	4
Middle Cambrian	517–505	4	5	76	6
Trilobite-bearing lower Cambrian	540–517	25	4	86	5
Pre-trilobite lower Cambrian	**545–540**	**33**	**5**	**66**	**8**
Post-Ediacaran Precambrian	550–545	11	4	36	11
Ediacaran	**570–550**	**31**	**10**	**35**	**29**
Varangerian	625–575	3	0	4	–
Riphaean	1,750–810	1	0	1	–

tropical forest ecosystems. Sometimes they refer to the asteroid that collided with the Earth about 65 million years ago and exterminated the dinosaurs (and a whole lot more). But they're all wrong, in my view. I would put my money on events much earlier, about two and a half billion years ago, when certain bacteria, blue-green algae specifically, learned how to use light to strip hydrogen off water and combine it with carbon dioxide to make sugars: in short, the origin of photosynthesis (Fig. 6.1).

Pointing to photosynthesis as an agent of catastrophe seems rather strange, since green plants are presently at the foundation of nearly all life on Earth. It has not always been so, however, and may be only partially true even today. For more than a billion years following life's origin, living things (all bacteria, then) relied on a diverse array of energy sources—such as atmospheric lightning, ultraviolet radiation, and complex molecules like ammonia, methane, and carbon dioxide—but left one unexploited, the virtually unlimited light streaming in from the sun. The origin of photosynthesis changed all that, because it gave the bacteria that were capable of it an immediate and insurmountable energetic advantage over their rela-

tively more plodding contemporaries. Consequently, photosynthetic bacteria soon became the dominant life form on Earth, and their abundance led directly to the second source of catastrophe. In stripping hydrogen atoms off water, photosynthesis produces oxygen gas (O_2) as a by-product. As photosynthesizers became abundant, they churned out ever increasing quantities of this waste gas. At first, the impact was not great, because the oxygen could be absorbed into solution in the oceans and bound to oxygen-hungry minerals, like iron and silicon. Once these reservoirs were full, however, oxygen began to accumulate in the atmosphere, fundamentally changing the chemistry of the planet. This was very bad news for most of the organisms then living.

The Oxidation-Reduction Potential

Oxygen is a curious substance. On the one hand, most organisms need it to live. On the other hand, it can be extremely toxic. It was oxygen's toxicity that posed the problem for the bacteria of a couple of billion years ago. If we are to understand how, we must delve into some basic chemistry.

Chemistry, at root, is the science of moving electrons around. A chemical bond between atoms is actually a pair of electrons shared by two atoms. Oxygen gas, for example, is O_2, two oxygen atoms bound together by a covalent bond. We usually represent a covalent bond with a connecting line to indicate the coupling between the atoms, O—O. It is truer to the nature of a chemical bond to write it this way:

O : O

where the two dots signify the two shared electrons.

Fundamentally, a chemical reaction is a movement of electrons from some bonds to others. In any chemical reaction, bonds between certain atoms, those in the reactants, are broken at the same time new bonds are formed in the products. This is said by chemists to be a coupled oxidation and reduction: the bond that gives up its electrons is oxidized, while the bond that receives the electrons is reduced (Fig. 6.2).

The art of chemistry is in knowing where the electrons will go if you put atoms or molecules together in such a way that electrons have a "choice" of bonds in which to reside. Chemists have a powerful tool to help them predict these preferences in the reduction potential, or redox potential. The redox potential is a voltage, a measure of the potential energy that can drive electrons from one bond to another. Electrons carry a negative charge, so electrons residing in a bond will tend to move toward other bonds with more positive redox potentials, just as electrons in an electrical current move through a wire from the negative pole of a battery to the positive pole.

Let us illustrate this concept with a simple example; the combining of oxygen and hydrogen to form water:

$O_2 + 2H_2 \rightarrow 2H_2O$

or, just to make it simple:

$\frac{1}{2}O_2 + H_2 \rightarrow H_2O$

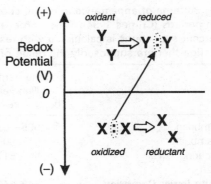

Figure 6.2 Movement of electrons in an oxidation-reduction reaction.

The movement of electrons can be tracked by splitting this reaction into two half-reactions. The electrons, symbolized as e^-, have to move from the chemical bond holding the two hydrogen atoms together (oxidizing the hydrogen):

$H : H \rightarrow 2H^+ + 2e^-$

to the oxygen to form a new chemical bond (reducing the oxygen):

$\frac{1}{2}O_2 + 2H^+ + 2e^- \rightarrow H_2O$

Where the electrons will go can be judged by comparing how strongly each bond, the hydrogen-hydrogen bond of H_2 or the oxygen-hydrogen bond of H_2O, attracts the electrons. This is quantified by the redox potential, symbolized as E'_o, of each of the half-reactions:

$$H : H \rightarrow 2H^+ + 2e^- \qquad E'_o = -0.42 \text{ V}$$
$$\frac{1}{2}O_2 + 2H^+ + 2e^- \rightarrow H_2O \qquad E'_o = +0.82 \text{ V}$$

The oxygen-hydrogen bond has a stronger affinity for electrons than the hydrogen-hydrogen bond.[2] To esti-

2. The negative redox potential for hydrogen-hydrogen bonds does not mean that electrons are repelled from hydrogen-

mate the potential energy driving the electrons from one bond to the other, calculate the difference in redox potential between the electron acceptor (in this case, the hydrogen-oxygen bond) and the electron donor (in this case, the hydrogen-hydrogen bond):

$$\Delta E'_o = +0.82 \text{ V} - (-0.42 \text{ V}) = +1.24 \text{ V}$$

So, there is a potential energy difference (1.24 volts, to be precise) that will drive electrons from H—H bonds to O—H bonds. The reaction will therefore proceed spontaneously: all that is needed is to bring the two molecules together closely enough and keep them there long enough for the electrons to make the move. The reverse reaction, namely getting hydrogen and oxygen away from water (essentially moving electrons from oxygen-hydrogen bonds to hydrogen-hydrogen bonds) will not be spontaneous. Here, the donor and acceptor bonds are switched around, and the potential energy difference is now $-0.42 - 0.92 = -1.24$ V. If the reaction is to proceed, work must be done to make it go. In photosynthesis, obviously, the energy to do this work comes from the capture of light.

The Role of Oxygen in Metabolism

The question remains, though: what is so bad about oxygen? Ironically, the very thing that makes it so toxic is what makes it so useful to organisms like us. To see why, let us look into the role oxygen plays in the metabolism of glucose. I have already outlined the basic reaction in Chapter 2. To refresh your memory, glucose is combined with oxygen to produce water, carbon dioxide, and energy:

$$C_6H_{12}O_6 + 6O_2 \rightarrow 6CO_2 + 6H_2O + 2.82 \text{ MJ [mol glucose]}^{-1}$$

---

hydrogen bonds. Redox potentials are referenced to a standard voltage, in the same way voltages in electronic devices are referenced to a ground. In the case of the oxidation of hydrogen, the negative value simply means the potential has a voltage lower than the agreed-upon reference value.

This simple equation masks a rather complicated set of chemical reactions (Fig. 6.3) that dismantles glucose piece by piece, if you will, and channels the electrons through mediators that capture the energy in them and stores it in ATP. Oxygen's part in all this is to serve as an ultimate acceptor of electrons released from the carbon-hydrogen bonds in glucose, incorporating them into the oxygen-hydrogen bonds of water. Along the way, the electrons released do chemical work for the cell. Because electrons are drawn so powerfully to oxygen, the work they can do is prodigious. The advantage is made clear by comparing how much energy can be captured from glucose when oxygen is present and when it is not. In anaerobic conditions (oxygen absent), only about 7 percent of the energy in glucose is converted to ATP. The remainder is left behind in the bonds of the lactate. When oxygen is present—that is, under aerobic conditions—the energy yield goes up nearly seven-fold, to roughly 40 percent (Fig. 6.3). Under the right conditions, it can exceed 95 percent. That kind of yield is not small potatoes.

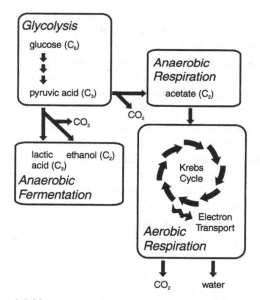

Figure 6.3 Movements of carbon in glucose metabolism.

The Anaerobic World

We're now in a position to explain why oxygen can be so toxic and at the same time so necessary. Both characteristics derive from oxygen's high redox potential. For the first billion years following the origin of life, oxygen gas was scarce, and bacteria living then could not take advantage of its high redox potential as we do presently. Without oxygen to accept electrons, other molecules have to serve that role, and organisms of the time were very creative in finding alternative electron acceptors to keep the juice flowing. The requirements for an alternative electron acceptor are simple: it simply must have a higher redox potential than the "food" molecule from which it draws electrons. When oxygen is not present, a broad array of different molecules may serve as electron acceptors and electron donors. For example, during the breakdown of glucose, with six carbons (a C_6 compound), one of the crucial reactions produces two molecules of a C_3 compound, pyruvic acid. Pyruvic acid, in turn, can accept the electrons released by glucose oxidation, forming lactic acid or ethanol. Pyruvic acid thus substitutes for oxygen as an electron acceptor (Fig. 6.4). "Internal" oxidation-reduction reactions like these are called fermentation, and they are common to both bacteria and eukaryotes. The problem for fermenters, of course, is all that energy left in the chemical bonds of lactate and ethanol, energy that cannot be used to do work because there is no electron acceptor powerful enough. Organisms can also use compounds outside the metabolic pathways as electron acceptors, in so-called anaerobic respiration. For example, nitrate (NO_3^-) is a common electron acceptor for anaerobes, in which it is reduced to nitrogen gas (N_2). Similarly, sulfate (SO_4^{-2}) can receive electrons, producing elemental sulfur (S), hydrogen sulfide (H_2S, the "rotten egg" smell), or other reduced sulfur compounds. Anaerobic respiration thus results in a net oxidation of the glucose, but in the absence of oxygen.

Alternative electron acceptors have their own redox potentials (Table 6.2), and there is no reason in principle why the electron acceptor of one organism cannot,

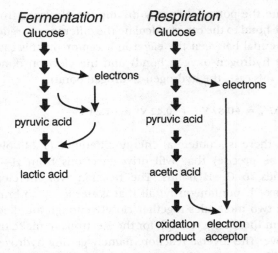

Figure 6.4 Fermentation versus respiration of glucose. Only respiration results in a net oxidation.

at the same time, serve as an electron donor for another. Essentially, the chemical "cast-offs" of organisms that cannot completely metabolize a food molecule may be taken up and used by another kind of organism (Fig. 6.5). This can lead to complicated and mutually dependent communities of microbes. Bacterial fermenters, for example, produce a host of compounds as by-products of the breakdown of large organic molecules. When the food molecule is glucose, the products typically are C_2 compounds like ethanol or acetic acid. When the food molecules are fats, larger compounds are produced, like propionic acid (C_3) or butyric acid (C_4), or even one-carbon acids, like formic acid. These compounds can, in turn, serve as electron donors, that is, food, for still other organisms. For example, a large class of bacteria, known as the acetogens, exploits these chemical "cast-offs." Acetogens come in three basic varieties: fermenters, which produce acetic acid (the acid in vinegar, usually designated by its anionic form, acetate) and carbon dioxide; the CO_2-reducing acetogens, which take formic acid and hydrogen gas and make acetate; and the H_2-producing acetogens, which use propionic and butyric

Table 6.2 Substrates and redox potentials of common biological reactions found in eukaryotes and bacteria.

Oxidation/reduction couple	E'_o (mV)	Comment
O_2 / H_2O	+815	Oxygen
Fe^{+3} / Fe^{+2}	+780	Iron bacteria
NO_2^- / NH_4^+	+440	Ammonium-oxidizing bacteria
NO_3^- / NO_2^-	+420	Nitrite-oxidizing bacteria
SO_3^{-2} / S	+50	Sulfur bacteria
Fumarate / succinate	+33	Intermediate in aerobic oxidation of glucose
Oxaloacetate / malate	−170	Intermediate in aerobic oxidation of glucose
Pyruvate / lactate	−185	Product of anaerobic fermentation of glucose
Acetaldehyde / ethanol	−197	Product of anaerobic respiration of glucose
S / H_2S	−270	Sulfur bacteria (anaerobic)
SO_4^{-2} / SO_3^{-2}	−280	Sulfate bacteria
H^+ / H_2	−410	Hydrogen bacteria
CO_2 / CO	−540	Carboxydobacteria

acids to produce CO_2, hydrogen gas, and acetate. The hand-me-down energy economy in anaerobic bacteria continues. The acetate produced by the acetogens itself acts as a feedstock for another type of bacteria, the methanogens, so called because they produce methane gas (CH_4) as an end product. One type of methanogen accepts acetate from the CO_2-reducing acetogens and fermenters, while a second type accepts acetate, hydrogen gas, and CO_2.

You now begin to see why oxygen gas constitutes a deadly threat to organisms such as these. These wonderfully diverse biochemistries operate on fairly small redox potential differences. If you introduce oxygen, with its very high redox potential, you change the balance of forces that move electrons through and between bacteria. Oxygen draws electrons away from the painstakingly crafted anaerobic pathways built for them, so they cannot do the work the biochemical pathway are intended to do. Before you know it, the organism that relies on this pathway is dead. Since all the organisms that existed prior to the evolution of photosynthesis relied on anaerobic biochemistries, you can now see why the evolution of photosynthesis was such an ecological catastrophe.

Faced with the aerobic onslaught that accompanied the evolution of photosynthesis, the bacteria of the anaerobic world could die, adapt, or retreat. The first was probably the most common, but that is evolutionarily uninteresting, of course. The second course was equivalent to yielding to the inevitable: since oxygen would get the electrons eventually anyway, why not try to make them do work along the way? The bacteria that adopted this strategy, in fact, found many new metabolic opportunities open to them. Remember, there is nothing magical about glucose as food—all that is needed to make a molecule serve as food is a sufficiently large redox potential difference between the food molecule (the electron donor) and the oxidant (the electron acceptor). Because oxygen is such a potent oxidant, microbes could now strip electrons away from many compounds that avidly held them in the presence of weaker electron acceptors. The result was a proliferation of all sorts of novel metabolic pathways among the bacteria, which exploited heretofore strange "food" molecules, such as ammonia, carbon monoxide, elemental sulfur and other sulfur compounds, hydrogen gas, and even iron and silica rocks (Table 6.2). One of these pathways, which we inherited from a class of bacteria that learned the trick of using the electrons to shuffle protons around and make

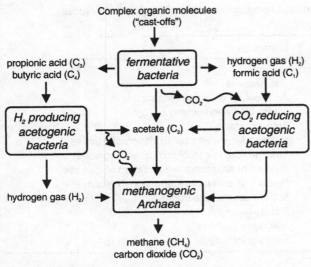

Figure 6.5 Ecological relationships among various types of anaerobic bacteria. [*After Ferry (1997)*]

just as the protozoans do in the oxygen transport "bucket-brigade" described in Chapter 4, and with the same result: a decline of oxygen concentration as one goes deeper into the sediments. In muds, though, the diffusion rate of oxygen is very much slower than the rate in a liquid culture broth. Consequently, oxygen concentration declines steeply with depth in sediments, so steeply that an abrupt discontinuity may be identified, usually at a depth of a few millimeters to a centimeter or so. This abrupt decline of oxygen concentration constitutes a sort of metabolic "Great Wall," behind which anaerobic bacteria could hide. And so, two and a half billion years ago, there they went, to escape the clouds of toxic gases swirling just a few millimeters above them. And there they hid, at least until about 600 million years ago, when this snug little world was rudely penetrated by those newly evolved Precambrian animal burrowers thrusting themselves forcefully through the Great Wall and bringing poison with them—oxygen.

them produce ATP, supports our own aerobic metabolism.

Retreat to Anaerobic Refuges

For the bacteria that were not so metabolically fortunate, retreat from the encroaching oxygen was the only option for them. Fortunately, they did not have to go very far, because most aquatic environments offer a variety of anaerobic refuges. In very deep oceans, for example, there is little oxygen, because there is no photosynthesis, and whatever oxygen does make it to great depths is quickly consumed by aerobic bacteria. Consequently, the abyssal depths are a very reducing environment, particularly where there is some input of reducing agents, as around hydrothermal vents. In shallower waters, where light can penetrate, though, photoautotrophs produce lots of oxygen, and aerobic bacteria can live in the water and in the upper layers of the shallow sediments of coastal margins (Fig. 6.6).

Below these upper layers, it is a very different world. Aerobic bacteria in the upper layers consume oxygen,

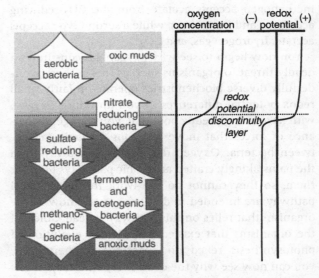

Figure 6.6 Distribution of bacteria along the gradient of redox potential typical in surface muds.

Redox Potential Gradients in Marine Sediments

When the first burrowing animals broke through the metabolic Great Wall into the anaerobic world that had long been hidden beneath it, they inadvertently tapped into one of the Earth's most potent sources of energy. Spanning the oxygen-depleted zone is a redox potential difference of about a volt. Any burrowing animal that could exploit this redox potential would be on easy street.

In a sediment, the redox potential has a different connotation than it does in a half-reaction. Sediments typically will contain a variety of electron donors and acceptors, and a sediment's redox potential is a kind of weighted average of these chemicals. For example, sediments with a lot of oxygen present will have very high redox potentials, but sediments with less oxygen will have redox potentials dominated by weaker oxidizing agents. Thus, the metabolic Great Wall, which appears where there is a decline in a sediment's oxygen concentration, exists in parallel with a gradient in redox potential (Fig. 6.6). In the top centimeter or so, where oxygen can readily move in from the water above to replace that consumed by aerobic bacteria, redox potentials are high, about $+600$ mV, and they decline gradually with depth to about $+400$ mV. At the metabolic Great Wall, redox potential drops sharply, forming a redox potential discontinuity (RPD) layer (Fig. 6.6). Below the RPD layer, redox potentials are typically about -200 mV or lower.

In an undisturbed sediment, this steep redox potential gradient is exploited by a rich microbial community (Fig. 6.6). For example, aerobic bacteria and other organisms that are obligately aerobic (like protozoans or small invertebrates) inhabit the top few millimeters of sediment, where oxygen is abundant and redox potentials are high. In this region glucose is typically metabolized to CO_2 and H_2O. Going deeper, as redox potential declines, the aerobic organisms disappear and fermenters take over. As the mud becomes more oxygen poor, anaerobic respiration comes to dominate. Around the RPD layer, because the gradients here are

so steep, the potential to do work is high, and conditions here support a diverse bacterial ecosystem. At and above the RPD layer, where oxygen is low or only intermittently available, nitrate is the favored electron acceptor, and nitrate reducers dominate. Below the RPD layer, where the likelihood of encountering oxygen is slight, the most common electron acceptor is sulfate. Still deeper, at redox potentials where even sulfate reduction cannot work, the acetogens and their clients, the methanogens, take over. These various bacteria sort themselves out neatly along the profile of redox potential, each group that functions well at one redox potential dependent upon the groups higher in the gradient for feedstock and for keeping the dreaded oxygen gas away.

Feeding Burrows

The animals that first breached the RPD layer 600 million years ago may have been driven there to escape predators. Once the metabolic Great Wall was breached, though, burrowers discovered entirely new ways to feed themselves. Many animals that now habitually live in muds in fact use their burrows as feeding burrows, and these seem to fall into two principal categories. In one type, constructed by so-called substrate feeders, the animal ingests nutrient-rich mud, leaving behind a tunnel as it eats its way through the mud. Alternatively, suspension feeders use the burrow as a conduit for water laden with planktonic organisms that can be captured and eaten.

Feeding burrows are usually simple in construction. The three simplest forms are named after the letters they resemble; I-burrows, J-burrows and U-burrows (Fig. 6.7). The I-burrow obviously points straight down, and the inhabitant (usually a polychaete worm) lies in it head down. The J-burrow is a natural extension of the I-burrow, with a straight vertical shaft terminated at the bottom by a horizontal cul-de-sac. Again, the inhabitant lies head down in the *J*. The common U-burrow, a further natural extension of the J-burrow, has two openings to the surface, connected, of course, by a U-shaped tube. The worm resides in the

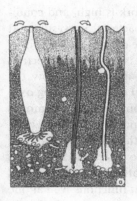 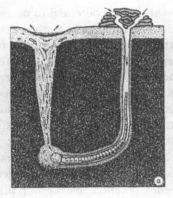 

a.   b.   c.

Figure 6.7 The different types of burrows. *a:* I-burrows constructed by *Molpadia (left)* and *Clymenella (right). b:* J-burrow constructed by the polychaete *Arenicola. c:* U-burrow constructed by *Corophium.* [*From Bromley (1990)*]

bottom of the tube and generates a current that, depending upon the animal, flows either from front to back, or from back to front.

Suspension versus Substrate Feeders

A feeding burrow in stinking mud hardly seems like a dream house—it smells, the food is bad, and it's stuffy. It is easy to look down upon the humble creatures that inhabit these burrows as backwaters of evolution, animals that are primitive and have been left behind in the race to become—us. Why else would they live in mud—or eat mud for that matter?

Simple digestive physiology seems, at first glance, to support this idea. Mud can be used as food, because it contains a lot of bacteria and other goodies, but, really, it is hardly a high-quality diet. Mud is mostly silica and other mineral: the nutrient value is high in a thin bacterial film coating the grains of silt. Bacterial films themselves are quite high-quality food, but what degrades the nutritional value of mud is the cost of getting at those films. Mud is viscous and heavy, and the costs of transporting it through the gut are high. Also, the more finely divided the clay, the more resistant the bacterial films are to digestion. Finally, an aerobic animal living in an anaerobic mud might not be able to avoid "doing as the Romans do": living in an anoxic layer means there will be little oxygen to support aero-

bic respiration, which forces the animal to use inefficient anaerobic pathways to extract energy from food. In short, the habitat of feeding burrows and the available food would seem to force on the inhabitants a sluggish and sedentary life style.

But anyone who visits an estuarine mud flat will be struck by just how abundant these creatures are. Population densities of a common burrowing worm, *Arenicola,* also known as the lugworm, can run as high as several hundred individuals per square meter. Other, smaller burrowers can exist in densities of thousands of individuals per square meter. So, despite their reputation for being sluggish and metabolically challenged, there seem to be an awful lot of them.

The quandary is deepened by taking a detailed look at what these animals are eating. This is easily done: you simply open up the gut and see what is inside. You would think that a substrate feeder living on bacteria on mud should have in its gut only mud and bacteria, but quite frequently they have lots of other things there, too. For example, substrate feeders' guts often contain abundant diatoms, small nematodes, and small arthropods that do not inhabit the mud they eat. Even more strange, the bacteria that are ingested frequently pass through substrate feeders' guts unscathed.

Things become even curiouser when you look at the

chemical composition of these animals' diets. An animal's meal is a package that contains some quantities of nutrients, like carbohydrates, fats, proteins, and minerals. As this package passes through the gut, some nutrients are absorbed. Consequently, the remains of the food, excreted as feces, should have fewer nutrients in them than the original food. Many substrate feeders, though, seem to defy this elementary demand of the First Law of Thermodynamics. Lugworms, for example, deposit fecal pellets that are actually nutritionally richer than the mud they are derived from. Obviously, there is some not very straightforward digestive physiology going on here. Key to this unusual physiology is the burrows the organisms build.

Conveyor-Belt Feeding in Arenicola Worms

The seemingly bizarre dietary physiology of lugworms has a rational explanation, but it requires looking beyond the worm itself to how the worm and the structure it builds interacts with the redox potential gradient in the sediments in which it burrows. The extra nutrients in the lugworm feces actually are produced using this source of potential energy.

Lugworms build J-burrows. The worm sinks a vertical shaft, the tail shaft, in the sediment, and then extends a short gallery horizontally from the bottom of the shaft. It lies head down in the burrow, with its head extending into the tail of the J. The worm ingests the sediment, extending the gallery as it feeds. The mud ingested passes through the gut and out through the worm's anus, located near the top of the tail shaft. There the feces piles up in coiled fecal casts. A lugworm will extend several horizontal galleries from a single tail shaft—viewed from above, the overall effect is a rosette of horizontal galleries extending from a single vertical shaft.

Animals that live in both I- and J-burrows commonly nourish themselves by a method known as conveyor-belt feeding. Among lugworms, it works in this way. The worm uses fleshy paddles on its body to drive a current through the burrow, drawing water in through the opening of the tail shaft and forcing it to percolate upward through the sediments above the worm's head. As the water percolates through the sediments, it stirs them up and causes them to collapse into a funnel-shaped head shaft. Meanwhile, the sediment particles in the developing head shaft sort by size, with the fine, and usually nutrient-rich, particles at the top filtering downward into the gallery. There they are ingested by the worm. As the worm feeds, the head shaft funnel deepens, until it is finally tapped out. Once this happens, the worm extends a gallery in another direction from the tail shaft and begins the process again with another column of mud.

For some time, the enrichment of lugworms' feces was explained away as a side effect of conveyor-belt feeding. The extra goodies, so the explanation went, were carried in with the water the animal pumped through its burrow. The animal's meals, therefore, were supposedly a mixture of both mud and the planktonic organisms drawn in by the feeding current. But, again, deeper analysis knocks the props out from under this explanation—protozoans and other small organisms do show up in substrate feeders' guts, but not the kinds that would be floating around as plankton. Rather, the most abundant protozoans in the guts are mud-dwelling types. This might seem perfectly logical—the lugworms are eating mud, so of course there would be mud-dwelling organisms in the diet. The truly puzzling observation is how much more abundant these mud-dwellers are in the guts than in the sediments that supposedly make up the meal.

Getting Something for Nothing?

Lugworms, as far as we know, have not received an exemption from the First Law, so the extra energy in the lugworm gut must come from somewhere. So it does: from the redox potential gradients that span the RPD layer. In the fine-grained sediments inhabited by lugworms, the RPD layer is very shallow, only about a few millimeters below the surface. The lugworm's burrow, on the other hand, is 5–8 cm deep. With the burrow breaching the metabolic Great Wall, the flow of electrons no longer must make their way through the convoluted cartels of bacteria that span the RPD layer. Now, because of the ventilatory flow of water through the burrow, there is a continual supply of highly oxidizing electron acceptors—specifically sulfate, nitrate, and oxygen—into the sediments below the RPD layer. The burrow, therefore, is analogous to a short-circuit across the RPD layer.

The shortcut sets in motion a complex series of events in the sediments surrounding the burrow (Fig. 6.8). In undisturbed sediment, the microbial community below the RPD layer is dominated by the acetogenic bacteria and the methanogens. These sediments are therefore rich in methane, as well as in the feedstocks for the methanogens: acetate, carbon dioxide, and hydrogen gas. More powerful electron acceptors used by some anaerobic bacteria, like sulfate, are relatively scarce, in part because they are not replaced as they are used up. Seawater is fairly rich in sulfate, though, and when lugworms ventilate their burrows, a stream of sulfate is reintroduced, providing anaerobic sulfate reducers the electron acceptors they need to grow. The increased growth of sulfate reducers thus diverts electrons away from the acetogenic fermenters and methanogens and toward sulfate (Fig. 6.8). Oxygen is also introduced with the ventilatory currents through the burrow, which supports the growth of aerobic bacteria, most notably those that use hydrogen, carbon monoxide, carbon dioxide and methane as their food molecules.

The end result is a flourishing microbial community that spans the short-circuited redox potential gradient around the burrow's walls. This community channels electrons away from the existing cartels of anaerobes to power a prolific growth of aerobic bacteria. The bacteria are in turn eaten by protozoans, mostly predatory diatoms, which are in turn consumed by other predatory diatoms and nematode worms (Fig. 6.8). These are ingested by the lugworm, along with the bacteria in the mud.

As the enriched mud passes through the lugworm's gut, some is digested, but there is enough residual energy in the mud to power the continued growth of this

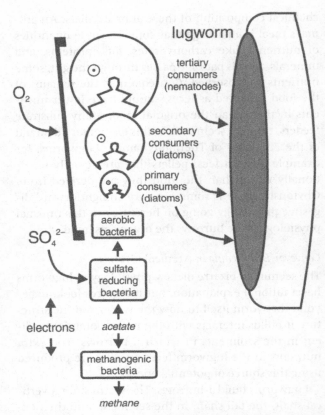

Figure 6.8 Stimulation of productivity in an anoxic mud following the introduction of sulfate by a lugworm.

newly energized community as it passes through, enriching the mud further before passing from the gut as a fecal cast. Metabolism continues even in the fecal casts—indeed, these are the most productive parts of the entire system. The point, though, is that the lugworm is not getting something for nothing. It is, rather, using its burrow as a parallel circuit for electron flow in the reducing environments of anoxic muds.

Burrow Linings as Metabolic Rectifiers

Thus, the burrow is a bit like the transistor-based structure described in Fig. 3.5. By directing a little metabolic energy into building and operating the burrow—namely, the cost of digging it and ventilating it—the worm activates a much larger flow of energy down an external potential energy gradient. This is not all the burrow does, though—it can also act as a rectifier (Fig. 3.5), a device that selectively impedes or allows a flux of matter one way but not the other.

Lugworms, when they construct their burrows, stabilize its walls, both by simple compaction and by secreting a coat of mucus that penetrates two or three centimeters into the sediments surrounding the burrow walls. The sticky mucus acts as a mortar, literally gluing grains of sediment together. Mucus has other peculiar properties, though, which make the burrow wall act as a metabolic rectifier. To understand why, you must first understand something about a chemical technique known as chromatography, one of a class of methods used by chemists to separate similar molecules from one another.

In chromatography, similar molecules are separated by differences in the interactions between the molecules in solution and a substrate that can bind them. We will use as an example the simplest type of chromatography, paper chromatography, which draws a solution through a piece of absorbent paper in just the same way water is drawn up into a paper towel. Imagine that we wish to use paper chromatography to separate a solution of different amino acids from one another. The twenty-three amino acids differ from one another by the identity of a so-called R-group. In glycine, for example, the R-group is a hydrogen atom. The R-group of alanine, on the other hand, is a methyl group (CH_3). The rest of the molecule is the same for all the other amino acids.

Remember, now, that paper is composed of numerous fibers of cellulose woven together into a tangled sheet, with spaces between the fibers. Imagine that we have a solution of amino acids being drawn through this array. The amino acid molecule, when it is in solution, is surrounded by a "jacket" of water molecules, forcibly held there by weak electrostatic forces between the water molecules and the amino acid mole-

cule. At the same time, there will be some attraction between the cellulose fibers and the amino acid molecules. The relative strengths of these interactions will depend upon the R-group. Some amino acids will make the water bind more avidly to them, some less so, and some will bind the cellulose fibers more avidly, others less so. When the solution moves past the fibers, there will be a competition between the water molecules and the cellulose fibers to see which can bind the amino acid molecule more strongly. If the attraction to cellulose is strong enough to overcome the forces holding the water molecules, the amino acid will be delayed slightly as the water carrying it flows past the fiber. If the attraction to cellulose is weaker, the amino acid will be less likely to tarry. Thus, molecules with a relatively weaker attraction for the cellulose will be carried ahead of those with a stronger attraction. Carried over many centimeters, the different amino acids will be separated from one another.

Cellulose can act as a medium for chromatography because it is a polysaccharide—that is, it is a polymer of glucose molecules. Glucose polymers are favorite tools for chromatography because of a curious distribution of charges that occurs in sugar molecules. Sugar molecules, while electrically neutral, nevertheless have some regions where the electrons spend more time than others. The charge distribution on a sugar molecule is therefore uneven: some regions, where the electrons are gathered, tend to be more negatively charged than regions where they are not. When a sugar is incorporated into a polysaccharide, and the polymer folds up into balls or sheets, and the polysaccharide presents an array of negative charges to the outside. The charged regions can then interact with charged molecules flowing by, slowing down some molecules and letting others past unimpeded.

The burrow lining secreted by a lugworm is, as I have said, mucus, which is mucopolysaccharide, a special type of protein-containing polysaccharide. Mucopolysaccharides also make good chromatographic media. The mere presence of a burrow lining impedes the flow of all materials across the burrow walls by 40 percent or so. Some molecules are impeded more than others, however. Negatively charged solutes, like bromide ion (Br^-), are impeded more than positively charged ions, like ammonium (NH_4^+) that are similar in size and strength (but not sign) of the charge they carry. The burrow lining seems to allow positively charged solutes through more readily than negatively charged solutes.

In the environment around the burrow, the burrow lining's chromatographic properties manage the flow of oxidants and nutrients around the burrow. They can have far-reaching effects on the microbial communities surrounding the burrow. Consider, for example, the fate of ammonia, produced as a waste product whenever animals use protein as food.

Ammonia itself is a neutrally charged molecule, but in solution it reacts strongly with water to form ammonium hydroxide, which further dissociates into ammonium ions (NH_4^+) and hydroxyl (OH^-):

$$NH_3 + H_2O \rightarrow NH_4OH \leftrightarrow NH_4^+ + OH^-$$

Ammonium is highly toxic to all animals, because it reacts with one of the intermediates in the aerobic oxidation of glucose, diverting it (and its electrons) away from the production of ATP. Consequently, animals go to great lengths to get rid of ammonium, usually by literally flushing it away. Lugworms usually flush ammonia away by ventilating their burrows with water, but they do not ventilate their burrows all the time. During the unventilated periods, ammonium can accumulate around the worm.

Because the burrow lining is relatively permeable to positive ions, the accumulating ammonium can diffuse easily out of the burrow into the surrounding sediments (Fig. 6.9). Once outside the burrow, it serves as a nutrient for ammonia-oxidizing bacteria, which produce the negatively charged nitrite ion (NO_2^-) as a waste product. You will recall that the burrow lining impedes the flow of negatively charged ions across it, so the nitrite will tend to stay in the sediment rather than diffusing back into the burrow. Thus, the burrow

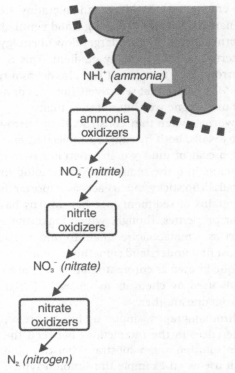

Figure 6.9 The nitrogen rectifier in the burrow lining of a lugworm.

lining acts as a nitrogen rectifier: it allows nitrogen to escape easily in its positively charged form, ammonium, but it impedes backflow in its negatively charged form, nitrite. The retention of nitrogen outside the burrow has other beneficial consequences. The accumulating nitrite is a feedstock for nitrite-oxidizing bacteria, which produce nitrate as an end product. This nitrate can then serve as an electron acceptor for anaerobic nitrate-oxidizing bacteria, in turn stimulating their growth (Fig. 6.9).

The chromatographic burrow lining also manages some interesting interactions with sulfur. In anoxic muds, hydrogen sulfide (H_2S) accumulates from the activities of sulfate-oxidizing bacteria. This is what gives anoxic muds their "rotten egg" smell. Hydrogen sulfide, like ammonia, is highly toxic to most animals. In water, it acts as a weak acid, dissociating into hydrogen ions and sulfide ions:

$$H_2S \leftrightarrow 2H^+ + S^{-2}$$

When hydrogen sulfide contacts oxygen, the sulfide is oxidized to sulfate (SO_4^{-2}). This reaction is spontaneous and very fast. Certain aerobic bacteria, known as sulfide oxidizers, can also oxidize sulfide to sulfate. Sulfide oxidizers, therefore, have a problem: they must compete with oxygen for the energy released by sulfide oxidation. If the electrons in sulfide can be channeled through the bacteria, these electrons can be made to do physiological work for them. If the electrons go directly to oxygen, of course, the bacteria get nothing. Consequently, sulfide oxidizers tend to inhabit the margins of the RPD layer, where there is enough oxygen to accept electrons but not so much that it outcompetes the bacteria for them.

The occasional ventilation of a lugworm burrow introduces oxygen into sediments only intermittently, and the oxygen concentrations in the muds around the burrow will tend to be low. These rise during ventilation, of course, but when ventilation stops, the oxygen is quickly consumed by aerobic bacteria and small animals residing around the burrow, and the local concentrations of oxygen will fall.

On average, the oxygen concentrations around the burrow are low enough to favor the sulfide-oxidizing bacteria. The burrow lining's low permeability to sulfide will ensure that it does not leak back into the burrow, and thus its concentration will remain high in the surrounding sediments. Thus, the growth of sulfide oxidizers will be stimulated, and their production of sulfate will provide oxidant to the anaerobic sulfate reducers, in turn stimulating their growth.

The Lugworm Feedlot

A lugworm and its burrow therefore mobilize the energy that exists in the redox potential gradient in undisturbed mud. This is accomplished partly by the straightforward introduction of oxidants into sediments below the RPD layer, where they are not normally present. The worm also biases the movements of material across the burrow lining, thereby altering the mix of oxidants and nutrients in the sediment around the burrow. The overall result is a stimulation of growth in the sediments that has been dubbed by some zoologists as "gardening," although I think "ranching" is probably a more appropriate term. The stimulation of growth is impressive: sediments with feeding burrows and active lugworms in them mobilize energy at nearly three times the rate that undisturbed sediments do.

It is worth asking just how the worm benefits from this enormous mobilization of energy. After all, the worm does all this work to construct a burrow and then pump oxidant into the sediment. Most of the energy that is mobilized in fact goes to benefit other organisms, not the worm. This is evident from a comparison or the energy consumed in respiration by the worm and by the various organisms in its surroundings. In feeding burrows of *Nereis*, another polychaete worm, the worm itself accounts for only about 10 percent of the total energy consumption. The rest is consumed by the vast community of other things mooching off the oxidant introduced by the worm: roughly 30 percent goes to things living in the lining of the burrow (mostly nitrogen fixers and sulfide oxidizers) and roughly 60 percent goes to things living in the sediments surrounding it (sulfate reducers mostly). So, what's in it for the worm?

The worm is benefiting by playing a game of leveraging. If a lugworm has a metabolic requirement of X joules per day, and its food has an intrinsic energy content of Y joules per kilogram, the worm's energy needs could conceivably be met by ingesting X/Y kilograms of food per day. In fact, it will have to ingest more, since there will always be some inefficiency in absorption and digestion. If its digestion is 10 percent efficient, for example, the worm will have to eat $10X/Y$ kilograms of food daily to meet its requirements.

The efficiency of digestion is related in part to the quality of the food. "Low-quality" food might require a lot of energy to obtain or process, or it might contain chemicals or other materials that impede its digestion. The muds that animals like lugworms inhabit are, as has already been noted, low-quality foods. Although a lot of energy is present in anaerobic muds, it is in a form that can fuel the growth mainly of bacteria—animals are not constructed to do all the strange metabolic things bacteria do. Fueling bacterial growth alone does substrate feeders no good at all, because most substrate feeders are incapable of digesting bacteria. All the extra bacterial productivity, provided courtesy of the lugworm, just passes right through the lugworm's gut. Stimulating bacterial growth does, however, benefit microbial predators, like predatory diatoms or nematodes, that *can* digest the bacteria. The lugworm leverages this secondary production to its own benefit because it is able to digest these types of organisms.

So lugworms essentially are using the redox potential gradient in an anaerobic mud to power the conversion of low-quality food (bacteria and mud) into the high-quality menu of diatoms and nematodes. This is just what ranchers do, of course. Ranchers feed low-quality grass, grain, and straw to livestock, which convert it to meat and milk. Just as ranching operates on an energy efficiency of about 10 percent (only 10 percent of the energy in cattle feed ends up as cow), so too, it seems, do worms like *Nereis*, which divert about the same portion of the energy flowing through the sediment to its own consumption. Even with this small profit margin, though, mud flats along with their substrate feeders rank as one of the richest ecosystems on the planet. Such is the power of mud.

Man is certainly stark mad; he cannot make a worm, and yet he will be making gods by the dozens.
—MICHEL DE MONTAIGNE

CHAPTER SEVEN

As the Worm Turns

If creationists are right and God did design the living world, then God must be a trickster. I say this because of the quandaries one is continually being led into by comparing what we believe God to be with the nature of His works. Let me offer a trite example: it is a common religious doctrine that God is perfection. It follows that God's works must reflect His perfection. Once we accept this, though, we soon are confronted with the manifest imperfections of the world, ranging from the profound (why is there famine and war?) to the trivial (why do so many kids need orthodontia?—a particular source of worry for me right now). This is an old, old conundrum, of course, and at least since Augustine, Christian doctrine has had an answer that satisfies *it*—our imperfection results from our original sin and our subsequent alienation from God.

Well, that explains nicely why people and their works are imperfect, sometimes even absurd. However, the rest of the living world is full of confounding creatures that fly in the face of any doctrine that the living world is a rational, well-designed place. Take, for example, this Zen-like riddle: why is an earthworm? Let me phrase it more prosaically: why are earthworms *earth*-worms? Earthworms, as we shall see, have no business living where they do, because they are physiologically quite unsuited for terrestrial life. Yet there they are, digging happily away, and flourishing in the bargain. How do they manage that?

In this chapter, I explore how the burrowing activities of earthworms benefit these physiological strangers in a strange land. This will require a new look at some commonly known facts about earthworms from an unusual perspective. For example, it has been known for ages that earthworms' burrowings build, aerate, and fertilize soils. Currently we are (finally!) reawakening to the fact that these activities are very advantageous for us, because earthworms are integral parts of productive agricultural ecosystems. But earthworms don't do all this work to promote *our* well-being (unless, given our common destiny as worm-meat, it is *they* who are cultivating *us*): they do it to benefit themselves. By burrowing as they do, I shall argue, earthworms co-opt the soils they inhabit and the tunnels they build to serve as accessory kidneys, ensuring their survival in an essentially uninhabitable environment.

The Rare Earthworm

Earthworms are annelid, or segmented, worms that have ventured onto land. Earthworms are so familiar to us that it is surprising to learn what a rare thing a terrestrial annelid is. Of the 15,000 or so species in the phylum Annelida, about 10,000 are polychaete worms inhabiting marine environments, like the lugworms discussed in Chapter 5. Another 4,000 or so are polychaete and oligochaete worms and leeches inhabiting fresh water. Less than a thousand species of oligochaetes (the earthworms) and a few species of leeches have moved out of water and onto land. So, despite their almost banal familiarity, there actually is something rare and wonderful about an *earth*worm.

But are they really earth-worms, that is, truly terrestrial annelids? Well, it depends upon how you look at it. One could argue, for example, that anything that does not live in water is terrestrial, and by that criterion earthworms are unequivocally terrestrial. Just as walking catfish are terrestrial, and diving spiders and whales are aquatic . . . There is another way to make the distinction, however, and that is to ask whether their physiology equips them for life on land. Now, it might just be my prejudice as a physiologist, but I think this definition offers a stronger criterion than a

judgement based simply on where an animal lives. I believe this because of what physiology is: doing work to maintain internal orderliness in a physical environment that is always pushing an organism toward disorder. Thus, the nature of the physical environment an animal is meant to inhabit should leave its mark vividly on that animal's physiology. Take, for example, the problems animals face in keeping the proper balance of water and salts within their bodies. It is almost absurdly obvious that animals living in water and animals that live on land will deal with this problem differently. It will be no surprise to find that the organs of water and salt balance of, say, a crayfish differ significantly from those of, say, a cockroach.

Physiological Attributes of Freshwater, Marine, and Terrestrial Animals

When we apply a physiological criterion to the problem of water balance, we find, surprisingly, that the physical world is divided not into just two environments, aquatic and terrestrial, but three: freshwater aquatic, marine or saltwater aquatic, and terrestrial. Each has associated with it a particular suite of physical challenges and physiological adaptations, which can mark an animal as "belonging" in one of those habitat types (Box 7A).

To illustrate, let us take a by now familiar example: kidneys. Generically, kidneys are organs that maintain the proper content of water and salts within the body. As we saw in Chapter 2, the water inside animals' bodies frequently differs in composition from the waters they live in. In accordance with the Second Law, a difference in solute concentration will drive fluxes of salts and water between an animal and its environment. To do their jobs, kidneys and the other water balance organs must do physiological work against these thermodynamically favored fluxes of salts and water. A freshwater animal must produce large quantities of very dilute urine, so that it may offset a large osmotic influx of water and diffusional loss of salts. An

Challenges and Opportunities of the Terrestrial Environment

Life originated in water, and for the most part it has stayed there. Beginning about 700 million years ago, however, living things began to move out of the water and onto land. The migration started with algae and other primitive plants, like mosses, but more complicated organisms, including animals, began moving onto land not long after, about 500 million years ago.

Like birth, the move onto land puts organisms into a radically different physical environment. Often, these differences impose severe challenges to survival. At the same time, the new environment must offer opportunities with high payoffs to the organisms successfully making the transition.

GRAVITY

Most organisms are about as dense as water, roughly $1,000 \, kg \, m^{-3}$. Consequently, organisms living in water are supported by buoyant forces and have little need for supporting structures like bones or shells. Where supports do exist in aquatic animals, they usually are for some other purpose, like protection (in the case of a shell) or locomotion.

In air, obviously, the buoyant forces are much weaker and structures capable of supporting the organism's weight are essential. These structures do not come free. Cellulose, which is the principal supporting structure of plants, is essentially glucose, which must be manufactured by photosynthesis. The chitins that support the bodies of arthropods likewise are largely made of sugars. The production of mineralized supporting structures, like bones or shells, also incurs energy costs in the gathering and transport of the minerals. Finally, the bodies of some organisms, like earthworms, are supported by the internal pressure of fluids, which forms a so-called hydrostatic skeleton. Maintaining the high internal pressures needed to operate a hydrostatic skeleton incurs the costs of powering the heart muscles.

WATER BALANCE

Since water balance is extensively discussed in the text of this chapter, suffice it to say here that terrestrial animals face a desiccating force that does not exist in aquatic environments—evaporation. One consequence of the move onto land, then, was facing the physiological demands this implies. At the same time, however, an interesting opportunity was opened. Because evaporation requires considerable heat, terrestrial animals are able to refrigerate their bodies in ways aquatic animals cannot. An animal that can support high rates of evaporation can cool its body considerably below the air temperatures commonly found in terrestrial environments.

OXYGEN

Oxygen offers an additional advantage to organisms not mentioned in the text. The easy availability of oxygen lifted constraints on maximum metabolic rate that limit the metabolism of aquatic animals. A "speeded up" metabolism, along with an ability now to refrigerate the body, made possible the metabolic "life style" of endothermy or "warm-bloodedness." A large portion of an endothermic animal's energy consumption is devoted strictly to the production of heat. Having an internal source of heat enabled animals to elevate and begin to regulate body temperature at the high and steady levels that characterize mammals and birds.

CARBON DIOXIDE

Animals produce carbon dioxide as a metabolic waste gas. The peculiar chemistry of carbon dioxide in water, discussed in Chapter 2, imposes a unique challenge to terrestrial animals. You will recall that carbon dioxide dissolves in water to form carbonic acid. Carbonic acid, in turn, dissociates to bicarbonate and hydrogen ions.

If an animal lives in water, getting rid of carbon dioxide is no problem: it simply leaves the body as bicarbonate ion. In fact, there are many physiological mechanisms for transporting bicarbonate across cell membranes or epithelial membranes. In air, though, bicarbonate must be converted back to carbon dioxide before the carbon can leave the body, and, as we saw in Chapter 2, this is difficult. Terrestrial animals, therefore, tend to retain respiratory carbon dioxide as carbonic acid, forcing the blood to be somewhat more acidic than the blood of aquatic animals. This has necessitated the evolution of new ways to cope with the greater acid load.

NITROGEN

Finally, terrestrial life presents the problem of nitrogen. Nitrogen is a problem because proteins contain nitrogen. When proteins are used as metabolic fuels, one of the waste products is ammonia, NH_3, which is highly toxic.

All animals, whether aquatic or terrestrial, have this problem, but it is rather more of a problem for terrestrial animals. The reason is that ammonia, like carbon dioxide, also reacts with water, but this reaction forms ammonium hydroxide:

$$NH_3 + H_2O \leftrightarrow NH_4OH \leftrightarrow NH_4^+ + OH^-$$

In water, ammonia can simply leave the body as ammonium hydroxide. In air, however, the ammonia must leave either as ammonia gas (which is hard to do) or in the urine (which requires a lot of water loss).

Many terrestrial animals have solved this problem by incorporating the ammonia into less toxic substances. Urea, for example, is a common component of the urine of mammals, while the less soluble uric acid is the favored substance for birds and reptiles.

Table 7.1 summarizes some of the physiological adaptations made by animals for solving the water balance and body waste problems in different environments. Earthworms seem to have more in common physiologically with freshwater animals than with "typical" terrestrial animals.

animal living in sea water faces an environment that is typically saltier than the body fluids. Osmotic loss of water and diffusional influx of salts is the problem faced here, and animals living in sea water must produce small quantities of relatively concentrated urine.

These different hydric environments leave their mark physiologically: each environment presents a particular water balance challenge that is reflected in the structures of the kidneys of animals living there. Compare, for example, the nephrons of freshwater versus marine fishes. The nephron of a freshwater fish has a large and well-perfused apparatus for filtration, which produces large quantities of filtrate. The tubules are long, indicating a high capacity for reabsorption, and the tubule walls are relatively impermeable to water, so that salt reabsorption is favored over water reabsorption. These structural features underlie the production of the voluminous and dilute urine we expect for creatures living in fresh water. The nephrons of marine fishes, on the other hand, frequently have small and poorly perfused filtration structures, which produce filtrate sparingly. The tubules are also relatively short, indicating lower capacities for reabsorption, and permeable to both salts and water. Consequently, marine fishes' kidneys are designed to keep filtration rates low and to reabsorb as much water and salts from the filtrate as possible. The result is the small quantities of relatively concentrated urine we expect for creatures living in marine waters.

Terrestrial animals, like marine animals, live in a desiccating environment. The major difference between the two involves the physical forces that draw water from the body. In the marine environment, water is drawn from the body by osmosis. In a terrestrial environment, it is lost by evaporation. Since both must conserve water, the kidneys of terrestrial animals are constructed very much like those found in animals inhabiting marine environments.[1]

---

1. The kidneys of terrestrial vertebrates—like frogs and, to a lesser extent, reptiles—are similar in design to those of marine fishes, that is, they produce small quantities of relatively con-

Table 7.1 Physiological characteristics of animals living in marine, freshwater, and terrestrial habitats compared with those of the earthworm.

Physiological activity	Animals' habitat			Earthworm
	Freshwater	Marine	Terrestrial	
Salt flux				
Diffusion flux *(TFF)*	−	+	∅	−
Filtration flux *(PF)*	−	− −	−	− −
Reabsorption flux *(PF)*	+++	+	+	+++
Water flux				
Osmotic flux *(TFF)*	+++	−	∅	
Evaporative flux *(TFF)*	∅	∅	−	−
Filtration flux *(PF)*	− −	−	−	− −
Reabsorption flux *(PF)*	+	+++	+	+
Excretion				
Of ammonia	Ammonia	Urea	Urea/uric acid	Ammonia/urea
Of carbon dioxide	Bicarbonate	Bicarbonate	Gaseous CO$_2$; bicarbonate	Calcium carbonate; bicarbonate; gaseous CO$_2$

Key: + = flux from environment to body; − = flux from body to environment; ∅ = no flux; *TFF* = thermodynamically favored flux; *PF* = physiological flux.

Thus, physiology gives us a useful criterion for deciding whether an animal "belongs" in a terrestrial, marine, or freshwater aquatic habitat. If the water balance organs are designed to meet the challenges of living in fresh water—that is, if they support high filtration rates and selective reabsorption of salts but not water—the animal that possesses them is properly a freshwater animal. Conversely, an animal with kidneys that are equipped to deal with the different challenges of the terrestrial environment is physiologically terrestrial.

The Earthworm's "Kidney"

What about the earthworm's "kidneys": do they reflect the workings of an aquatic or a terrestrial animal?

centrated urine (having about the same concentration as blood). The mammals and birds have taken kidney design a step further and are able to combine both high filtration rates and high degrees of water conservation; they produce urine that is more concentrated than the blood.

When we apply physiological criteria to deciding what the proper habitat of an earthworm might be, it is hard to escape the conclusion that they are not really terrestrial. Rather, earthworms seem to belong in a freshwater habitat. Let us see why.

Annelids do not have kidneys as such, but rather a series of small water balance organs called nephridia, one pair per segment (although not every segment has them). Each nephridium is essentially a coiled hollow tube connecting the water space inside the worm, the coelom, with the outside (Fig. 7.1). At one end of a nephridium, a nephridiostome (literally "nephridium mouth") opens into the coelom and thence into the hollow tube, the tubule. The tubule opens at its other end to the outside through a nephridiopore, through which urine is voided.

Earthworm nephridia, like the nephrons of the vertebrate kidney (Fig. 2.5), produce urine through filtration, reabsorption, and secretion, but with a few differences. In the vertebrate nephron, for example,

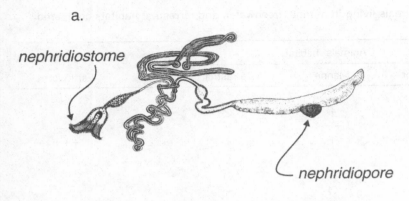

a.

nephridiostome

nephridiopore

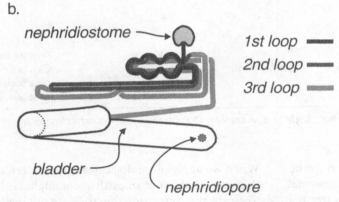

b.

nephridiostome

1st loop ▬▬
2nd loop ▬▬
3rd loop ▬▬

bladder

nephridiopore

Figure 7.1 "High filtration" nephridia of annelids. *a:* Sketch of the nephridium of *Pontoscolex corethrurus,* showing open nephridiostome, long and extensively folded tubules, and capacious bladder, opening to the exterior through a nephridiopore. [*From Goodrich (1945)*] *b:* Schematic diagram of the nephridium of the common earthworm, *Lumbricus terrestris,* showing multiple loops that promote the reabsorption of salts. [*After Boroffka (1965)*]

filtration occurs in one step, when the filtrate is formed from blood at high pressure in the glomerulus (Fig. 2.2). Filtration in earthworms, in contrast, is a two-step process. First, coelomic fluid is produced by filtration directly from the blood, analogously to the deposition of filtrate in the glomerulus. Then coelomic fluid is filtered across the nephridiostome into the tubule. Both filtration steps are pressure driven.[2] The first step is driven by the worm's blood pressure. The annelid heart can sustain blood pressures that, while modest compared with ours (roughly 20–50 percent of the average systolic pressure of humans), are high by invertebrate standards. The second step is driven by

coelomic pressure, which is imparted to the coelomic fluid by the tunic of locomotory muscles that invest the worm's body. Once in the tubule, salts and water are recovered from the filtrate by reabsorption, and the filtrate's composition is otherwise modified by secretion, just as it is in the vertebrate nephron. The urine then passes out the tubule through the nephridiopore to the outside.

The structures of annelid nephridia correlate with particular habitats, just as the fish nephron reflects the animal's "proper" physiological place. Marine polychaetes, for example, have very small and poorly vascularized nephridiostomes, and the nephridiostomes are closed, not open, imposing a barrier to the production of filtrate by coelomic pressure (Fig. 7.2). The overall result, as one would expect, is a

2. Among some annelid worms, differences in osmotic pressure between the coelom and blood also drive filtration.

nephridium that produces only small quantities of filtrate. Nephridiostomes of freshwater polychaetes and freshwater oligochaetes, on the other hand, are large, well perfused, and open, apparently "designed" to produce large quantities of filtrate (Fig 7.1).

Freshwater annelids take this pattern a couple of steps further. In most annelids, the nephridiostome sits in the same segment as the rest of nephridium, and the septa between segments are incomplete, so that coelomic fluid can flow from one segment to another (Fig. 7.3). This arrangements limits the extent to which coelomic pressure in any one segment can rise above that in an adjacent segment. Freshwater oligochaetes, however, typically have complete septa, a design that allows large pressure differences to develop between segments, as high as a few hundred pascals (Fig. 7.3). Furthermore, the nephridiostome for one nephridium pokes through the septum into the adjacent segment. The pressure differences between segments provide an additional motivating force for production of filtrate. The bottom line, though, is that nephridia of freshwater oligochaete worms are constructed to produce large quantities of filtrate, as one would expect.

What about the nephridia of earthworms? They, in fact, have much in common with the nephridia of freshwater oligochaetes, and very little in the way of the structural adaptations one would expect to see in an animal living on land. Functionally, earthworms produce urine as if they were freshwater animals, with losses of water ranging from 60 to 90 percent of body weight per day (in contrast, our own urinary losses of water range from 5 to 10 percent of body weight per day). There appears to be considerable recovery of salts as well, on the order of 90 percent of the salts in the original filtrate. Marine polychaete worms, with their more appropriately constructed nephridia, have urine production rates that are only about 5–35 percent of body weight per day, with little recovery of salts.

It is hard to escape the conclusion that earthworms are essentially aquatic oligochaetes, poorly equipped physiologically for life on land. Yet there they are. Earthworms still face the water balance problems of a terrestrial animal, and their successful habitation on land implies *something* is doing the physiological work necessary for a terrestrial existence. This raises an obvious question: if earthworms' *internal* physiology equips them so poorly for this task, then what *is* doing the work? You probably will have guessed by now that

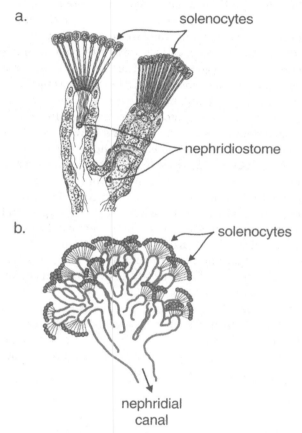

Figure 7.2 "Low filtration" nephridia of annelids. *a:* A "closed" nephridiostome from the polychaete *Phyllodoce paretti.* The end of the nephridiostome is a closed tubule, in which filtration is driven by currents set up by the solenocytes (flagellated cells that drive fluid down the tubule). *b:* The nephridium of *Phyllodoce paretti* is formed from a cluster of these tubules, each topped with a cluster of solenocytes. [*From Goodrich (1945)*]

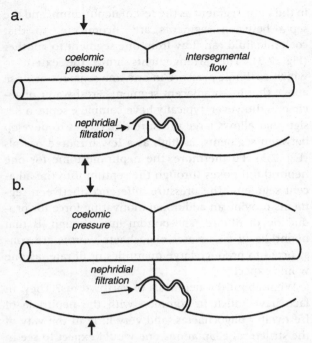

Figure 7.3 The effect of the septum on filtration across the nephridiostome. *a:* When the septum is incomplete, pressure in one segment can do work moving coelomic fluid from one segment to another. Consequently, little energy is left to drive filtration. *b:* When the septum is complete, more energy is available to drive filtration.

my answer to this question lies in the structures built by earthworms as they tunnel through soils.

Before I am able to convince you of this, I must first develop two arguments. First (and fairly straightforward), I need to show how respiratory gases like oxygen behave in air and in water, because this characteristic illuminates why animals go to the trouble of coming onto land in the first place. Second (and somewhat more involved), you need to know something about the physical forces that move water around in the porous medium of soils. Once those pieces are in place, we will be ready to look at how the burrows and other structural modifications earthworms make to soil help do their physiological work for them.

Life on Land: Why Bother?

If moving onto land is so challenging physiologically (Box 7A), it is tempting to ask why organisms bothered to make the move in the first place. This is one of biology's Big Questions, and, not surprisingly, several possible answers have been proposed, some of which make more sense than others. Explanations have included: easier access to light for powering photosynthesis, escape from aquatic predators, more readily exploitable food stores, and so on. Among the more sensible answers, in my opinion, is simply that oxygen from water is scarce and expensive to obtain, much more so than oxygen obtained from air. Just how scarce and expensive it can be is illustrated with a simple calculation. Suppose an animal wants to metabolize and extract the energy from one gram of sugar. How much oxygen does it need to do this? And how much air, or water, is required to provide that much oxygen?

The first question is easy to answer. We have already seen (Chapter 2) that six molecules of oxygen are required to oxidize one molecule of glucose. If you work out the respective molecular weights of oxygen and glucose, you will see that about 1.1 grams of oxygen are required to oxidize one gram of glucose.[3] The amount of air you need to get 1.1 g of oxygen is also easy to calculate: you need a little over 3.6 liters of air. Remember this number.

The availability of oxygen in water is limited by its solubility, which describes how much oxygen can be dissolved in water before it starts coming out as gas bubbles. The quantity of oxygen in solution is governed by Henry's law, which states:

$$[O_2]_s = \alpha_{O_2} p_{O_2} \qquad [7.1]$$

3. The molecular weight of glucose is 180 g mol^{-1}, so 1 g of glucose is 5.56 mmol. Glucose oxidation will require six times that molar quantity of oxygen, or 33.3 mmol, or (since the molecular weight of oxygen gas is 32 g mol^{-1}) 1.07 g. With gases, one mole will occupy roughly 22.4 l. The volume occupied by 33.3 mmol of oxygen will therefore be 746 ml. Air is a mixture of gases, of which 21 percent is oxygen. The volume of air required to hold 746 ml of oxygen is therefore 3.56 l.

where $[O_2]_s$ is the concentration of oxygen in solution (in moles per liter), pO_2 is the oxygen partial pressure (in kilopascals, or kPa), and α_{O_2} is oxygen's Bunsen solubility coefficient (mol l^{-1} kPa^{-1}). In water at 20°C oxygen's solubility is 13.7×10^{-6} mol l^{-1} kPa^{-1}. At sea level, where atmospheric pressure is about 101 kPa, the oxygen partial pressure will be about 21 percent of that, or about 21.3 kPa. Water exposed to atmospheric air will therefore have an oxygen concentration of about 292 μmol l^{-1} (1 μmol = 10^{-6} mol). To extract 1.1 grams of oxygen (which is about 0.037 moles, or 37,000 μmol), you will need 126 liters of water. This is roughly 35 times the volume needed (3.6 liters) of air. This is only the best case: it gets worse if the water is warmer than 20°C, or if its access to the air is somehow impeded, or if the altitude is higher than sea level, or if you are sharing the water with lots of other things that are consuming oxygen.

Not only is oxygen in water scarce, it is expensive to extract. Whenever an animal extracts oxygen from either water or air, it must move the fluid past a gas exchange organ, either a gill or lung. This means doing work on the fluid to pump it past the gas exchanger: obviously, the less fluid that has to be pumped, the lower the energy costs for pumping will be. Air is the clear winner here, because only about 3 percent as much of it must be pumped to extract the same quantity of oxygen from water. Air is also easier to pump, because it is about a thousand times less dense than water, and less viscous. The bottom line is that the costs of breathing are considerably less for animals that breathe air (roughly 0.5–0.8 percent of total energy expenditure) than it is for animals that breathe water (5–20 percent of total energy expenditures). The much lower "overhead" means that air breathers have more energy left over to make babies.

So that—breathing easy—is the first piece of our puzzle. The aquatic ancestors of earthworms probably made the transition from water to terrestrial soils because it enabled them to tap into the more abundant and easily exploitable oxygen there.

Water in Soils

The second piece of the puzzle is how water moves in soils. This is rather more involved than the straightforward matter of oxygen, but at root it is really fairly simple. You just have to remember two things. First, for a parcel of water to move anywhere, it must be pushed or pulled there—that is, some force must act on it to start it moving and keep it moving. Second, if a parcel of water is standing still, this does not mean there are no forces acting on it. It does mean that the *net* force acting on it is *nil*—that is, every force acting to push it one way is balanced by an equal force pushing it the other way.

Soil scientists have been concerned for a long time with how water moves in soils, and they have developed a comprehensive body of theory to help them with this problem. We need to understand the elements of this theory, which is centered around a quantity known as the water potential. As the name implies, the water potential is a measure of the potential energy that can do work on water (namely move it around). The water potential, symbolized by the Greek letter Ψ (psi), is equivalent to a pressure, conveniently enumerated with units of pascals (equivalent to newtons per square meter).

There are several kinds of forces that can act to move water, and the water potential accounts for all these forces. These forces are: the pressure potential, Ψ_p; the gravity potential, Ψ_g; the osmotic potential, Ψ_o; and the matric potential, Ψ_m. These add to give the water potential, Ψ:

$$\Psi = \Psi_p + \Psi_g + \Psi_o + \Psi_m \qquad [7.2]$$

Of the four components of the water potential, the pressure potential is the easiest to grasp: it is simply the hydrostatic pressure acting on a parcel of water. Water in a hose pipe has a pressure potential, for example. Pressure potentials can be positive (forcing water away) or negative (drawing water in, that is, a suction pressure). For an earthworm, pressure potential is important because the heart and muscles of the body can elevate the pressure of its internal water by several thousand pascals (kilopascals).

The gravity potential is also straightforward.[4] Water has mass, and therefore gravity will pull it downward. Gravity potential is proportional to the height of a parcel of water, h, with respect to some reference level, h_0.

$$\Psi_g = \rho g(h - h_0) \qquad [7.3]$$

where ρ is the water's density (roughly 1,000 kg m^{-3}) and g is gravitational acceleration (9.8 m s^{-2}). The reference level, h_0, usually is the height (in meters) at which water would come to rest independent of any other force acting on it. For example, if a parcel of water resting on the ground was lifted 1 m, it will have a gravity potential of 9,800 Pa, 9.8 kPa, with respect to ground level. Gravity potential is also handy for determining such practical things as pumping requirements for ground water. For example, if the water table is 10 m below the ground ($h - h_0 = -10$ m), getting it up to the surface where it can water a crop requires a pump capable of developing 1,000 kg m^{-3} × 9.8 m s^{-2} × −10 m = −98,000 Pa or −98 kPa of suction pressure.

The osmotic potential reflects the effect of solute concentration on movement of water. The osmotic potential represents a pressure that draws water toward parcels with higher concentrations of solute. By convention, the osmotic potential is always negative. In other words, it is always represented as a suction pressure, the magnitude of which depends upon solute concentration:

$$\Psi_o = -CRT \qquad [7.4]$$

where C is the molar concentration of solute (mol m^{-3}), R is the universal gas constant (8.314 J mol^{-1} K^{-1}), and T is the absolute temperature (in kelvins).[5]

The matric potential is the trickiest of the components of the water potential, and it cannot be encapsulated in a neat formula the way the other components can be. It is also the most important to the story I am building, so we must give it a fair bit of attention.

Anyone who has ever soaked up a spill with a paper towel has seen matric potentials at work. Water is drawn up because the towel exerts a force on the water, the so-called matric force, sometimes called capillary action. The matric force arises from an interaction between electrical charges on surfaces, in this example charges on the cellulose threads of the paper towel and similar but opposite charges on water molecules. Just as opposite poles of a magnet exert a mutual attractive force on one another, so too do these opposite charges attract one another.

This forceful interaction is visible to the naked eye in a glass containing water. Most of the water's surface is flat, but near the sides of the glass the water creeps up slightly onto the glass wall, forming a curved surface called a meniscus. The meniscus arises from a balance of three forces. Pulling upwards are the electrostatic

---

4. Some of this discussion may be confusing without some explanation of the "convenient" units of pascals. The pascal is a so-called derived unit, which distinguishes it from the fundamental units of length, time, mass, and temperature. The pascal is a measure of pressure, the quotient of a force, in newtons (N), acting over an area, in square meters. Thus, 1 Pa = (1 N) ÷ (1 m^2) = 1 N m^{-2}. The newton itself is a derived unit, the product of a mass (in kg) and an acceleration (in m s^{-2}), which in fundamental units would be written as 1 N = (1 kg) × (1 m s^{-2}) = 1 kg m s^{-2}. The pascal in its fundamental units would therefore be 1 Pa = (1 kg m s^{-2}) × (1 m^{-2}) = 1 kg m^{-1} s^{-2}. This should clarify how a water potential arises from the formulas given in the text. Taking the gravity potential as an example (equation 7.3), Ψ_g is the product of a water density, ρ (in kg m^{-3}), a gravitational acceleration, g (in m s^{-2}), and a height, h (in m). The units of this product are (kg m^{-3}) × (m s^{-2}) × (m) = kg m^{-1} s^{-2}, the same as the derived unit of the pascal.

5. Similarly, expressing osmotic potential as pascals requires some legerdemain with units. The product of concentration (in mol m^{-3}), gas constant (in J mol^{-1} K^{-1}) and temperature (in K) results in units of (mol m^{-3}) × (J mol^{-1} K^{-1}) × (K) = J m^{-3} (equation 7.4). The joule is also a derived unit, being the product of force (in newtons) and distance (in meters), that is 1 J = 1 N m. In fundamental units, the joule is therefore 1 J = (1 kg m s^{-2}) × (1 m) = 1 kg m^2 s^{-2}. The osmotic potential is therefore equivalent to 1 J m^{-3} = (1 kg m^2 s^{-2}) × (1 m^{-3}) = 1 kg m^{-1} s^{-2}, which, as was shown in footnote 4, is equivalent to the pascal.

forces I described just above, but this time between the glass surface and the water molecules. The stronger this attraction is, the more wettable the surface is said to be. The meniscus is formed by the interaction of this force with the other two. Gravity obviously pulls the water down with a force proportional to its density. In addition, the water's surface tension exerts a force parallel to the water surface that pulls inwardly toward the center of the glass. The balance of all three gives the meniscus its gracefully curved surface.

When the multiple forces that give rise to a meniscus are translated to a porous medium containing a matrix of small pores and channels, the result is the matric force. Commonly, matric forces are much stronger than those that draw water up the side of a glass: even a cheap paper towel can lift water ten or fifteen centimeters. This indicates that the paper towel exerts a more forceful upward pull than the glass, which lifts water only by a few millimeters. The stronger force in the paper towel arises, in part, because the very small pores force the menisci to be more tightly curved there. Forcing curvature onto a water surface magnifies the surface tension forces: it follows that the smaller the pores in the matrix, the stronger the matric forces will be. Different grades of paper towels prove my point: "better" paper towels are woven more tightly and will draw water into them more strongly than will the more loosely woven "inferior" brands. Soils are like paper towels in that they are a matrix of small pores, small air spaces bounded by the mineral grains and small organic particles that make up the soil. Consequently, water in these small spaces between the soil grains, or micropores, will experience strong matric forces.

Understanding matric potentials in soils is fairly straightforward as long as you remember two simple rules. First, matric potentials require that a boundary, or interface, exist between water and air. It is the interaction between the wetting forces and surface tension that makes up the matric potential. If there is no air-water interface, there is no surface tension, and hence no matric potential. Second, creation of new surface

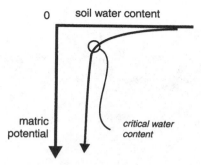

Figure 7.4 Typical relationship between a soil's matric potential and its water content (quantity of water held in a particular quantity of soil—say kilograms of water per kilogram of dry soil). As the soil water content decreases (moves toward zero), the matric potential becomes more and more negative (that is, the suction pressure becomes greater and greater).

area requires work to be done, because surface tension opposes the increase of surface area. Just think of a stretched rubber sheet: the elastic forces in the sheet act similarly to surface tension, and you cannot stretch a rubber sheet without doing work on it.

Let us now apply these rules to understanding a particular behavior of the matric potential: its variation with soil water content, the quantity of water held by a particular quantity of soil (Fig. 7.4). When soil is saturated—that is, when all the pores spaces in a soil are completely filled with water—the matric potential is *nil*. In wet but less than saturated soil, matric potential is weak, and it strengthens only slightly as the soil dries. At a critical water content, however, matric potential strengthens considerably, dropping rapidly to hundreds of thousands or millions of pascals below zero.

Let us now use our rules to explain this behavior. The total volume of the micropores in a soil forms a void space, which can be occupied either by air or water (Fig. 7.5). The *nil* matric potential in saturated soil is easily explained by our first rule. Water completely fills the void space, there is no air-water interface anywhere within the void space, no surface tension forces there, and hence no matric potential. As a soil dries,

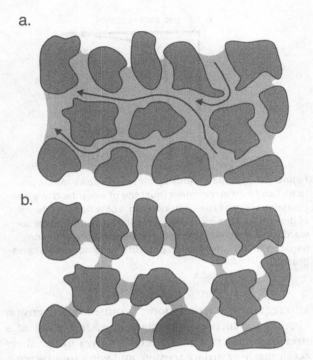

a.

b.

Figure 7.5 Different regimes of matric forces in wet and dry soils. *a:* In very wet soil, many of the micropores are filled with water, and matric forces are weak. *b:* In dry soils, the small amount of water is distributed among many micropores, and matric forces are strong.

though, some fraction of the void space comes to be occupied by air, and we now need to fall back on our second rule.

The existence of a critical water content implies that there are two mechanisms for the creation of new interfacial surface, one of which dominates in soils wetter than the critical water content, and the other dominating in drier soils. When soil is wetter than the critical water content, water is distributed through the void space in fairly large parcels. The pores that do contain an air-water interface are relatively few in number and the interfaces gravitate toward pores of larger diameter. As the wet soil dries, the water retreats deeper into these pores, and the interfaces become more tightly curved as they retreat (Fig. 7.5).

The increase of the interfacial surface area results mostly from this increased curvature of already existing interfaces. Hence, matric potential changes slowly as soil dries, because interfacial surface area varies only slightly as the water retreats deeper into the pores. Below the critical water content, though, these large parcels of water break up into many smaller parcels. This creates new interfacial surface at a much greater rate than the simple shrinkage of a few large parcels.[6] The rapid increase of interfacial surface is reflected in the dramatic strengthening of the matric potential (Fig. 7.4) as the soil dries.

We can also use our rules to explain how matric potentials behave in different soil types (Fig. 7.6). Clays have higher critical water contents than do more coarsely divided soils, like loams or sands, and their water potentials drop to much lower values. The reason, obviously, is that pore spaces are smaller across the board in finely divided clays than in the coarser sands and loams. A parcel of water will break up when the forces holding it together (its cohesiveness) are weaker than the forces pulling it apart (the surface tension). When water is distributed into small pores, as in clays, surface tension forces pull more strongly than they will in more coarsely divided soils, and parcels of water will break up at higher water contents than they do in sands. Clays, therefore, can hold a lot of water, but this abundant water will be difficult to obtain because it is held by very strong matric forces.

Handy Features of the Water Potential

The water potential is such a powerful tool because it enables one to analyze the movements of water and their energetic consequences. Suppose, for example,

---

6. You can convince yourself that this argument is true with an easy calculation. Suppose you have one liter of water occupying a single spherical mass. One liter occupies a volume of 1,000 cm^3, so the sphere has a radius $r = \sqrt[3]{3V/4\pi} = 6.2$ cm. The surface area of this sphere is $4\pi r^2 = 484$ cm^2. Now divide this 1,000 cm^3 of water into four equal parcels of 250 cm^3 each. Each parcel will have a radius of 3.91 cm and a surface area of 191 cm^2. All four will have a combined surface area of 4×191 $cm^2 = 767$ cm^2, roughly 60 percent greater than the surface area of the single, larger sphere.

you want to know how high above the water table a soil's matric forces will draw a column of water. Just to simplify the calculation, let us suppose the matric potential of the soil is −100 kPa. A parcel of water will rise to the height where the gravity potential pulling it down is matched by the matric potential sucking it up. In equation form:

$$\Psi_g + \Psi_m = 0 \qquad [7.5]$$

or:

$$\Psi_g = -\Psi_m$$

Since $\Psi_g = \rho g (h - h_o)$,

$$\rho g (h - h_o) = -\Psi_m$$

If we set the reference height, h_o, to be the height of the water table, the height we want to find is:

$$(h - h_o) = -\Psi_m/\rho g \qquad [7.6]$$

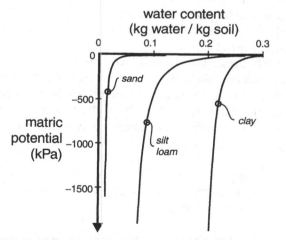

Figure 7.6 Typical soil water profiles for three different types of soils, ranging from very fine grained (clay) to coarse (sand). [*After Campbell (1977)*]

Remember that $\Psi_m = -100\,kPa = -100 \times 10^3\,Pa$, $\rho = 10^3\,kg\,m^{-2}$, and $g = 9.8\,m\,s^{-2}$. Plugging in these values, we see that the soil can draw water up 10.2 m above the water table. Soils with stronger matric potentials will draw the water up further.

The water potential also allows one to estimate the energy costs of transporting water, which is very important for understanding how organisms interact physiologically with the soil. Suppose, for example, that a plant is living in a soil with a matric potential of −101 kPa and that the water inside the plant has a water potential of only −1 kPa. There will therefore be a water potential gradient of −100 kPa drawing water out of the plant. Unless the plant actively moves water into its "body," it will dry up. It would be useful to know how much energy the plant will have to expend transporting water in at a rate sufficient to keep its water content steady. We know that the soil is doing work on the plant when it draws water out, and the rate this work is being done is the product of the water flux rate and the potential energy gradient doing the work:

$$P = V\Delta\Psi \qquad [7.7]$$

where P = work rate (joules per second = watts), V = volume flow rate of water out of the plant ($m^3\,s^{-1}$), and $\Delta\Psi$ = the potential energy gradient doing the work (Pa). If we make our calculation simple and assume we know that water is being drawn out of the plant at a rate of 1 ml ($10^{-6}\,m^3$) per second, work is being done on the plant at the rate of ($10^{-6}\,m^3\,s^{-1} \times 10^5\,Pa$) = 0.1 J s^{-1}, or 100 mW. To stay in water balance, the plant must use metabolic energy to transport water in at the same rate it is being lost. This means doing work against the thermodynamic gradient in water potential, essentially doing physiological work on the environment at a rate of 100 mW. If the plant cannot mobilize energy at least at this rate, its inward rate of water transport will not match the water loss, and the plant will wilt. In fact, agronomists use just these sorts of calculations to estimate water requirements of various crop plants in various soils.

Water Potentials of Worms and Soils

We are now ready to begin assembling the pieces of the puzzle into a picture of how essentially aquatic earthworms manage to live in the terrestrial environment of the soil, and how the structures they build help them do it.

Let us begin with what governs the water balance between an earthworm and the soil it inhabits. This is easily understood in light of their water potentials: let us call them Ψ_w and Ψ_s, for the worm and soil respectively. The worm's water potential, like that of the soil, comprises the sum of the water potentials arising from the various forces that can move water:

$$\Psi_w = \Psi_{w,p} + \Psi_{w,g} + \Psi_{w,o} + \Psi_{w,m} \qquad [7.8]$$

where the subscripts p, g, o, and m refer respectively to the pressure, gravity, osmotic, and matric potentials.[7]

The water inside a worm will have a pressure potential arising from two sources: its blood pressure and the pressure imparted to the body water when its locomotory muscles are active. These have actually been measured: we know the blood pressure of an earthworm varies from +2.5 kPa at rest to about +6.5 kPa when active. Coelomic pressures vary from roughly 600 Pa when resting to as high as 1.5 kPa when active.

Opposing this outward pressure will be the worm's osmotic potential. Worms generally regulate the concentrations of solutes in their body water to between about 100 millimolar (100 millimoles per liter) at the most dilute and 300 millimolar at maximum. This translates into a range of osmotic potentials from −250 kPa to about −750 kPa. Thus, the osmotic and pressure potentials do not balance: the net water potential of the worm will be negative by a few hundred kilopascals, and osmotic forces will draw water into the

---

7. I am ignoring losses of water to evaporation because the humidities of the soil environment are so high that evaporative water loss is virtually nil.

earthworm faster than the worm's internal pressure squeezes it out.

The worm's matric and gravity potentials are best considered in light of the soil's water potential. Like the worm, the soil will have a water potential that is a composite of the water potentials for pressure, osmosis, gravity, and matric forces:

$$\Psi_s = \Psi_{s,p} + \Psi_{s,g} + \Psi_{s,o} + \Psi_{s,m} \qquad [7.9]$$

Pressure potentials in soils usually are small, and we can safely estimate these to be less than 1 kPa. Unless the soil is very salty, we can also assume that osmotic potentials of soil waters will be small. Let us assume the water in soil contains solutes at roughly 10 percent the concentration of the body water inside the worms: this gives soil osmotic potentials of roughly −25 to −75 kPa. If we use a simple trick, we can safely ignore gravity potentials of both the soil and worm. Gravity potentials are only important for vertical translocations of water. Earthworms are only 2–3 mm in diameter, and any vertical movement of water between a soil and worm will be correspondingly small, with gravity potentials on the order of about 10 Pa, too small to be of any significance. We can similarly ignore the worm's matric potential: most of a worm's pore spaces will be filled with water, and so will contribute little to their matric potentials—let us say they are roughly −10 Pa.

With these assumptions, we can write a simple equation describing the forces moving water between soil and worm. The important quantity is the net water potential difference, $\Delta\Psi = \Psi_w - \Psi_s$. This can be expanded to reflect all the components of the worm's and soil's water potential:

$$\begin{aligned}
\Delta\Psi &= \Psi_w - \Psi_s \qquad [7.10] \\
&= (\Psi_{w,p} - \Psi_{s,p}) + (\Psi_{w,o} - \Psi_{s,o}) + (\Psi_{w,g} - \Psi_{s,g}) \\
&\quad + (\Psi_{w,m} - \Psi_{s,m})
\end{aligned}$$

Okay, that's not so simple. But if we neglect the small terms, the simple equation we are looking for falls out:

Table 7.2 Water balance and its energetic costs for a hypothetical 500 mg earthworm living in soils of various matric potentials. Water potential difference, $\Delta\Psi$, is calculated as $\Psi_w - \Psi_s$. Negative water losses indicate a loss from the worm to soil, and positive "losses" indicate a gain to the worm from soil. Metabolic costs are calculated with respect to an earthworm's average metabolic rate of 35 J g^{-1} d^{-1}. Soil water is assumed to have an osmotic potential of 10 percent of the earthworm's osmotic potential. Earthworms' urinary losses of water are assumed to decline with soil matric potential, from 60 percent of body weight per day at maximum to 10 percent of body weight per day at minimum. Similarly, osmotic potential of the earthworm's body is assumed to increase with soil matric potential, from −250 kPa at minimum to −750 kPa at maximum.

Soil matric potential Ψ_m (kPa)	Water potential difference $\Delta\Psi$ (kPa)	Urinary water loss (mg/h)	Skin water loss (mg/h)	Net water loss (mg/h)	Metabolic cost (mW)	Percent of metabolic cost
0	224	−12.5	70.0	57.5	3.6	2%
−10	264	−11.5	82.4	71.0	5.2	3%
−20	304	−10.4	94.9	84.5	7.1	4%
−40	334	−9.4	104.2	94.8	8.8	4%
−80	344	−8.3	107.3	99.0	9.5	5%
−160	314	−7.3	97.9	90.6	7.9	4%
−320	204	−6.3	63.6	57.3	3.2	2%
−640	−66	−5.2	−20.7	−25.9	0.5	0%
−1,280	−656	−4.2	−204.7	−208.9	38.1	19%
−2,560	−1,887	−3.1	−588.4	−591.5	310.0	153%
−5,120	−4,397	−2.1	−1,371.2	−1,373.3	1,677.2	828%

$$\Delta\Psi = \Psi_{w,p} + \Psi_{w,o} - \Psi_{s,m} \qquad [7.11]$$

Equation 7.11 lets us do some simple calculations about the consequences of living in soils with various water contents, and I have tabulated some of these in Table 7.2. It is a very complicated table, but its contents are important, so please bear with me as I go through it with you.

At the heart of the table is an imposed variation of soil matric potential, listed in the leftmost column, each entry being double the one above it. These values span the range of soil matric potentials in which one most frequently finds earthworms, from a matric potential of zero, as would occur in a soil saturated with water, to a very dry soil with matric potential of about −5 MPa. In the second column is the difference in water potential between the soil and worm, calculated with equation 7.11. As you can see, in very wet soils, the water potential difference is positive, dominated by the worm's osmotic potential, which draws water

into the worm. At or around a soil matric potential of −500 kPa, the worm and soil will be in water balance: net water potential is null, and there is no net force moving water one way or the other. In very dry soils, of course, the water potential difference points the other way, drawing water out of the worm and into the soil, because the soil's matric potential is so strong. The last three lines in Table 7.2 represent desiccating environments.

The really interesting numbers are the estimates of the energetic costs of living in these environments. We have already seen how such a calculation is done (equation 7.7): if the net water flux can be estimated, the work required to oppose it is the product of the water flux and the water potential.[8] Calculating the

---

8. One has to make a few assumptions about the biology of the worm, of course, to make this model realistic. For example, we know that an earthworm's urine production varies from a maximum of about 60 percent of the body weight per day to a minimum of about 10 percent of the body weight per day (when

net water flux requires estimates of water loss in urine and the movement of water across the worm's skin. Again, net water flux (tabulated in column 5) is inward at low matric potentials and becomes outward once the soil matric potential exceeds about −500 kPa.

In fairly damp soils, then, matric potentials in the soil are not sufficiently powerful to overcome the net potential difference drawing water into the worm. Even in fairly dry soil environments, matric water is sufficiently available to offset the voluminous losses of water through the worm's high-filtration nephridia. Over this range, there are finite metabolic costs, mostly associated with the cost of transporting water out, but these are small—the worm need invest at most about 5 percent of its resting energy consumption to stay in water balance.

Once soil matric potentials fall below −500 kPa, however, a very different picture emerges. The costs of transporting water at sufficiently high rates dramatically increase. A doubling of matric potential from 1.28 MPa to 2.56 MPa, for example, increases metabolic costs of water balance about eight-fold (from 38.1 mW to 310.0 mW). Worms can survive in soils with matric potentials as strong as −5 MPa, but the energy costs are steep: roughly eight times the resting metabolic rate (expressed as 828 percent of metabolic cost). Earthworms are capable of mobilizing energy at this rate—most aerobic animals are capable of about a tenfold increase in energy consumption above their resting values—but the costs cut deeply into other uses, like locomotion or most important, reproduction. How deeply they cut becomes clear when you realize that an earthworm can only divert about 10 percent of its total energy production rate to growth and reproduction—when more than 90 percent of its energy is going to maintaining water balance, as it will in soils

with matric potentials stronger than about −5 MPa, the worm's capacity to grow and reproduce will inevitably decline.

It seems, therefore, that the most effective water balance strategy of the worm will be to stay in soils with matric potentials of a few hundred kilopascals. This will at least give the worm a fighting chance against the torrents of water being sucked out of it by the matric forces in drier soils.

Oxygen and Water Profiles in Soils

Where does a worm go to find the best soils? To some degree, locating a suitable soil environment is simply a matter of how deep the worm goes. Soils are commonly divided into horizontal layers, called horizons, that have different features at different depths. The trick for the worm is finding the right horizon.

Soil horizons reflect the processes that go into making a soil. The organization of different types of horizons are a soil's profile, essentially a record of the history of that soil. For example, a soil's lower horizons might be made up mostly of material derived from the bedrock below. Over this, flowing water might deposit sediments eroded from distant bedrock. Drying or waterlogging of the soil also might change the distribution of minerals in the horizon, and the activities of microbes might modify it further, as they do in marine muds (Chapter 6). Water filtering downward through a soil and the settling action of gravity may sort soil particles by grain size, with the smallest particles falling to the deepest levels. Organic material, like leaf litter, may accumulate at the top. The end result will be a characteristic pattern of horizons that differentiates between soil types and tells something about how they formed.

The diversity of soils and soil horizons is bewilderingly complex, and soil scientists have developed numerous classification schemes to make sense of them. One convenient way is to divide a soil column into three basic horizons that reflect the degrees to which water or air occupies the porous spaces of the soil.

---

the worm is in a drying environment). We also know that a worm will let its concentration of solutes in the body vary: in very wet environments, it is about 150 milliosmols, but in a drying environment it can rise to about 300 milliosmols.

Starting from the top and working down, these are the aerial, edaphic, and aquatic horizons.

The aerial horizon, as the name implies, is fairly dry, and most of the soil's void space is filled with air. The pore spaces tend to be large, but the water that does exist there is located mostly in the horizon's smallest pores. Consequently, matric potentials in the aerial horizon are very strong, well into the megapascal range. Air also diffuses easily through the large air-filled pores. Consequently, oxygen concentrations are typically greater than 19 percent, not much less than the 21 percent oxygen in the atmosphere. So, while oxygen is readily available in the aerial zone, water isn't. An earthworm living in the aerial horizon would face serious desiccation problems.

A worm would have no desiccation problems in the aquatic horizon. Gravity will draw water downward through the horizons above, and void spaces are filled nearly completely with water. Although pore spaces in the aquatic layer are usually very small, matric potentials are also very weak, almost always weaker than -30 kPa or so. The only trouble with the aquatic horizon is its limited availability of oxygen. Oxygen can diffuse down from the air and through the horizons above, but because it must travel a long way it is slow getting there. Furthermore, biological activities in the upper horizons consume oxygen as it diffuses down, exacting the kind of "oxygen tax" described in Chapters 4 and 6. Consequently, oxygen concentrations in the aquatic horizon are commonly low, ranging from 16 percent down to nearly zero. So, while water is readily available, oxygen isn't.

A worm in soil seems to have a dilemma. If an earthworm descends to a soil horizon that is wet enough for it to stay in water balance cheaply, it will suffocate. Ascending to the horizon where oxygen is abundant puts the worm in jeopardy of drying out or incurring large energy costs for water balance.

Sandwiched in between the aquatic and aerial horizons, however, is an intermediate zone, the edaphic horizon, which is not too dry, as in the aerial horizon, and not too stuffy, as in the aquatic horizon. As I like to tell my students, it meets the "Goldilocks criterion": it is ju-u-u-ust right. In fact, earthworms most commonly inhabit the edaphic horizon, and if they happen to be displaced from it, into either aerial or aquatic horizons, they will migrate back to the soil environment they favor. So the earthworms' secret—how they are able to live in a terrestrial environment even though their water balance system is physiologically suited to aquatic life—is to live in the best of both worlds, the edaphic horizon of soils.

Soil Building by Earthworms

Up to this point, the water balance of earthworms seems fairly mundane. Clearly, earthworms need to seek out environments that suit their physiology, and these environments clearly exist in soils. What is interesting about this?

Many people would view the specialized habitat requirements of an earthworm as a limitation: sure, they've found their little niche, but their limited physiology traps them there. Owing to the fairly limited set of environmental circumstances to which their physiology suits them, they really are at the mercy of what the environment presents to them. If a soil offers no edaphic horizon, or if the edaphic horizon is thin, then opportunities for earthworms to live there will be commensurably limited. This viewpoint, unfortunately, reflects the attitude I described in Chapter 1, that animals always adapt to their environments, and never the other way round.

Few people who are not familiar with the history of biology know this about Charles Darwin, but his last great work (and the work that he claimed to have derived the most satisfaction from) was on earthworms and what they do to soils. Entitled *The Formation of Vegetable Mould through the Action of Worms*, it was published just a year before his death in 1882. On the face of it, it seems to be one of those curious anomalies so characteristic of the British eccentric. In fact, nearly all Darwin's work had a common theme, of which his brilliant synthesis of natural selection was but one dimension: his work on earthworms was another. Put simply, Charles Darwin was not so much an evolutionist as he was a uniformitarian, meaning that he believed that small, seemingly insignificant processes could be enormously powerful agents for change if they were allowed to operate over a sufficiently long time. Just as small changes in body shape from generation to generation in a line of evolutionary descent could, given enough time, generate a new species, and just as small accretions of calcite by corals could produce mighty atolls and reefs, so too could the seemingly insignificant activities of earthworms have enormous effects on the structure and functioning of soils. Indeed, soil scientists can claim Charles Darwin as the founder of their discipline as confidently as evolutionary biologists do.

Just as Darwin's thoughts on evolution aroused controversy, so too, remarkably, did his work on earthworms. Before Darwin, earthworms, if they came to mind at all, were considered agricultural pests, because they were thought to eat the roots of crop plants. One Mr. Fish, for example, wrote an article to the *Gardener's Chronicle* of 1869 criticizing Darwin's claim that earthworms were major forces in the origins and modifications of soils. "Considering their weakness and size, the work [earthworms] are represented to have accomplished is stupendous," he sniffed.

Well, we all remember Mr. Darwin and not Mr. Fish. In fact, earthworms are major agents for change in soil ecosystems, and they have three basic *modus operandi*. First, earthworms open up large persistent tunnels in the soil. As they tunnel, the worms press sideways on the soil, compacting and slightly stabilizing the walls of the burrow. Earthworms also leave behind a mucus coating on the burrow wall, which further stabilizes it. Second, a worm ingests the soil as it burrows and passes it through the gut, digesting the organic matter and bacteria and finally extruding the remnants as fecal pellets. The fecal pellets are permeated with mucus and other secretions from the gut. When these dry, they form large, surprisingly durable casts. Finally, earthworms come to the soil surface, gather large bits of decomposing organic matter, usually rotting leaves or grass, and drag them back into their tunnels. These items are either ingested immediately or are cached in a storage tunnel to be eaten later.

The end result of this incessant mixing and churning is to build soil. When multiplied by the numbers of earthworms in a typical field, the maneuvers described above produce enormous results. Darwin himself estimated the rate of soil-building by measuring how quickly large, immobile objects, like large stones or foundations, appeared to "sink" into the soil—what he was really measuring, of course, was how fast soil newly created by earthworms built up around the object. In the Kentish countryside where Darwin lived, for example, new soil accumulated at a rate of about 5 mm per year, which corresponds to an annual addition of new soil of roughly 18 tons per acre (40 metric tons per hectare). Earthworms clearly are doing a lot of work out there.

Soils, Earthworms, and the Second Law

Soils are dynamic structures, and like all dynamic structures, they exist by resisting the inexorable slide toward disorder demanded by the Second Law. When earthworms build soils, they are working to oppose, even reverse, this slide toward disorder.

One important source of disorderliness in soils is weathering. Soils start as particles of rock: silica or some other mineral. A newly formed soil will consist predominantly of larger particles, a millimeter or so in diameter. During the course of weathering, water flowing through soil will jostle particles against one another, gravity will push them against one another, and so forth. The relatively large particles of newly weathered rock are in this way broken into smaller particles, and the process continues as long as weathering continues. If a soil's inexorable slouch toward entropy is not opposed, its particle size will decline and it will follow a progression from sand to silt and loam, ending finally at the lower limit of clay.

As you would expect from what I have said about

the matric potential, the course of weathering in a soil will affect the interactions between soils and water. As the average particle size of a weathered soil declines, the soil's pore size likewise will decline. Because matric potential depends upon pore size, the soil will gradually become more and more clay-like in its interactions with water: it will hold lots of water, but will hold it with stronger and stronger matric forces (Fig. 7.6).

Opposing this tendency are processes that prevent or even reverse this decline in pore size. One obvious strategy is the aggregation of small soil particles into larger pieces. Aggregation can be a simple chemical process: sand grains sometimes become glued together into larger composites when minerals like calcite precipitate from soil water. Often, though, aggregation is a biological process. For example, numerous microorganisms in soil produce long-chain carbohydrate polymers (like the dextrans often used as thickening agents in processed foods). These literally glue soil particles together, making the average particle size larger, and slow the rate of weathering because the particles in the aggregate are coated and protected.

Earthworms aggregate soils in several ways. First, as they open a tunnel, they secrete mucus from the surface of their bodies, which infiltrates the soil around the tunnel. This aggregates the soil particles, keeps the tunnel open, and, because mucus is a great absorber of water, helps retain water around the tunnel. Second, as soil particles pass soil through the worms' guts, they also pick up some mucus. The fecal casts so formed obviously glue many small soil particles together into larger pieces. Third, one of the peculiar features of annelid biochemistry is their ability to eliminate bicarbonate from the body as crystalline calcium carbonate or calcite. This comes out in the worm's feces, and the combination of digested soil particles, mucus, and calcite makes the cast very resistant to erosion. Throw into the mix the incorporation of large pieces of leaf litter from the surface into feces and tunnels, and you can see that earthworms are powerful forces for the aggregation of soils, working against the disordering

processes of weathering. Of course, by keeping soil particles large, they also keep matric potentials weak, which, as we have seen, makes it easier for worms to draw water into their bodies.

Earthworms change the soils they inhabit in another way. I have already spoken at length about the behavior of water in the soil's micropores, defined as spaces smaller than about a millimeter or so. There exist in soils, however, another class of pores, larger ones designated macropores, which, as the name implies, are larger than a millimeter. Macropores can arise from physical processes—cracks in blocks of bedrock or erosion features, for example—but most frequently, macropores arise biologically. One of their common origins, for example, is the penetration of roots into soil. While the root is alive, it occupies a space, but once the root dies and decays, it leaves behind a tunnel that can range from about a millimeter in diameter (for finely divided roots) to several centimeters in diameter (for main roots). Obviously, earthworms form macropores when they burrow through soils.

The macropore-micropore dichotomy is important because the movement of water through each type of pore is dominated by different physical forces. Movement of water in micropores tends to be dominated by the matric forces discussed extensively in this chapter. Water movement through macropores, on the other hand, is dominated by gravitational forces pulling water down these relatively large "drain pipes" into the soil. Thus, the degree to which water infiltrates into soil, as it would after a rain, is determined by the relative abundance of micropores and macropores. If rain falls on a dry clay soil, where micropores dominate, the surface layers of soil absorb a large quantity of water but will hold it there tightly, which means that the water will not percolate downward, nor be available to plants or animals that might have use for it. In the language of the water potential, the clay's matric potential is strong enough to overcome the water's gravity potential, and the water does not sink, at least not until enough of the pore spaces in the surface layer are filled and the local matric potential declines. After a rain, therefore, clay soils tend to hold water only in a thin layer at the surface. Because it is these surface layers that are warmed the most by the sun, this water will most likely evaporate right back to the atmosphere rather than sink deeper into the soil. Although clays are most susceptible to this phenomenon, even sands have strong enough matric potentials to keep rainwater confined to the top few centimeters of soil.

In a soil with abundant macropores, water flows rapidly downward, rather than being trapped by strong matric potentials in the upper layers of the soil. As it sinks down the macropore drain pipes, it interacts with soil along the way, "spreading the wealth" so to speak, deep into the soil. Technically, the sinking rate is measured by two important soil properties: the infiltration rate (how fast water poured onto the surface sinks into the soil) and the hydraulic capacity (how much water a particular volume of soil can hold).

Earthworms, in building their tunnels, obviously increase the abundance of macropores in the soil, and by doing so they commensurately alter the properties of the soils they occupy. Infiltration rates and hydraulic capacities of fields occupied by earthworms are several hundred percent higher than the rates in fields that do not have earthworms in them. Fields in which earthworms have been killed or their populations reduced substantially undergo a marked degradation in their abilities to absorb rainwater or to hold it weakly enough so that plants can use it. This is one of the reasons why no-till and low-till agricultural methods are so effective—deep tilling kills worms, and when worms are killed, more energy and money must be spent in delivering water to soils in quantities sufficient to support plant growth. The greater infiltration rates of "wormy" soils also make for warmer surface temperatures, and hence faster growth rates of roots and the plants that are supported by them.

The end "goal" of all this squishing and chewing and crawling and tunneling is to develop, maintain and expand a soil horizon where infiltration rates of water

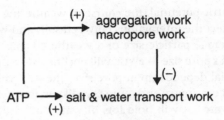

Figure 7.7 The physiological "choices" made by an earthworm.

are high (but not too high), where water is held capaciously (but not too tenaciously), and where there is sufficient air space so that oxygen is abundantly available (but humidities are high). In short, the building activities of earthworms expand the range in the soil horizon where the Goldilocks criterion is met. It is, in other words, the soil environment adapting to the earthworm, not the other way around.

The adaptation of the environment is occurring because earthworms are making a physiological "choice" (Fig. 7.7). They can use ATP energy to do physiological work against the physical forces in the environment drying them up. Or they can use ATP energy to do burrowing work, to synthesize secretion of chemicals—mucus and calcite—to stabilize their tunnels, and otherwise to modify their environment to expand the range of soil habitats in which a physiologically aquatic animal is able to live.

Something seems to have driven earthworms to the latter strategy, and it is interesting to speculate why. As I outlined at the beginning of this chapter, doing physiological work involves having an infrastructure—an engine, if you will, that can use ATP as a fuel to do the work expected of it. These engines we call organs. Moving from one environment to another in which the physiological demands might change necessarily involves a "retooling" process, converting the physiological engines to meet the new demands placed on them.

Clearly, this is something that animals have done repeatedly throughout their evolutionary history, so there is no intrinsic reason why earthworms could not have done the same when their ancestors moved onto land. Remember, however, that unbreakable connection between embryological development and body plans: retooling of physiological organs involves radical modifications of the body plan and the developmental programs that realize it. This means that retooling internal physiology is a laborious and time-consuming process. For example, the conversion of a gill-breathing fish to a lung-breathing mammal involved a massive plumbing renovation to modify the heart, major blood vessels, intestine, gills, and lungs, a process that took roughly 200 million years to complete. Vertebrates undertook the task and succeeded, obviously, but this success is no reason to suppose our ancestors followed the best, or even a particularly good, strategy for adapting to a new environment.

Earthworms, when they came onto land, seem to have opted out of the retooling option and pursued a strategy of using ATP energy to work against soil weathering. Along the way, they essentially co-opted the soil as an accessory organ of water balance. The advantage of adopting this strategy is clear: it is accomplished much more rapidly than a retooling of internal physiology. One can follow the "evolution" of the changing physiology of the soil by monitoring changes in the soils' physical properties in fields newly inoculated with earthworms. Prior to inoculation, soils are compact, they don't retain water well, and variations of surface temperature are extreme. Following the introduction of earthworms, one can see, within about a decade, marked changes in the infiltration rate, water holding capacity, and temperatures, all of which make the soil environment more equable for its annelid inhabitants. Even given the fairly short generation time of earthworms, you have to agree that a decade for remaking the soil environment is clearly preferable to the megadecades needed for a retooling of internal physiology by evolutionary adaptation.

CHAPTER EIGHT

Arachne's Aqualungs

You may know the Greek legend of Arachne, the young Lydian weaver who was so skilled at her craft that she threatened Athena's stature as the source of all greatness in such arts. To punish Arachne both for her superior skill and her hubris, Athena turned her into a spider. As they say, it's not nice to fool Mother Nature.

We humans like to consider ourselves the master weavers of the planet, and indeed, we have a strong claim to that title. Just look around you and reflect on all the versatile things we weave: textiles, automobile tires, high-pressure hoses, duct tape, to name a few. Arachne's skill has also spread widely among animals, though—consider the fantastic array of woven structures to be found among birds' nests—and animal weavers are at least as versatile in the products they weave. Weaving seems particularly widespread among the insects and spiders, which produce their own fiber, silk, from glands in the body. The woven structures produced by these terrestrial arthropods include: nests, cocoons, egg cases, and other types of housings; nets, bolos, snares, booby traps (some devilishly clever), and other devices for capturing prey; parachutes and drag lines to aid in flight; signaling devices and accessory sense organs. In this chapter, I wish to look at woven structures as external organs of respiratory gas exchange. It takes up the theme begun in Chapter 7, in which we explored how the physiologically aquatic earthworms live in a terrestrial environment where they really have no business being. In this chapter, I turn the tables and examine how certain essentially terrestrial arthropods, mostly beetles, but including some spiders, manage to live underwater. The physiological problem faced by these animals is that they are irrevocably committed to breathing air, and this, of course, makes them completely unsuited physiologically for life underwater (Box 8A). Yet they live and prosper there: in some instances, woven structures are key to their ability to do so.

Most commonly, the woven structures built by underwater creatures are used to contain bubbles of air. A well-known example is the "diving bell" woven by spiders of the genus *Argyroneta* (Fig. 8.1). Diving bell spiders weave a dome-shaped web underwater, which contains a bubble held in place by the web's finely woven sheet of threads. The spider captures air at the surface and carries it down to its web as a bubble held to the body by a dense pile of water-repellent hairs, called hydrofuge hairs. Once it has filled its web with air, there the spider sits in its little diving bell, venturing out from time to time to hunt; like a spearfisherman, it returns to the bell after each foray to breathe. Occasionally, the spider ventures to the surface, where it gathers a fresh bubble and carries it down to replenish the air in its web.

Many aquatic insects, most commonly beetles, share this habit of carrying bubbles around with them. Usually, the bubbles are visible as a silvery coating on some portion of the animal's surface and are held in place by patches of hydrofuge hairs. Sometimes the bubbles are tucked away out of sight, under the wings or wing covers. Like diving bell spiders, many of these beetles come to the surface periodically to replenish their bubbles. Some bubble-carrying beetles do not surface, however; even if confined underwater indefinitely, they swim happily about, seemingly oblivious to the fact that they are supposed to be air-breathing animals. And herein lies a mystery. The ability of aquatic insects and spiders to breathe underwater is not a straightforward matter of carrying a bubble of air around and breathing from it. In fact, some marvelously subtle physics is at work, which enables these animals to use bubbles as accessory gills. We shall spend some time exploring the physics of bubbles and bubble gills, in part because it is intrinsically fascinating, and in part because understanding the physics at

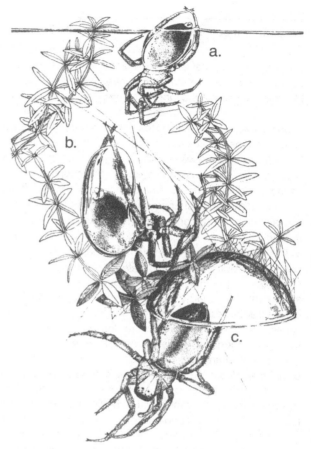

Figure 8.1 *Argyroneta aquatica* in its web. [*From Preston-Mafham (1984)*]

work pays a bonus: it gives us the tools to look for other animal-built structures that serve the same function but less obviously.

Simplicity and Complexity: Occam's Razor vs. Goldberg's Lever

The bubble-carrying beetles posed a mystery to biologists in the early part of the twentieth century in part because of a human failing: we are prone to confuse simplicity with truthfulness or authenticity. One

Why Must Insects and Spiders Breathe Air?

Among most animals, gas exchange systems are constructed to bring together two convective flows—air and blood in the case of the lung—at a diffusion exchange surface that is very thin. In the human lung, for example, about a micrometer separates the alveolar air from the blood, and the surface area of the entire exchange surface in the lungs is about 80 m^2. This design is known as a coupled convection-diffusion exchanger. The coupling of convection and diffusion is found in many configurations among many gas exchange organs, and its ubiquity suggests that the limitations on diffusion exchange are widespread.

The gas exchangers of terrestrial spiders and insects, in contrast, operate by diffusion alone, with minimal or no requirement for convection. Despite this limitation, insects can grow to large body sizes and engage in vigorous physical activities such as flight. They are able to do so because their respiratory organs are designed to use air, not blood, as a medium for distribution of oxygen.

Insects exchange gases through small breathing holes, called spiracles, piercing the animals' flanks at regular intervals. The spiracles open up into large air-filled tubes, the tracheae, which ramify into a network of tubes reminiscent of the bronchi and bronchioles of the lung. The tracheae branch and penetrate throughout the body of the animal, ending in blind-ended tubes, also filled with air, called tracheoles. The spiracles, tracheae, and tracheoles together form the gas exchange organs of terrestrial arthropods, the tracheal system.

Spiders' gas exchange organs are similar in principle to insects', although they differ in some details of their construction. Spiders and other arachnids have a voluminous gas exchange organ known as a book lung, which consists of a series of chitinous sheets, known as lamellae, bounded on one side by air and on the other by the spider's body fluids, or hemolymph. In most spiders, the book lungs have elaborated into a tracheal system that is constructed and works very much like the tracheal systems of insects.

Gas exchange in the tracheal system takes place solely by diffusion between the air in the tracheole and the cell the tracheole serves, along the length of the tracheoles, and subsequently between the tracheolar air and outside air. The importance of using air as a distribution medium may be understood by examining Fick's law:

$$J = -DA(C_I - C_{II})/x \qquad [8A.1]$$

(See equation 5.4 for a review of Fick's law.) Generally, diffusion coefficients for oxygen diffusing through air are about 10,000 times larger than those for oxygen diffusing through water. By using air as the medium for distributing gases through the body, insects and arachnids can exploit these very high diffusion coefficients in a way vertebrates have not. Vertebrates, of course, distribute gases via the blood, and the very low diffusion rates for gases in blood have forced vertebrates into adopting convection-driven gas exchange systems.

It is important to remember at this stage that convection systems do offer significant advantages over strictly diffusion-based systems. For one thing, they allow animals possessing them to attain much larger body sizes. Vertebrates, with their complicated convective exchange systems, have attained body sizes much larger than any insect, living or extinct, of which we are aware. Also, convection systems offer a greater degree of control over gas exchange. Diffusion-based gas exchangers can alter exchange rates only by altering the partial pressure gradients that drive gases. Because insects exchange gases with the very large sources and sinks of the atmosphere, these partial pressure gradients can change only through variation of the gas partial pressures in the insect. An insect that increases its oxygen consumption rate, for example, does so by becoming hypoxic, which can have undesirable consequences for the insect's metabolism. In a convection exchanger, gas exchange rates can be altered by changing convection rates rather than partial pressure gradients. When we exercise, for example, the higher exchange rates for oxygen and carbon dioxide are accommodated mostly by increased ventilation of the lungs and increased heart rate. The blood concentrations of oxygen and carbon dioxide, in contrast to those in insects, remain relatively steady. Indeed, many highly active insects have developed ventilatory schemes to take advantage of the benefits afforded by convection exchange.

BREATHING AIR UNDERWATER

For insects and spiders, reliance on diffusion for gas distribution has imposed a stringent limitation: it commits them irrevocably to breathing air. Should the tracheoles become flooded with water, for example, the insect could no longer enjoy the very high rates of diffusion that prevail in air. Consequently, insects and spiders that have returned to water have had to cobble together devices and structures that enable them to continue breathing air.

Despite being committed to breathing air, insects have evolved many clever ways of living underwater. Mosquito larvae, for example, use their colons as snorkels, hanging suspended from the surface of the water, held in place by surface tension. There, the larvae open their anuses to the air, breathing through their bums, so to speak. Another clever solution is to use plants as snorkels. Many "aquatic" plants are in fact physiologically terrestrial, with the same requirements for "breathing" air, mostly to supply oxygen to the roots. These plants modify some of their vascular tissues, which normally carry fluid between the roots and leaves, and fill them with air. This gives the roots easy access to oxygen in the atmosphere. In a nice double-cross, some types of beetle larvae tap into these air-filled tubes and use them for snorkels. Still others have "gone native," so to speak, developing so-called spiracular gills, in which the spiracles are permanently closed and the air-filled tracheal system is separated from the water by a thin membrane across which diffusion can move gases quickly. In some instances, the spiracle's cover is elaborated into a gas exchange organ consisting of air-filled tracheae. One of the more "deep-tech" (no pun intended) solutions is exemplified by diving bell spiders and aquatic beetles—these creatures simply take a bit of air down with them underwater.

would expect that scientists, with characteristic hard-nosed insistence upon evidence and proof, would be immune to this failing. They are not. In fact, there is among scientists a sort of cult of simplicity, which traces its origins to the fourteenth-century writings of a Franciscan monk, William of Occam. Brother William's legacy to us is a philosophical rule—which has come to be known as Occam's razor—which supposedly differentiates the one "true" explanation from many possible explanations. Put simply, Occam's razor asserts that the explanation most likely to be true is the simplest one, the one that requires the least in terms of special assumptions, rules, or exceptions.[1] Take, for example, the apparent motion of the planets in the sky. There are two competing explanations for these motions, each corresponding to models of the solar system that place either the Earth or the Sun at the center. The geocentric, or Earth-centered, model explains the apparent motion of the planets just fine, but it requires the planets actually to move in some rather complicated ways. For example, it depends on various combinations of rotational and circular movements, unexplained accelerations or decelerations of the planets, and other special rules. A heliocentric, or Sun-centered, solar system requires only that the planets move in elliptical paths, obviously the simpler explanation. Occam's razor would lead us to select the heliocentric model of the solar system as the one most likely to be "true."

Like any useful tool, Occam's razor is prone to abuse. One of the more common abuses is the assumption that it should apply equally to all scientific endeavors, whether they be in the realms of physics, chemistry, or biology. This is probably a pretty safe assumption for sciences like chemistry and physics: I am less certain of its usefulness for biology. Among the assumptions one must make to use Occam's razor is that the universe is a simple place, for which simple

---

1. Occam's razor, from the horse's mouth, as it were, is stated thus: *Pluralitas non est ponenda sine necessitate.* This translates into English as "Plurality must not be posited without necessity."

explanations suffice. Biology, mixed up as it is with a complicated evolutionary history and the often commented upon opportunistic nature of natural selection, offers every reason to believe that the simplest explanation will, in fact, often be the wrong one. Indeed, biology is so full of seemingly claptrap solutions to rather simple problems that I would not simply reject Occam's razor as a useful tool in biology. I would go further and pose a new philosophical rule, that the more complicated the explanation, the more likely it is to be true. We can call it, for lack of a better name, Goldberg's lever, after the cartoonist Rube Goldberg and his ingeniously complicated solutions to simple problems.[2]

The Case of the Bubble-Carrying Beetles

Let us apply Occam's razor to the diverse uses of bubbles by beetles. Do we require two explanations for the two kinds of aquatic beetles, one for those that periodically come to the surface and one for those that do not? In the late nineteenth century, Occam's razor provided a nice way of judging this question: beetles used bubbles for buoyancy. Certainly, so the story went, the bubbles might play some role in respiration for the beetles that regularly come to the surface, but it does not explain why beetles that *never* come to the surface should carry them. If all beetles use bubbles for buoyancy, though, two explanations may be rejected in favor of one: if some beetles use their bubbles for buoyancy, then all must.

I sometimes ask my graduate students what they think moves science forward. Because most of them spend enormous effort chasing down financial support for their work, it is not surprising that most answer

"money." Often, I find I have to agree with them, even though my idealistic side doesn't want to—I still want to believe that science, even in our careerist age, is still primarily an intellectual endeavor divorced from the crass scramble for money and prestige. I usually am succored in my faith by examples of how an individual with a good idea, a healthy dose of skepticism, and only sparse resources can change the course of a science.

One of my favorite examples of this, even if it is a rather prosaic one, concerns the problem of the bubble-carrying beetles. The hero of the story is a German biologist, Richard Ege, who took on this problem in the years just prior to World War I. Two things made Richard Ege's work remarkable. First, the problem he chose exemplified nicely how Occam's razor got in the way of a simple experiment that would have provided a good answer. Why do an experiment, after all, when the answer is obvious (how many times have I seen *that* on my grant proposal reviews?). Second, his work showed how the relentless rejection of the simple can open up marvelous vistas of new biology to explore.

Richard Ege used in his work several species of aquatic beetles, including the main character in the story to follow, *Notonecta*. These beetles normally can stay submerged for five or six hours, after which they must come to the surface or drown. Ege had the ingeniously simple idea of testing whether the bubble had a respiratory function by altering the composition of the bubble gas, confining the beetle underwater, and seeing how long it survived. If the bubble was carried strictly for buoyancy, its gas composition would have no effect: a bubble of nitrogen will float a beetle as well as a bubble of air. If, however, the bubble had some respiratory function, then replacing its air with nitrogen should shorten the beetles' survival times. Sure enough, that was what Ege observed. When beetles were confined underwater with air-filled bubbles, they survived on average about six hours, but when they were confined with only nitrogen in the bubble, they suffocated in only five minutes.

So, the bubbles carried by *Notonecta* clearly were

---

2. Occam's razor has been expressed as various aphorisms, such as "The simpler the explanation, the better the explanation" and so forth. In the hope that Goldberg's lever might catch on, I offer the following aphorisms rendered into Latin by my colleague Jim Nakas. The essence of Goldberg's lever is *Quanto implicatius, tanto verius est* ("The more complicated it is, the truer it is"). A more succinct aphorism might be *Mirationem meam nihil moveat* ("Nothing would surprise me").

acting as a source of oxygen. There the matter would have rested for most people, and Ege's work would have been uninspiring, certainly to me and perhaps others. However, one of the marks of a great experimental biologist is the unwillingness to take a simple "yes" for an answer. Ege happened to be a great experimental biologist, and so he went on to do a second experiment, from which an extraordinarily puzzling result emerged. Let us go through the reasoning behind Ege's second experiment.

- Perhaps the beetle is using the bubble as an oxygen store: that is, it captures in the bubble a certain quantity of oxygen (it would be about 21 percent of the bubble's total volume), uses it up, and then returns to the surface to get a fresh supply.
- A beetle confined with a bubble of nitrogen should not survive as long as a beetle carrying a bubble of air, simply because the quantity of oxygen in the bubble's initial store is smaller. This was shown by experiment to be true.
- The converse should also be true: a beetle using a bubble enriched in oxygen should survive longer than a beetle carrying a bubble of air.

Now here is the remarkable part. Ege confined his beetles with bubbles of pure oxygen. By all rights, they should have survived longer than the beetles carrying bubbles of air, because they had more oxygen to start with. Beetles carrying bubbles of pure oxygen survived for only 35 minutes, however, longer than the 5–6 minutes beetles carrying nitrogen bubbles lived but *one-tenth* the survival time for beetles carrying bubbles of air.

The Ege Effect

Clearly, something strange was going on with the bubbles carried by *Notonecta*. Ege went on to document this curious phenomenon in several different species of aquatic beetles, so extensively that it came to be known as the Ege effect. Understanding the Ege effect is the key to understanding how insects and spiders use bubbles as accessory gas exchangers. It is also the looking glass that helps us find novel structures that might also serve as gas exchangers.

The first step is to understand how gases move between a bubble and the water surrounding it. To do so, we must be able to explain the following phenomenon: an ordinary bubble, drawn from air, will gradually shrink until it disappears. The shrinkage indicates that the gases contained in the bubble are leaving it and dissolving in the water. This movement can only occur if the gases in the bubble are at higher partial pressure than the gases in solution, that is:

$$pO_{2(b)} > pO_{2(s)} \qquad\qquad [8.1a]$$
$$pN_{2(b)} > pN_{2(s)} \qquad\qquad [8.1b]$$

where the subscripts b and s refer to the gases in the bubble and in solution, respectively.

To understand why a bubble shrinks, we must explain where this difference in partial pressure comes from (Fig. 8.2). The partial pressures of the gases in solution are easy to understand: at equilibrium, a gas in solution is at the same partial pressure as the gas in the atmosphere above it. To illustrate, let us assume the atmosphere is similar to air, a mixture of 79 percent nitrogen and 21 percent oxygen:[3] for dry air at sea level (atmospheric pressure = 101 kPa), $pO_{2(s)}$ is about 21.2 kPa and $pN_{2(s)}$ is about 79.8 kPa. Most importantly, these partial pressures are *independent of depth*: they are the same at the surface as they would be at 10 cm or 10 m depth.[4]

Pressures in the bubble are another matter, because

3. I am ignoring the trace gases that collectively make up about 1 percent of atmospheric air, as well as the water vapor it contains.

4. Strictly speaking, this is only the case if there are no sources or sinks for oxygen in the water. If the water contained microorganisms that consumed oxygen, the oxygen partial pressures would decline with depth.

they are affected by how deep the bubble is and by how big the bubble is. A bubble held at depth is squeezed by the water's hydrostatic pressure, which increases at a rate of about 100 pascals per centimeter of depth: this is the pressure you feel in your ears when you dive to the bottom of a swimming pool. If a beetle takes down a bubble of air from the surface at atmospheric pressure, the bubble will be squeezed, and as a result the partial pressures of all the gases in the bubble will be elevated. For example, a bubble derived from air at atmospheric pressure of 101 kPa and carried down to 5 cm depth would have a total pressure of 101.5 kPa, an increase of roughly 0.5 percent. The partial pressures of all the constituent gases would

increase commensurably. Thus, $pO_{2(b)}$ would now be roughly 21.3 kPa (100.5 percent of 21.2 kPa), and $pN_{2(b)}$ would be roughly 80.2 kPa (100.5 percent of 79.8 kPa). Both are higher than their respective partial pressures in the water, and so both gases will be driven from the bubble to dissolve in the water. As the gases leave, the bubble shrinks until the partial pressures in the bubble and water equilibrate (Fig. 8.3). At this point, the bubble's size should stabilize, because there are no longer any partial pressure differences to drive gases out of the bubble. Remember, though, that the bubble does not stabilize—it continues to shrink until it collapses. Some additional force therefore must be compressing the bubble and keeping the gas partial

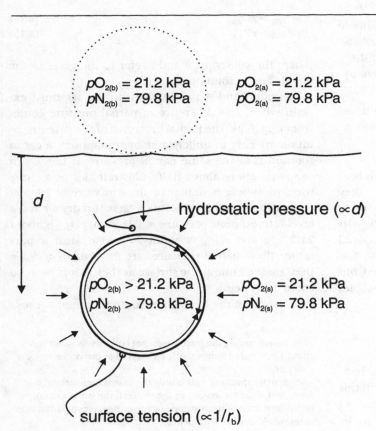

$pO_{2(b)} = 21.2$ kPa
$pN_{2(b)} = 79.8$ kPa

$pO_{2(a)} = 21.2$ kPa
$pO_{2(a)} = 79.8$ kPa

d

hydrostatic pressure ($\propto d$)

$pO_{2(b)} > 21.2$ kPa
$pN_{2(b)} > 79.8$ kPa

$pO_{2(s)} = 21.2$ kPa
$pN_{2(s)} = 79.8$ kPa

surface tension ($\propto 1/r_b$)

Figure 8.2 Forces acting on a bubble of air at depth d and their effects on the partial pressures of the various gases contained within it. Partial pressures are given for gases in the air (a), in the bubble (b), and in solution in the water (s).

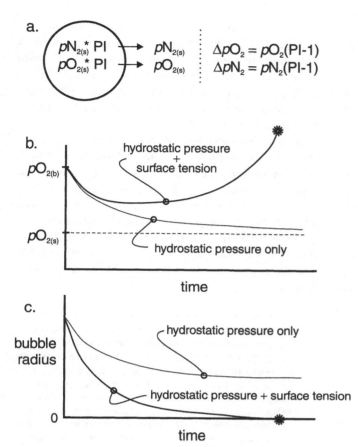

a.

$$pN_{2(s)} * PI \rightarrow pN_{2(s)}$$
$$pO_{2(s)} * PI \rightarrow pO_{2(s)}$$

$$\Delta pO_2 = pO_2(PI-1)$$
$$\Delta pN_2 = pN_2(PI-1)$$

b.

$pO_{2(b)}$

hydrostatic pressure
+
surface tension

$pO_{2(s)}$

hydrostatic pressure only

time

c.

bubble
radius

hydrostatic pressure only

0

hydrostatic pressure + surface tension

time

Figure 8.3 Variation of oxygen partial pressure and radius in a bubble held at a depth. *a:* Partial pressures inside the bubble are equal to partial pressures in solution that has been elevated by a pressure increment, *PI,* determined by both hydrostatic pressure and surface tension at the bubble. *b:* Oxygen partial pressures, pO_2, in the bubble (subscript *b*) and in solution (subscript *s*) with respect to time, under a hypothetical regime of hydrostatic pressure only (thin line) or combined hydrostatic pressure and bubble surface tension (heavy line). *c:* Bubble radius under a hypothetical regime of hydrostatic pressure only (thin line) or combined hydrostatic pressure and bubble surface tension (heavy line).

pressures in it high. This additional force comes from surface tension.

You will remember surface tension from the last chapter. At any interface between air and water, a surface tension force pulls on the interface, in parallel to it (Fig. 8.2). When surface tension acts at the spherical air-water interface of a bubble, it compresses the bubble, just as the stretched wall of a rubber balloon compresses the air inside. What keeps the bubble from stabilizing is the way the surface tension force depends upon bubble size. This relation is expressed by the law of LaPlace, which states:

$$\Delta p = 2\gamma/r \qquad\qquad [8.2]$$

where Δp is the increased pressure, γ is the water's surface tension (about 73 mN m^{-1}), and r is the bubble's radius (m). A bubble 1 cm in diameter ($r = .5$ cm $= .005$ m), therefore, would have an internal pressure elevated by about 29 Pa from surface tension forces alone. The bubble collapses because these surface tension forces increase as the bubble shrinks. If the bubble shrinks to one-tenth its original diameter ($r = .0005$ m), for example, the pressure inside the bubble will increase tenfold. This increased pressure will keep the partial pressures of gases in the bubble perpetually higher than those in the water (Fig. 8.3). Consequently, the bubble never equilibrates and must collapse itself out of existence.

The Ege Effect and Bubble Aqualungs

We're now ready to revisit the Ege effect. So far, we have been dealing with bubbles in which the only possible movement of gas is between the bubble and water. The physical forces operating in a bubble in this circumstance always drive both nitrogen and oxygen one way and one way only—out of the bubble and into solution. If an insect is breathing from the bubble, however, the picture changes dramatically (Fig. 8.4). We now have an additional avenue for the movement of gas—namely, the movement of oxygen from the bubble to the insect. As the insect consumes oxygen from the bubble, the bubble pO_2 will drop, faster than it would if the bubble was simply losing oxygen to the water. If the beetle draws down the oxygen fast enough, the bubble pO_2 eventually falls below the pO_2 in the water. Now the partial pressure difference for

oxygen is reversed, and oxygen will diffuse *from* the water *into* the bubble, where it can be consumed by the insect. In short, the bubble carried by an aquatic insect is not simply an oxygen store; it is also serving as a gill, removing oxygen from solution and conveying it through the bubble to the insect. Bubbles used in this way are known as bubble gills.

We now see why Ege got his seemingly bizarre result with oxygen-filled bubbles. When the bubble is filled with air, the insect need not remove too much oxygen before the ΔpO_2 reverses and starts driving oxygen from the water into the bubble. With a bubble of pure oxygen, however, pO_2 inside the bubble never falls below the pO_2 in solution. Oxygen flux is always outward, the bubble shrinks quickly, and the animal will die if it is not able to come to the surface.

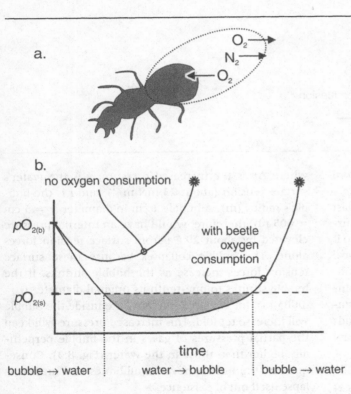

a.

b.

no oxygen consumption

$pO_{2(b)}$

with beetle oxygen consumption

$pO_{2(s)}$

time

bubble → water   water → bubble   bubble → water

Figure 8.4 Conversion of a bubble to a gill. *a:* Fluxes of oxygen and nitrogen in a bubble attached to an insect that is consuming oxygen. *b:* For oxygen in a bubble attached to an insect, the characteristic curve of $pO_{2(b)}$ is not as described in Figure 8.3. At first $pO_{2(b)}$ $pO_{2(s)}$ and oxygen diffuses into the water. But for a short while $pO_{2(b)} < pO_{2(s)}$, as indicated by the shaded area (where oxygen diffuses from the water into the bubble). Eventually, $pO_{2(b)}$ rises again, oxygen diffuses out into the water, and the bubble collapses.

Occam's Razor and the Replenishing of Bubble Gills

A bubble gill, useful as it is, suffers from a serious limitation: it is temporary. The insect might be able to extract more oxygen from the bubble than was originally contained in it, but invariably the bubble will shrink, and the animal will have to return to the surface to replenish it. But air is a mixture of oxygen and nitrogen. When a beetle replenishes its bubble, what, exactly, is being replenished? The simplest answer to this question is that, obviously, it is the oxygen that is being replenished—it is oxygen that is consumed, and so it is the oxygen that must be replaced. The simplest answer is, in this case, not quite the correct answer—illustrating, again, the dangers of relying too closely on Occam's razor.

A bubble's effectiveness as a gill is enhanced by slowing the bubble's shrinkage rate. The longer the bubble endures, the longer the beetle will be able to extract oxygen from the water, and the more oxygen it will extract. In fact, it is fairly easy to predict how effective a bubble gill will be from a number known as the gill factor, G. The gill factor simply compares the quantity of oxygen extracted from the water with the quantity of oxygen initially contained within the bubble. A gill factor of five, for example, means that a bubble gill initially containing one milliliter of oxygen will extract five milliliters of oxygen from the water.

The gill factor is determined by the relative ease with which nitrogen and oxygen cross the boundary between the bubble and water. This is quantified by the invasion coefficient, i, which is the product of two properties: the diffusion coefficient of the particular gas in water, D, and the gas's Bunsen solubility coefficient, α. The invasion coefficients for oxygen, nitrogen, and carbon dioxide are as follows:

$$i_{O_2} = \alpha_{O_2} D_{O_2} = 7.05 \times 10^{-9} \text{ cm}^2 \text{ s}^{-1} \text{ kPa}^{-1} \qquad [8.3a]$$
$$i_{N_2} = \alpha_{N_2} D_{N_2} = 3.22 \times 10^{-9} \text{ cm}^2 \text{ s}^{-1} \text{ kPa}^{-1} \qquad [8.3b]$$
$$i_{CO_2} = \alpha_{CO_2} D_{CO_2} = 1.56 \times 10^{-7} \text{ cm}^2 \text{ s}^{-1} \text{ kPa}^{-1} \qquad [8.3c]$$

Note that carbon dioxide has the largest invasion coefficient, nearly two orders of magnitude larger than the coefficients for oxygen or nitrogen. Carbon dioxide, therefore, crosses the boundary between bubble and water most rapidly of the three. Paradoxically, this means that CO_2 will have little role to play in the behavior of a bubble gill: it leaves the bubble so rapidly once it is released from the insect's body that it contributes only slightly to the bubble's total pressure. However, the invasion coefficients of oxygen and nitrogen are closer in value, with nitrogen's being about half that for oxygen. This similarity has important implications for the performance of bubble gills. Oxygen's larger invasion coefficient means it will cross the boundary between the bubble and water more rapidly than nitrogen. Consequently, in a bubble standing alone, the air contained in it will become richer in nitrogen as the bubble shrinks. A bubble gill turns this disparity to advantage. As long as the animal keeps the oxygen's partial pressure lower in the bubble than in the water, oxygen's larger invasion coefficient will ensure that oxygen flows into the bubble more rapidly than nitrogen flows out. Thus, bubble volume is maintained mostly by the *nitrogen*: oxygen is mainly flowing *through* the bubble rather than *into* it. It is the presence of the nitrogen, therefore, that retards the shrinkage rate of the bubble and permits it to exist as a gill for a longer time. How long is made explicit by calculating the gill factor for a bubble gill. Gill factor is:

$$G = (i_{O_2}/f_{O_2}) \times (f_{N_2}/i_{N_2}) \qquad [8.4]$$

where f_{O_2} and f_{N_2} are the fractional concentrations of oxygen and nitrogen, roughly 0.21 (21 percent) and 0.79 (79 percent), respectively. Plugging in the numbers reveals the gill factor for a bubble gill made from air to be about 8.3. A bubble gill therefore delivers to the beetle 830 percent of the oxygen initially contained in the bubble. But look what happens if the quantity of nitrogen in the bubble (expressed as f_{N_2}) is allowed to drop, say, to 60 percent ($f_{N_2} = 0.6$) and oxygen is elevated to 40 percent ($f_{O_2} = 0.4$): the gill factor drops to 3.28, and the bubble extracts less oxygen.

Clearly, what is being replenished by the insect during its trips to the surface is not the oxygen in the bubble but the nitrogen. The oxygen initially contained in the bubble is quickly used up: the insect will benefit only if this oxygen can be replaced with oxygen from the water. These benefits accrue only as long as the bubble endures, and this is ensured by the continuing presence of nitrogen: it is the maintenance of nitrogen volume in the bubble rather than the extraction of oxygen that is the secret of the bubble gill.

Clearly, the simplest explanation for bubbles as diving bells is wrong on two counts. First, the bubble is not a buoyancy device, it is a respiratory structure even if it is used in different ways by different types of diving beetles. Second, it functions as a respiratory structure not because of the oxygen contained within it but because of the nitrogen. Thus, diving bell spiders and some beetles are solving the problem of being aquatic air breathers in a manner similar to how physiologically aquatic earthworms inhabit soil. Rather than retooling their bodies to form gills, aquatic beetles and spiders have solved their gas exchange problems by co-opting their locomotory muscles to power a bulk flow of nitrogen from air into a submerged structure that does the work of a gill for them.

The Plastron Gill

We still don't have a good explanation for bubble-carrying beetles that don't need to surface. However, our more complex, but more correct, understanding of bubble gills now lets us use Goldberg's lever to pry out the answer.

Bubbles normally collapse because the forces acting on them—hydrostatic pressure, surface tension, gas partial pressures—never balance. Bubble gills work because the insects using them intervene and manipulate two of these forces—partial pressures of oxygen and nitrogen—to delay the bubble's inevitable demise. What would happen, though, if an insect could somehow stop the bubble's march to self-destruction? Let us explore this question with a thought experiment.

We begin, as we did before, with a bubble formed from air and submerged to a certain depth. As the bubble is submerged, the increased hydrostatic pressure compresses the bubble, raising the partial pressures of all the gases in it. Oxygen and nitrogen therefore begin to diffuse out. So far, I am describing the behavior of a conventional bubble. As part of our thought experiment, though, let us do something novel. A real bubble is flexible, and its size can change depending upon the balance of forces acting on it: increase the hydrostatic pressure squeezing the bubble and it shrinks, for example. Let us now suppose that our imaginary bubble is stiff, so that it no longer can change size (Fig. 8.5). The gases in the stiff bubble can now come into equilibrium with the gases in the water in a way that they could not in the flexible bubble. Specifically, nitrogen and oxygen will diffuse out of the bubble, driven by their temporarily elevated partial pressures, until these partial pressures equilibrate with the pressures of the gases in solution.

Let's now complicate our thought experiment a bit: allow a beetle to breathe from the stiffened bubble and see what happens to the gas concentrations in it (Fig. 8.5). For two of the gases, nitrogen and carbon dioxide, the beetle will have little or no effect. The beetle neither consumes nor produces nitrogen, and because the bubble nitrogen is already in equilibrium with the water ($\Delta p N_2 = 0$), the bubble $p N_2$ will be unchanged. As the beetle releases carbon dioxide into the bubble, its very high invasion coefficient will ensure that it dissolves rapidly in the water, keeping the bubble $p CO_2$ near zero. But with oxygen, the picture is different. As the beetle consumes oxygen, the $p O_2$ in the bubble obviously must fall, which drives a flux of oxygen in. The magnitude of this flux, as we well know by now, will be proportional to the $\Delta p O_2$, and this will increase until it is sufficiently large to deliver oxygen from the water as fast as the beetle consumes it. There it will come to equilibrium, allowing the beetle to sit and extract oxygen from the water indefinitely. The only reason a beetle with a stiff bubble ever need come to the surface

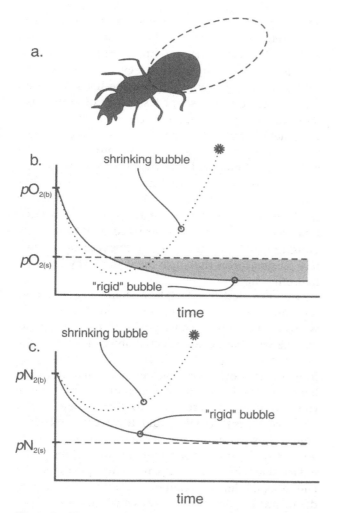

a.

b.

$pO_{2(b)}$

shrinking bubble

$pO_{2(s)}$

"rigid" bubble

time

c.

$pN_{2(b)}$

shrinking bubble

"rigid" bubble

$pN_{2(s)}$

time

Figure 8.5 The development of a plastron gill. *a:* A hypothetical "rigid bubble" is attached to an insect consuming oxygen. *b:* Comparisons of the course of $pO_{2(b)}$ in a "rigid" bubble and a bubble that is able to shrink. The $pO_{2(b)}$ in a rigid bubble doesn't increase, so oxygen continues to enter the bubble from the water. *c:* Comparisons of the course of $pN_{2(b)}$ in a "rigid" bubble and a bubble that is able to shrink. Eventually, $pN_{2(b)}$ reaches equilibrium with $pN_{2(s)}$.

would be depletion of the oxygen in the water (which sometimes happens if the beetle is sharing its home with lots of other things that consume oxygen) or for needs unrelated to respiration (like having to lay eggs out of the water).

We now see how a bubble-carrying beetle could live underwater indefinitely. If a bubble gill's shrinkage could be opposed somehow, it could extract oxygen from the water indefinitely, and the beetle carrying it would never have to replenish it. The kind of gill I have just described in fact is widely employed by diving beetles, and it is called a plastron gill.[5] Our little thought experiment has shown us how plastron gills work—if anything, they are simpler than conventional bubble gills. Most importantly, though, the thought experiment has given us a design principle for plastron gills: "stiffen" the bubble. In looking for novel types of plastron gills, one should look for mechanisms that make bubbles stiff, that is, resistant to collapse.

Let us apply this principle to a conventional plastron gill, formed from a permanent bubble carried on the surface of an insect's body. The internal pressures of a stiff bubble will come to a steady state when its internal pressure is ΔpO_2 below the hydrostatic pressure acting on the bubble. In a formula:

$$P_b = P_h - \Delta pO_2 \qquad [8.5]$$

where P_b and P_h are, respectively, the total pressure in the bubble and the hydrostatic pressure on the bubble (Fig. 8.6). The bubble is, therefore, operating under an imbalance of forces: water is squeezing down on it more powerfully than the gas in the bubble is pushing out. A real bubble, in contrast to our imaginary stiff bubble, would collapse under this imbalance. The fact that the bubble of a plastron gill does not can only mean that there is some force, not accounted for in

---

5. *Plastron* is from the Greek for "breastplate," so called because the bubble is usually carried on the beetle's ventral surface (on its "chest," as it were).

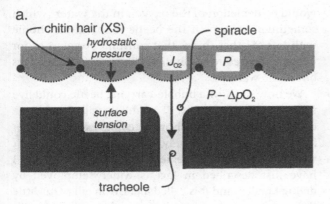

a.

chitin hair (XS)
hydrostatic pressure
spiracle

J_{O_2} P

$P - \Delta pO_2$

surface tension

tracheole

b.

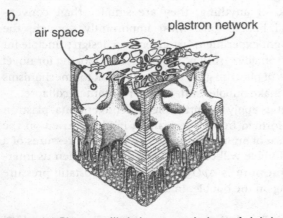

air space   plastron network

Figure 8.6 Plastron gills in insects. *a:* A sheet of air is held in place by a mat of hydrofuge hairs (shown in cross-section as filled circles), which span a space between the insect's body (dark shading) and the water (light shading). *b:* Detail of a plastron gill of a blowfly egg. [*From Hinton (1963)*]

equation 8.5, that is pushing outward at a force equal and opposite to the ΔpO_2. To give you an idea of the magnitude of this force, the ΔpO_2 in a typical plastron gill is about 5 kPa.

The most common way beetles stiffen the bubble involves using the hydrofuge to manipulate the surface tension forces. The hydrofuge consists of very closely spaced hairs; individual hairs are separated from their neighbors by distances of a micrometer or less. Because chitin is not wettable (that is, it resists absorbing water), these mats of hair resist penetration of water into them, by a sort of "reverse matric potential." In this case, water between the hairs form tiny meniscuses—crescent-shaped surfaces of water like those discussed in Chapter 7. The surface tension of each meniscus pushes outward on the water, resisting the excess hydrostatic pressure that ordinarily should collapse the bubble. This outward force can be quite large—the plastron can hold an air space against a hydrostatic pressure of 500 to 600 kPa, roughly five to six times atmospheric pressure and roughly a hundred times more forceful than the usual ΔpO_2 of 5 kPa or so. Thus, the bubble is "stiffened," in this case, by the surface tension forces that push the water away from the hydrofuge hairs.

With this design principle in hand, we can now recognize plastron gills popping up in circumstances we would never have considered before. For example, anyone who has been neglectful about taking out the household trash (as I have been on occasion) has confronted the problem of maggots, writhing around in liquefying rotting meat. Maggots, of course, have to come from somewhere, and this means there had to have been eggs laid in that fluid. If you can choke back your disgust for a moment, a question might occur to you—how do those eggs in there breathe? It turns out, in fact, that many insect eggs, particularly eggs that are laid in liquid, like a stream or a lake or a rotting puddle of carrion, are coated with woven mats of fibers, supported by complicated arrangements of pillars (Fig. 8.6). The woven mats exert outward "reverse matric" forces, just as the hairs in a hydrofuge do, and the pillars resist the collapsing force of the excess hydrostatic pressure. Clearly, the eggs are structured to be a plastron gill.

The Dynamic Bubble Gill of the Water Beetle Potamodytes
A particularly interesting example of a novel plastron gill is found in an aquatic beetle common to western

Africa, *Potamodytes*. Living in swift-flowing streams, *Potamodytes* faces into the current, clinging to rocks or sitting just behind them. These beetles often carry with them a prominent bubble, which encloses their legs and streams out behind them as it is pulled by the current (Fig. 8.7). In still water, the bubble carried by *Potamodytes* is a conventional bubble gill: it eventually shrinks and disappears and the beetle must come to the surface to capture a new one. Yet, in flowing water, the beetle is capable of maintaining its bubble and staying underwater apparently indefinitely, just as it could if the bubble were acting as a plastron gill. Under certain conditions, the beetle can even grow a bubble directly out of the water. Yet *Potamodytes* has none of the specialized structures, like hydrofuge hairs, that other plastron breathers use to stabilize their bubbles. Rather, these beetles maintain the plastron gill by using kinetic energy in the environment (or, as Mr. Spock might say, "from pure energy").

The beetle maintains its bubble by exploiting a behavior of flowing fluids known as the Bernoulli principle, named for the eighteenth-century physicist Daniel Bernoulli.[6] The Bernoulli principle expresses how different types of energy in a fluid interact. For example, flowing water has mass and velocity, which imparts to it kinetic energy. This can be precisely quantified:

$$KE = \tfrac{1}{2}\rho v^2 \qquad\qquad [8.6]$$

where KE is the kinetic energy (joules), ρ is water's density, about $1,000$ kg m^{-3}, and v is velocity (m s^{-1}). If this fluid is brought to a halt, as it would be if it encountered a solid wall, this kinetic energy would not disappear, it would be converted to a potential energy, which in fluids is expressed as a pressure. Bernoulli's principle simply states that the relationship between a fluid's kinetic energy and its potential energy are constrained by the First Law:

$$P + \tfrac{1}{2}\rho v^2 = k \qquad\qquad [8.7]$$

where P is the fluid's pressure and k is a constant. Thus, the total energy in a parcel of water can exist either as potential energy or kinetic energy, and these can interchange, as long as their sum remains constant.

To illustrate, suppose, for example, a parcel of water is abruptly brought to a dead stop. This means the kinetic energy in the parcel falls to zero, because now $v = 0$. But the First Law tells us that the energy cannot simply disappear. Rather, it is converted to potential energy, in other words, a pressure. Indeed, the pressure will be equal to the kinetic energy in the parcel when it was flowing at its unobstructed velocity; that is, $P = \tfrac{1}{2}\rho v^2$. Such pressures, called dynamic pressures, are a part of common experience—it is harder to run

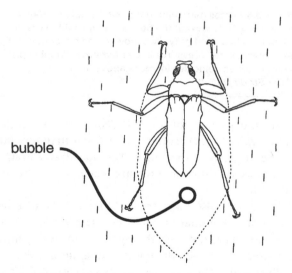

Figure 8.7 A *Potamodytes* beetle with its dynamic bubble gill. The outline of the bubble is indicated by the dotted line. [*From Stride (1955)*]

bubble

---

6. For the purposes of this discussion, I am presenting a simplified version of Bernoulli's principle, which in full is expressed as $P + \rho gh + \tfrac{1}{2}\rho v^2 = k$. The terms are designated the *head pressure*, P, the *hydrostatic pressure*, ρgh, and the dynamic pressure, $\tfrac{1}{2}\rho v^2$. What I have left out, of course, is the hydrostatic component of the pressure.

into the wind than with it, and you can even be knocked off your feet when a sufficiently vigorous wind is brought to a halt by your body.

Dynamic pressures are important in helping *Potamodytes* maintain an ordinary bubble as a plastron gill. In this case, the dynamic pressures arise from water being accelerated around the beetle and its bubble. Equation 8.6 tells us that an increase in velocity must result in an increase in the water's kinetic energy. The pesky demands of the First Law insist that the energy come from somewhere, and the only place it can come from is from pressure, that is, the pressure must drop to *less* than *P*. In other words, a dynamic *suction* pressure must arise that will pull outward on the obstacle to the flow. If a *Potamodytes* beetle carrying a bubble faces upstream, water must inevitably be accelerated around it. In accordance with Bernoulli's principle, there should be a suction pressure acting on the bubble, and this suction pressure should increase with the square of the velocity of the water. This is indeed the case (Fig. 8.8).

It is clear now how swift currents can make the bubble behave like a plastron gill. The outward force of the dynamic suction pressure opposes the forces that would normally collapse the bubble. Remember that the suction force need not be very great, on the order of a few hundred to a few thousand pascals or so. What normal plastron breathers accomplish by structure—hydrofuge hairs that use surface tension to resist the bubble's collapse—*Potamodytes* accomplishes by clever use of kinetic energy in the physical energy stream.

The mere presence of the beetle in a flowing stream is sufficient for the Bernoulli inflation of its bubble, but *Potamodytes* takes the principle a couple of steps further. For example, one curious anatomical feature of *Potamodytes* is the flattening of the proximal segments of the limbs, those closest to the body (Fig. 8.9). The beetle holds these limbs out to its sides, and the flattened surfaces are held at an angle pointing downward toward the substratum. The flat segments therefore act as hydrofoils, which accelerate the water up-

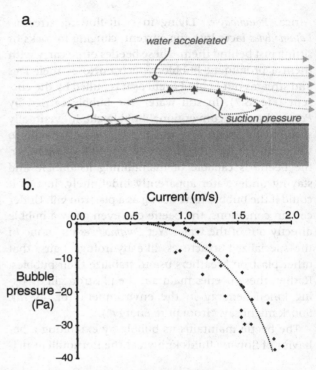

Figure 8.8 Forces maintaining the bubble gill of *Potamodytes. a:* Side view of the beetle with its bubble drawn out behind it. Dotted gray lines indicate the streamlines of flow. *b:* Measured pressures in a *Potamodytes* bubble go negative when velocities increase above about 0.5 m s^{-1}. [*After Stride (1955)*]

ward and inward toward the center of the beetle's body. Water flows past the bubble faster than it would in the absence of the limbs. The added acceleration magnifies the suction pressure keeping the bubble inflated.

Potamodytes also uses structural features of its environment to enhance plastron respiration. Most streambeds are not smooth: especially in swiftly flowing streams, where silt that might settle in between pebbles and rocks will have been swept away, the bottom can be quite rough. *Potamodytes* frequently position themselves just downstream from pebbles or other obstructions to flow, as if trying to find shelter

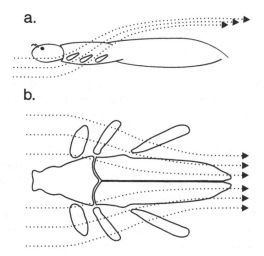

Figure 8.9 Use of the limbs of *Potamodytes* as hydrofoils to accelerate water above and to the center of the bubble. *a:* As viewed from the side, the tibiae of the beetle are in position to impart an upward acceleration to water flowing past the beetle. *b:* As viewed from the top, the tibiae are held at acute angles to the body. In this position they impart a centripetal acceleration to water flowing past the beetle.

in the lee of the obstruction. Fluids are accelerated around pebbles too, though, and suction pressures on the beetles' bubbles will be magnified by the enhanced velocity over the pebble.

The Winter Web of Argyroneta

We began this chapter with the diving bell spider, and the use of woven structures as gills. Let us now return to this problem, starting with the aquatic spider, *Argyroneta*, which, as noted above, uses its web as a conventional bubble gill. In fact, the spider only uses its web in this way during its active season, when it must get about and forage in its territory. In the winter, however, *Argyroneta* weaves a very different web. Where the summer web is thin and lacy, the winter web is a thickly woven mat of silk, and sometimes it encloses a rigid bracing structure like a leaf, a twig, or a snail shell. Sealed within its cocoon, the spider sits and waits out the winter. We can now see clearly that the winter web of *Argyroneta* is a candidate for a structural plastron gill. The thickly woven web and its bracing are structurally more resistant to collapse than the spider's rather diaphanous summer web. If the winter web sustains the pressure differences needed for a plastron gill to operate, it will, in accordance with the design principle, allow the spider to sit out the winter in its little house, underwater but dry and snug, breathing the oxygen that seeps in across the walls.

The Cocoon of the Ichneumon Wasp Agriotypus

One of the more bizarre examples of a woven plastron gill is built by a wasp that parasitizes caddis fly larvae. Caddis flies belong to an order of insects (the Trichoptera) whose larvae are predominantly aquatic. Caddis larvae themselves are architects of sorts—they construct little tubular houses for themselves from which they capture whatever prey wanders close enough. Typically, the caddis larva weaves a silken tube, open at both ends, which it then decorates with bits of leaves, tiny pebbles, even pieces of its prey. Sometimes, these decorations add structural strength to the caddis house (Fig 8.10).[7]

As every homeowner with a mortgage will understand, having a house restricts the movements of the caddis larva away from it. Being a "homebody" makes a caddis larva vulnerable to predators, just as a homeowner is more frequently harassed by telemarketers. Among the predatory dangers faced by caddis larvae are nasty little wasps known as ichneumons: the parasite in the movie *Aliens* was loosely based on an ichneumon wasp larva. The ichneumon lays its eggs on a caddis larva. The wasp larva, when it hatches, has the endearing habit of consuming its living host, eventually leaving only an empty husk.

---

7. Caddis houses have been extensively studied by behavioral and evolutionary biologists because their variation provides a useful indicator of a behavior and its evolution. Their construction is an example of "frozen behavior" alluded to earlier in the book, a durable record of an instinctual behavioral program. See von Frisch and von Frisch (1974) and Dawkins (1982).

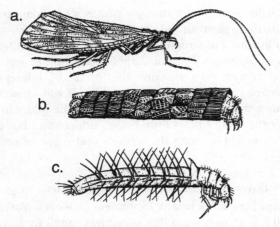

Figure 8.10 *a:* Adult of a caddis fly, about 2 cm long. *b:* Caddis house, built from flakes of plant material glued to a silken sheath. *c:* Caddis larva, showing spines it uses to anchor its house in place and sensory bristles around the thorax and head. [*From von Frisch and von Frisch (1974)*]

Our example concerns pupae of the caddis flies *Silo,* which are parasitized by larvae of the ichneumon wasp *Agriotypus. Silo* construct fairly rigid houses coated with tiny bits of gravel. During the initial stages of its parasitic life, the *Agriotypus* larva breathes through spiracular gills and so lives quite happily in the water-filled house of its host. During its last larval instar, though, just before pupation, the spiracles open and the *Agriotypus* larva becomes an air breather. Its habit of parasitizing aquatic insects now becomes a bit of a liability, because if it remains in water, it will drown. Unfortunately, it is not really in a position to leave the water for air, either. The newly air-breathing wasp larva therefore weaves itself a heavy web, which closes the open ends of the caddis house. By some process still unknown, the wasp larva then evacuates the water from the now-closed cocoon, opening up an air space inside.

This is a perfect situation for a plastron gill, and in all likelihood, the simple closed structure would behave as one, just as the winter web of *Argyroneta* does. The *Agriotypus* cocoon has an unusual feature, how-ever; a long ribbon, woven from silk by the wasp, extrudes four or five centimeters from the end of the larval case (Fig. 8.11). Remarkably, this ribbon appears to be a plastron gill. The ribbon's threads resist wetting, so that the ribbon encloses an air space that is continuous with the air space in the cocoon. Furthermore, the ribbon is crucial to the wasp larva's survival—if it is pinched off, the larva suffocates.

Even though the *Agriotypus* ribbon meets these criteria for a plastron gill, it seems to fall short in another. Consider a "conventional" plastron gill, held in place by a mat of hydrofuge hairs. These bubbles seem to be well-designed gas exchangers, according to the dictates of Fick's law. They are thin (x is small, a few micrometers) and they cover a fairly large area of the beetle (A is large, a few square millimeters). The ratio A/x is therefore very large, and Fick's law dictates a rapid oxygen flux. Well and good. Now consider the air-filled ribbon of an *Agriotypus* larva and try to fit *that* into the good design principles of Fick's law. It fails on two counts. Oxygen must travel a great distance along the ribbon before it gets to the larva, 5 cm or so (x is large). The ribbon is also narrow in cross-section (A is small). If *Agriotypus* is using Fick's law as a guiding

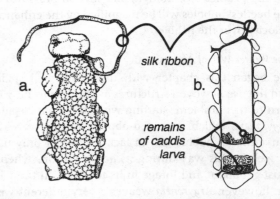

Figure 8.11 Parasitism of caddis larvae by *Agriotypus. a:* External view of a caddis house parasitized by an *Agriotypus* larva, showing the characteristic silk ribbon. [*From Thorpe (1950)*] *b:* Cross-sectional view of a parasitized caddis house. The *Agriotypus* larva is not shown, and only a portion of the ribbon is shown. [*From Clausen (1931)*]

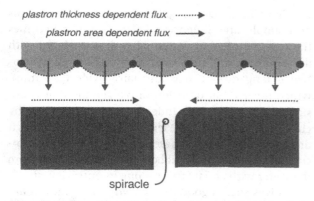

plastron thickness dependent flux ········▶

plastron area dependent flux ──────▶

spiracle

Figure 8.12 Two-dimensional flow of oxygen through a flat plastron bubble.

principle for its plastron gill, it seems to have developed a case of physiological dyslexia.

Occam's razor would, in this case, guide us to reject the hypothesis that the *Agriotypus* ribbon is a plastron gill. But let's see what Goldberg's lever can do for us. First, let us concede what is right: good design of a plastron gill must account for the limitations imposed by Fick's law, and indeed, we find that most plastron bubbles are both thin and capacious in area. Insect plastron breathers must fit this principle into their evolutionary legacy as air breathers, however, of which spiracles are an important feature. Thus, by dint of its ancestry, a plastron breather must convey oxygen from a bubble surface that covers a broad area of the insect and channel it to the very localized point of the spiracle (Fig. 8.12). This is rather more complexity than can be handled by the simple version of Fick's Law used so far in this book. It would do perfectly well if the entire surface of the beetle covered by the bubble could absorb oxygen—if, in the jargon of such things, the flow of oxygen across the bubble was one-dimensional. However, the flow of oxygen through a real plastron gill is not: it is two-dimensional. For oxygen to move from the water to the spiracle, it must flow across the bubble, perpendicular to the beetle's surface (one dimension), and then parallel along the surface to the spiracle (a second dimension).

If we are prepared to embrace complexity for its own sake, rather than avoid it lest it lead us astray, then some new principles for "good design" of plastron gills emerge. To illustrate, let us compare a "well-designed" plastron gill with a "poorly-designed" one. The plastron gill of an aquatic beetle, like *Aphelocheirus*, serves as an example. If we measure the oxygen partial pressures in the *Aphelocheirus* plastron, the ΔpO_2 between the bubble and water varies only little between the edge of the bubble and the spiracle it serves (Fig. 8.13, solid curve). There is a slight dip right at the opening to the spiracle, but in general, the ΔpO_2 changes little, even at points in the bubble far removed from the spiracle. What makes this a "well-designed" gill is the uniformity of oxygen partial pressure across the entire surface of the bubble. Because oxygen flow from the water into the bubble is proportional to the ΔpO_2, it is sensible to keep this difference as large as

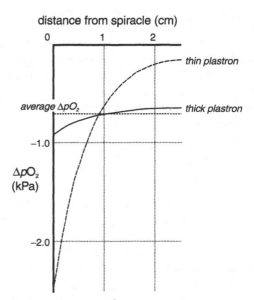

distance from spiracle (cm)

average ΔpO_2

thin plastron

thick plastron

−1.0

ΔpO_2 (kPa)

−2.0

Figure 8.13 The difference in oxygen partial pressure in water and in a poorly designed and a well-designed plastron bubble. [*After Crisp and Thorpe (1948)*]

possible over the entire outer surface of the bubble. This ensures that oxygen flows into the bubble over its entire surface. The flat distribution of ΔpO_2 simply expresses this steady level of difference graphically.

Now let us modify the bubble in a way that would seem to make sense according to Fick's law: make the bubble thinner (reduce x). If oxygen moved through the bubble in only one dimension (from the surface of the bubble to the beetle's skin), thinning it should enhance the flow of oxygen, making it a better gill. A thinner bubble actually makes for a poorer plastron gill, however. To see why, look at the distribution of ΔpO_2 in a plastron bubble one-tenth as thick as the normal *Aphelocheirus* plastron (Fig. 8.13, dashed curve). While the ΔpO_2 is quite steep near the spiracle, it is much shallower at the bubble's margins. This occurs because thinning the bubble impedes the flow of oxygen in the second dimension, that is from the margins of the bubble to the spiracle (Fig. 8.12). As a result, oxygen flowing into the bubble accumulates at its margins, and this reduces the ΔpO_2 that can develop there. Consequently, little oxygen enters the bubble along its margins because the concentration gradients that could drive a flux there are small. Put simply, the bubble's margins are "useless machinery," like the fins on those 1950s cars. Take it away, and the performance changes little.

These design principles are in fact encapsulated into a neat formula:

$$\Delta pO_2 \propto \sqrt{iO_2 s^2 / Dh}$$

where h is the thickness of the bubble, iO_2 is oxygen's invasion coefficient, s is the distance to the spiracle, and D is oxygen's diffusion coefficient in air. "Good design" is evidenced by making the thickness of the bubble large compared to the distance from the spiracle. Practically, the ratio on the right side of this equation should be less than 1, although the smaller it is, the better the plastron gill will perform.

Let us now return to the problem of the *Agriotypus* ribbon. This structure, which seemed such a poor candidate for a plastron gill when considered in terms of a simple one-dimensional flow, actually becomes the epitome of good design when it is analyzed with due appreciation for its complexity. Although the ribbon is long (5 cm), it is also fairly wide (2–3 mm), which means that diffusion of oxygen along it is not significantly impeded. Indeed, a calculation of the change of ΔpO_2 along the length of the ribbon shows it drops by only about 1 pascal from the tip to the cocoon. Thus, the entire ribbon can draw oxygen from the water with equal facility along its entire length.

If this is such a good design, why then is it not applied more often? In fact it may be used more often than we might think. For example, many insect eggs sport long projections from their egg cases. Often, these are used to attach the egg to some surface, but not always—in some eggs, the stalks are air-filled and project directly to the spiracles of the developing larva. In one of the more interesting cases, eggs of a blood parasite use a stalk to extract oxygen directly from a host's blood, essentially co-opting the respiratory machinery of its host to support its own gas exchange.

"Anti-Gills" in the Bubble Nests of Spittlebugs

Finally, we turn to an interesting example of a bubble gill that lets an aquatic insect live in the air: the bubble nests of spittlebugs. Spittlebugs are the nymphal stages of a group of insects, the Cercopidae, that are included among the insect family Homoptera. Among the more familiar Homoptera are the leafhoppers, aphids, and cicadas. Spittlebugs are especially common on meadowland crops like grass and lucerne (alfalfa), and most people who have spent any time walking around such habitats in the spring will have seen their nests: wads of white, bubbly "spittle" attached to stalks or stems (Fig. 8.14).

The spittle itself has some interesting folklore attached to it. Probably the strangest idea is the oldest, proposed by Isidorus in the sixth century, who believed the spittle to be the spit of cuckoos. Indeed, a common name for spittle nests in Europe is some variant of "cuckoo-spit." Among the slaves held in North America, spittle nests were thought to be constructed

Figure 8.14 Photograph of a spittle nest.

fall into two broad categories. On the one hand, the "smart" phloem feeders, mostly aphids, tap into a nice rich sap of sugars and amino acids. On the other hand, the "stupid" xylem feeders tap into the nutritionally impoverished sap in the xylem. Spittlebugs are "stupid" xylem parasites.

The spittle nest is produced from xylem sap mixed with secretions from glands in the bug's anus and from intestinal glands commonly found in insects, the Malpighian tubules. The spittle starts as a liquid drop extruded from the anus, which is taken up by the hind legs and folded so that it incorporates a bubble of air. The bubble is then passed forward into a tubular fold of the body wall along the bottom of the insect, called the ventral tube. The insect's spiracles open up into the ventral tube. The insect continually secretes raw spittle, and the repeated formation of bubbles from raw spittle forms the froth, which is eventually forced out the front end of the ventral tube. The overflow covers the outside of the insect and the stem it sits on, forming the spittle nest.

The spittle nest, despite looking like an ordinary wad of foam, is actually a woven structure, like the bubble webs of *Argyroneta*. The fluid of the raw spittle, which ends up as the bubble walls, includes lots of fibrous proteins, including short fibers of silk. Because the bubbles are impregnated with these silk fibers, they are very long-lived—the bubbles in a spittle nest will last for weeks, as long as the nest is not allowed to dry out. The large quantity of liquid produced—as much as thirty times the weight of the bug per hour—and its rich protein content makes the spittle nest a very energetically expensive structure. Some estimates place the energy investment in spittle production as high as 90 percent of the nymph's total energy budget. So, it is reasonable to suppose that the nest does something important for the animal. But what?

As many theories for the spittle nest's function have been proposed as there have been for its origin. Some of the more easily dismissed theories: it is a trap for insects (but spittlebugs eat only plant sap), it is an antimicrobial or anti-fungal shield (it actually is a quite good culture medium), it is a deterrent against preda-

by horseflies, perhaps reflecting some African folklore concerning their origins (and a story closer to the truth than the legends of their European masters). The great eighteenth-century naturalist John Ray got it right, though, by identifying the spittlebug as the source of the spittle. Surprisingly, how the bug produces the spittle, or even which end of the bug it comes from, the mouth or the anus, was not really settled until the early twentieth century (it's the anus, by the way).

Many homopterans are parasites of plants' vascular tissues that live by tapping into a stem or leaf and sucking out the fluid within. Among the vascular plants, there are two types of vascular tissue, phloem and xylem, each of which transport different types of fluid in the plant. Phloem usually carries in it a dilute mixture of water and sugars produced by photosynthesis—when maple sap flows at the end of winter, it is being transported by phloem from the roots, where it was stored the previous summer, to the branches, where it can fuel growth of a new leaf crop. Xylem, on the other hand, carries in it a very dilute solution of salts and amino acids and virtually no sugar. The homopteran parasites of plant vascular tissue likewise

tors (I can say from personal experience it doesn't taste bad). Current conventional wisdom, though, is that the spittle nest helps keep the insect from drying out. There is some justice to this claim: spittlebug nymphs do have thin skins, and they dry out quickly and die if they are removed from their spittle nests. Because spittlebugs probably were aquatic insects that secondarily returned to the air, it seems reasonable that the spittle nest might provide the bug an aquatic microenvironment, the same way amniotic fluid supposedly provides a "uterine pond" for a mammalian fetus.

Watch out for Occam's razor, though: the role of the spittle nest is actually not so simple. First, we have to ask: is there any possibility that spittlebugs will ever face a water deficit? The answer is probably not, since they feed from an extremely rich source of water, the xylem sap. How could they ever dry out while sitting in the middle of a veritable spittlebug Niagara? Indeed, the enormous production rate of spittle—thirty times the nymph's body weight per hour—is only possible because of the abundant supply of water provided through the xylem tissues of the plant host. So we are forced to ask another question: why should the spittlebug spend such enormous amounts of energy to construct a spittle nest to protect it against water loss when it has such an abundant supply of water to begin with?

The story gets even stranger. Spittle, being wet, obviously loses water by evaporation. The rate of evaporation is about 30 percent less per square centimeter of surface than it is from a plain water surface, a phenomenon that, again, fits into the supposed role of the spittle as protection against desiccation. Probably, the fibrous web of the bubble walls helps retard evaporation, in the same way a thin layer of oil on a pond reduces evaporation from it. But then you have to ask still another question: why should the nymph go to the trouble of blowing up its anal secretion with bubbles? If reducing water loss was the object, it would make more sense to just let the anal secretion ooze over the insect as a smooth sheet. By blowing the liquid up with bubbles, the insect increases the surface area from which water can evaporate. Indeed, spittle nests lose water at a total rate roughly 10 percent faster than the bug would if it simply let water, without all the added costly proteins, ooze over the smaller surface area of its body. So the spittle nest does not provide much protection against water loss, even supposing such protection is needed.

So, what is going on? It may be that the spittle nest is part of a plastron gill that has been carried up into air when spittlebugs made the transition from water back to land. For example, the ventral tube of spittle bugs is coated with a waxy cuticle that elaborates into a very finely reticulated network, similar in appearance and function to the plastron hairs of other aquatic insects. It seems a bit strange, though, that a plastron gill would be retained in the rich oxygen reservoir of the air. A plastron makes more sense, though, if it is intended to exchange some gas other than oxygen. Probably, that other gas is ammonia.

When any animal uses protein for energy, one of the inevitable by-products is ammonia. Ammonia is highly toxic, and animals that produce it must keep it at fairly low concentrations in the fluids of the body. For an animal that lives in fresh water, this is no problem. Ammonia diffuses across the body wall or gills and is diluted in the water of the environment. Indeed, as we saw in Chapter 7, the large flow of water through the bodies of freshwater animals helps keep ammonia concentrations low. In marine or terrestrial environments, however, animals face an ever-present danger that ammonia will build to unpleasantly high concentrations (Box 7A). This is an especially serious problem for terrestrial animals: unless the ammonia can be converted to something less toxic, like urea, it can only leave the body by volatilization—moving from solution into the gaseous phase.[8] Ammonia is

---

8. You will remember from Box 7A that many terrestrial animals use alternative nitrogen-containing compounds, like urea or uric acid, to remove ammonia wastes from the body. Spittlebugs seem not to use this strategy—neither urea nor uric acid is detectable in spittle nests. The implication is that these animals lose ammonia as ammonia gas.

not easily volatilized, however, because of the way it reacts with water to form the weak base ammonium hydroxide:

$$NH_3(g) \leftrightarrow NH_3(s) + H_2O \leftrightarrow NH_4OH \leftrightarrow NH_4^+ + OH^-$$

Getting ammonia to evolve away from a solution as a gas involves somehow biasing the reaction to the left.

Spittlebugs face an unusually serious challenge in this regard. Their diet of xylem sap is virtually all amino acid. Thus, virtually everything they consume for energy produces ammonia as a waste product. Furthermore, spittlebugs, as far as is known, are not able to detoxify the ammonia to urea or uric acid, as other terrestrial animals commonly do. Consequently, one of their major physiological challenges is how to handle the high ammonia load that is, for them, an unavoidable cost of living.

It just may be that the spittle nest is the solution. Ammonia loss is facilitated by providing a large surface area for it to volatilize from. Blowing up the spittle nest may be the spittlebug's way of increasing the volatilization rate of ammonia, in just the same way that blowing up the nest increases its rates of evaporative water loss. Furthermore, by filling the liquid portion of the spittle with bubbles of air, it increases the diffusion rate of ammonia away from the bug, for the same reason that diffusion is so effective in the air-filled tracheoles of insects. Indeed, ammonia moves through a layer of bubbles about twice as fast as it does through a parcel of water with no bubbles. So it just may be that the spittle nest of spittlebugs serves as an accessory gas exchange organ—one that promotes the exchange of ammonia rather than oxygen.

CHAPTER NINE

Manipulative Midges and Mites

The image captured in this epigraph—the arrogance of kings mockingly ground to rubble and dust—reflected the poet's radical (for the time) antipathy to monarchy and monarchs. Still, you have to give Ozymandias (known to us as Ramses II) some credit. It was, after all, *his* image and *his* words that sat there disintegrating in the sands, not those of some other fellow who lived four thousand years ago. *Something* about Ozymandias made him capable of leaving behind such a formidable monument.

What made Pharaoh's legacy an enduring one had nothing to do with his own physical prowess: it wasn't Ozymandias who carved out all those stones and hoisted them into place. Rather, the monuments were built because he had a talent for getting others to build them for him. Whether this talent derived from terror, persuasion, or inspiration, we really cannot say, but the magnitude of his legacy is undeniable.

The ability to appropriate the energy of others to build structures for you is not limited to human societies. In fact, many aphids, mites, flies, and wasps engage in this type of construction by proxy. These insects hijack the physiology of organisms much larger than they and force them to build structures that feed, shelter, and protect them. These structures are known collectively as plant galls, and it is to them that we shall turn in this chapter.

My thesis is that a particular type of gall, specifically leaf galls, are structures that a parasite tricks a tree into building for it. The gall is built for the "purpose" of altering the heat budgets of leaves afflicted by them. A parasite that instigates the construction of a gall in effect activates an adaptation normally employed by a leaf to cope with abnormal temperatures for its own benefit. To develop this idea, I shall first describe in general what galls are, with special attention to what leaf galls are and how they develop. I shall then explain how leaf temperature normally affects the energy budgets of leaves and how leaves use shape to control or otherwise manipulate their own temperatures. At that point, we will be ready to explore the consequences of galls for leaves' heat budgets and temperatures, and how these galls might affect the exchange of energy between leaves and their parasites.

Growth and Development of Galls

Galls are a developmental anomaly found in a wide variety of plants and in nearly every type of plant tissue, including stems, leaf blades, leaf veins, and the vascular and woody tissues of trunks, roots, and buds. They are usually induced by an arthropod parasite, although fungal, bacterial, and viral infections also can induce them. Frequently, the gall grows so that the parasite is enclosed in it. In some cases, the gall houses only one stage in the parasite's life cycle, like the larva, but in others, such as in some galls induced by aphids, the gall is the animal's permanent home—it may even house entire societies or communities of aphids. Some galls have economic benefits for humans—in medieval times, certain galls of oaks, for example, were an important source of ink pigments. More often, galls are deleterious to the health of the plant, as evidenced by the common designation of galls as "plant cancers." Indeed, many trunk and stem galls of trees do show the type of uncontrolled growth that is typical of cancerous tumors in animals (Fig. 9.1). Plants afflicted with these tumors face problems similar to those suffered by animals with cancer: a diversion of energy toward the growing mass that saps energy away from the host, weakening it and eventually killing it.

Some galls, however, particularly those that affect leaves or buds, are highly organized structures, and characterizing them as "plant cancer" seems decidedly

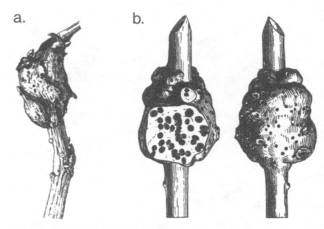

Figure 9.1 Two examples of "plant cancers," stem galls of oaks. *a:* The noxious oak gall, *Neuroterus noxiosus. b:* The gouty oak gall, *Andricus punctatus.* [*From Felt (1940)*]

inapt—I have selected a sampler of examples for Figure 9.2. Consider just one of these remarkable galls, the spruce cone gall. Cone galls arise at the leaf buds and bases of leaves of spruce (*Picea* spp.) and their relatives, hemlock (*Tsuga* spp.), and fir (*Abies* spp.). They are induced by a small fly, about a millimeter long, known as a gall midge, which lays its eggs at the bases of buds or leaves. Once the larvae hatch, the bud begins to grow abnormally into an anomalous woody structure with a shape very reminiscent of a spruce cone (Fig. 9.2a). The remarkable thing about these galls, of course, is their highly organized and most *un*cancer-like growth and development. It is as if the midge, in depositing its offspring at the base of the leaf bud, is commanding the tree (like Pharaoh to his slaves), "build a cone here!"

On the face of it, this is an extraordinary achievement. The formation of complex structures like spruce cones would seem to require a complex developmental program that controls rates and timing of growth and differentiation, as similar programs do in animals (Chapter 5). How does a gall midge make the tree build such a structure? Two possibilities come to mind. First, the midge may stimulate the plant's cells to begin

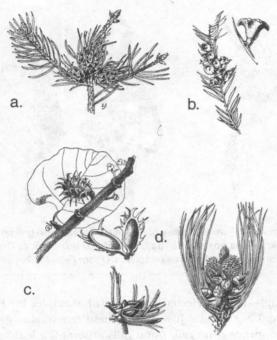

Figure 9.2 More highly organized galls develop in leaves and buds. *a:* The spruce cone gall, *Adelges abietis. b:* The cypress flower gall, *Itonida anthici,* showing a cluster of flower galls and one flower gall enlarged. *c:* The spiny witch-hazel gall, *Hamamelistes spinosus. d:* The pine bud gall, *Contarinia coloradoensis,* showing a cluster and a single gall enlarged. [*From Felt (1940)*]

to grow and, as they grow, direct the cells' rates of growth and differentiation. If the midge can force the developing plant tissue to mimic the patterns and rates of growth that characterize the development of a normal cone, the gall will resemble a cone. On the other hand, the midge could hijack the developmental program already used by the plant to produce a cone; that is, it may activate the plant's own developmental programs inappropriately, the way a hacker might break into a company's computer system and make it write checks. In this case, the hacker does not write the code to make the victim's printer cut the checks—he or she simply activates software already written by the company and normally used by it to cut its checks.

Plant Homeosis and Galls of Buds and Leaves

Cone galls, or other galls that likewise mimic normal plant structures, probably are induced by parasites acting as genetic hackers. This is surprisingly simple to do. Take, for example, galls that mimic structures derived from leaf buds. The leaf bud is a sort of foundation structure for a variety of specialized structures. Leaf buds can develop into (of course) leaves, but they also develop into flowers, or thorns, or hairs. A leaf bud's developmental fate can be mapped out as a series of contingent pathways. The early stages of development are the same whether the structure is destined to become a leaf or a flower. At some point, though, the developmental pathways for the different structures will diverge. Follow one pathway, and the bud develops into a leaf. Switch pathways at some point, and a flower might be the result.

The points of divergence in developmental programs, both in animals and plants, are controlled by so-called homeotic genes, which act essentially as genetic switch points. For example, one homeotic gene might initiate the developmental pathway for a leaf-derived structure like a thorn. Keep the gene switched off, and development will proceed toward leaves or flowers. Switch the gene on, and development will proceed toward a thorn. Another homeotic gene, later in the sequence, might control the point of divergence for leaves versus flowers. Keep the gene switched off, and development proceeds toward a leaf. Switch it on, and development proceeds toward a flower.

In normal development, homeotic genes are useful because they provide simple ways of controlling complex patterns of development. You don't need to regulate an entire complex program, all you need to do is regulate the switch that activates it. Having these developmental switches, however, makes the plant vulnerable to attack by organisms that have learned how

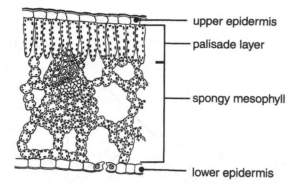

Figure 9.3 Cross-section through the lamina of a leaf, indicating the four principal layers of cells. [*From Bridgewater (1950)*]

to switch the genes on themselves. This is probably what gall midges do: they initiate anomalous production of cones by switching on the homeotic gene that controls cone development. Once activated, the program runs whether doing so suits the needs of the plant or not, and a cone gall results.

Growth and Differentiation in Normal and Galled Leaves

Gall inducers can also work the hard way—by intervening in and controlling the growth and differentiation of cells, rather than the homeotic switch points. These types of galls are common in growing or already mature leaves. Leaf galls commonly form relatively late in development, after the leaf bud is already developmentally committed to the formation of a leaf, rather than another structure like a flower. By controlling the leaf's normal patterns of growth, the gall-inducer can significantly alter the leaf's shape as it develops. To see how, you must first understand something about how leaves normally develop.

A mature leaf consists principally of two parts: the leaf ribs and associated vascular tissue, and the leaf lamina, the flat sheet of photosynthetic tissue that fills in between the vascular tissue. In a typical leaf, the lamina itself consists of four layers of cells (Fig. 9.3).

The upper epidermis and lower epidermis, located on the upper and lower surfaces of the leaf, are composed of cells that are typically flattened and tough. Below the upper epidermis is a layer of columnar palisade cells that together form the sheet-like palisade layer. Sandwiched between the palisade layer and lower epidermis are the loosely organized spongy cells, which form the spongy mesophyll. The cells in the spongy mesophyll are rather loosely packed so that oxygen and carbon dioxide can percolate through the layer.

A leaf begins as a small protuberance from the stem known as a foliar buttress (Fig. 9.4). Within the foliar buttress, two types of primordial tissues are evident. Running along the midrib of the buttress are the cells of the primordial vascular tissue, which will become the midribs and vascular tissues of the leaf. Flanking this are two bodies of marginal or lateral meristem, which will grow and differentiate into the leaf lamina.

The shape of a leaf is determined by the basic patterns of growth of these primordial tissues. The primordial vascular tissue grows by elongation along the long axis of the foliar buttress, while the lateral meristem grows outward from the stem. If unmodified, the result of these combined patterns of growth would be an elongate leaf, similar in shape to a tobacco

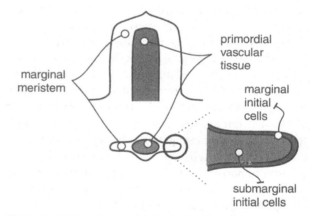

Figure 9.4 Cell types in the foliar buttress.

leaf, consisting of a single midrib flanked by the leaf lamina.

Leaf shape can be modified in various ways by modifying these basic patterns of growth. One common modification is the branching of the primordial vascular tissue. For example, if the primordial vascular tissue elongates for a time, sends branches off to each side, and then continues to elongate, the leaf will take on the shape, say, of a maple leaf. Leaf shape can also be altered by modifying the growth of the lateral meristem relative to the vascular tissue. If vascular tissue and lateral meristem grow equally rapidly, for example, the leaf will have a simple outline—that is, its margin will not be cut with lobes, teeth, or other complicating shapes (Fig. 9.5a). But if vascular tissue grows relatively faster than the lamina, the leaf might have a lobed or toothed appearance, like a maple or oak leaf (Fig. 9.5b). If the initiation of growth of the lateral meristem is delayed with respect to the vascular tissue, the leaf might even develop as a compound leaf, with numerous leaflets arrayed along "stems" of vascular tissue, as is the case in leaves of ash or locust (Fig. 9.5c).

The internal structure of a leaf also depends upon characteristic patterns of growth and differentiation in the marginal meristem. Early on, the marginal meristem differentiates into two cell types. The marginal initial cells are arrayed along the top and bottom surfaces of the marginal meristem: these cells will eventually form the upper and lower epidermis of the leaf. The submarginal initial cells, sandwiched between the two layers of marginal initial cells, will form the palisade and spongy mesophyll layers. Normally, the marginal initial cells will go through several cycles of cell division and then cease dividing and continue to grow by lateral elongation. This pattern of growth both flattens the cells and extends the space between the two epidermal layers. This space is filled as the palisade cells and spongy mesophyll continue to divide and grow (Fig. 9.6, *third line from top*). Commonly, the spongy mesophyll cells cease dividing first, and the continued lateral elongation of the leaf causes these

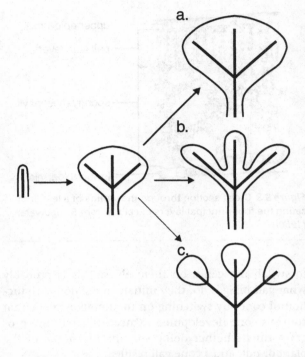

Figure 9.5 Development of leaf shape and the relative developmental rates of the primordial vascular tissue and the marginal meristem. *a:* The development of the simple leaf. *b:* The development of a lobed leaf. *c:* The development of a compound leaf.

cells to separate from one another, opening up the air spaces of the spongy mesophyll (Fig. 9.6, *bottom*). The palisade cells continue to divide as long as the epidermal cells continue to elongate, and they consequently pack themselves into the compact palisade layer.

Many leaf galls arise through modifying these basic patterns of growth. For example, leaf-rolling is a common gall disease (Fig. 9.7a). As the name implies, the leaf grows into a curled-up cylinder rather than a flat blade as it normally does. Leaf rolling is a direct result of the parasite forcing a mismatch in the growth rates of the upper and lower epidermes. If the upper epidermis grows more slowly than normally, for example, the leaf lamina will roll upward. The mismatch in growth rates comes about in a number of ways, all of which disrupt the temporal synchronization of the growth of the marginal initial cells.

Leaf crinkling is another common example, but this time the mismatch is between the growth rates of the primordial vascular tissue and the marginal meristem (Fig. 9.7b). If the primordial vascular tissue prematurely stops growing, or if the leaf lamina continues to grow past the time it would normally stop, the lamina will buckle and crinkle along its surface. Both leaf rolling and leaf crinkling can come early in a leaf's development, in which case the entire leaf will roll or crinkle. In some gall diseases, the abnormal growth sets in later in leaf development, in which case the malformation is limited to the leaf's margin or focal points on the leaf blade.

Galls can also arise in mature leaves, as a result of the insect forcing a dedifferentiation of the leaf tissue; in effect, the insect causes a portion of the leaf's cells to revert to their primordial and undifferentiated states. Dedifferentiated cells are more prone to grow, and as

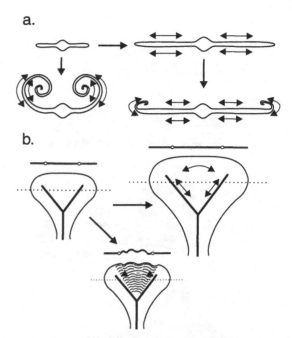

Figure 9.7 Crinkling and curling of leaves induced by developmental anomalies. *a:* Leaf curling. *b:* Leaf crinkling. The dotted line indicates the level of the cross-sectional view shown above each top-down view of the leaf.

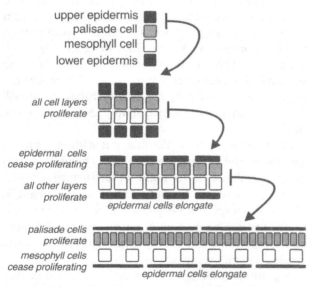

Figure 9.6 Sequence of development in the growth of the leaf lamina.

they grow the gall-inducer directs their subsequent development. A simple example of this type of gall is the so-called bladder gall commonly found on the leaves of hackberry (*Celtis* spp.), induced by a gall midge (Fig. 9.8). The formation of a bladder gall begins with the parasite puncturing some of the palisade cells, injuring them so they and the surrounding cells begin to dedifferentiate and divide. The locally high rate of growth causes the leaf surface to buckle, producing a small pit. As the larva matures, the palisade cell layer and epidermis are then induced to dedifferentiate and grow, forming an upward-growing cup that surrounds and eventually encloses the larva. Such cups are not in any way a normal part of the leaf, but they are characteristic of the species of midge inducing it.

Needle galls, which afflict the leaves of the sugar

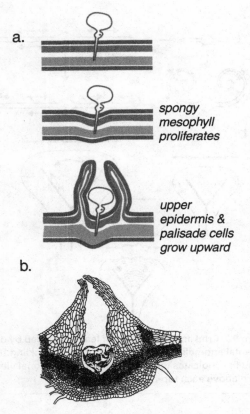

a.

spongy mesophyll proliferates

upper epidermis & palisade cells grow upward

b.

Figure 9.8 Development of a bladder gall on the leaf of hackberry *(Celtis)* by the growth of a gall midge *(Pachypsylla). a:* Schematic development of the bladder gall. *b:* Cross-section through a *Pachypsylla* gall on a hackberry leaf. [*From Wells (1964)*]

maple, are formed in a similar way, but by a mite, not a midge. The mite dedifferentiates the epidermal tissues of the leaf, and the ensuing growth produces a thin column of epidermal tissue, which grows up around the mite, enclosing it. The thin needle-like column of tissue gives the gall its name.

Leaf Temperature and Photosynthesis

Let us now shift gears a bit. Leaves, of course, are a plant's principal engines of energy acquisition. How well they perform this task affects all other aspects of a plant's biology. One would expect, therefore, that natural selection should refine leaves to work as well as possible. Natural selection can only do so much, however. Leaves function in a fickle physical environment—just consider the wild and difficult-to-predict fluctuations in so seemingly simple a thing as the air temperature. No matter how refined the functioning of a leaf is, natural selection's perfecting power will always be subverted by the chaotic variation of the physical environment in which the leaf exists. Often, this inescapable conflict is resolved by some compromise between efficient photosynthesis and variable leaf temperature.

The compromise is best understood by invoking a crass materialist analogy—the tree as an industrial corporation. A corporation works by taking money and doing something with it that adds value. The money put in is the investment, or capital, which is used to develop and operate the infrastructure that generates the added value. A corporation does best when it sustains the production of added value. This does not necessarily mean maximizing profit: sometimes it means diverting some of the added value away from profit and plowing it back into more investment. A healthy corporation is one that looks soberly at all the costs and returns that accrue to its operations and sustains the maximum sustainable difference between them. Generally, corporations that do so will fare better than those which simply seek to maximize profit.

Plants do something like this, but with energy, not money, as the currency. Plants take chemical energy (sugar) and use it to build infrastructure (leaves) whose purpose it is to capture the energy of light in chemical energy (photosynthesis). A plant does best when it sustains a maximum return on its investment, and temperature has a very important role in determining a plant's returns.

A plant cell, like any other cell, consumes energy both to maintain itself and to grow. Having a leaf commits a tree to meeting these metabolic costs, which are measured by how rapidly the leaf releases carbon dioxide. Similarly, benefits are measured by how rapidly the leaf consumes carbon dioxide, essentially, the rate at which the carbon in carbon dioxide is incorporated into sugar, also known as the gross photosynthesis (Fig. 9.9, top graph). A leaf returns added value on the investment in it if, in its lifetime, CO_2 consumption exceeds CO_2 production. In other words, the leaf succeeds if it has produced more sugar than it has consumed and its net photosynthesis (gross photosynthesis-metabolic cost) is maximized (Fig. 9.9, bottom graph).

It is important to realize that maximizing net photosynthesis does not necessarily mean maximizing gross photosynthesis. If we plot rates of CO_2 consumption (by photosynthesis) or production (by metabolism) against leaf temperature, for example, we see a charac-

teristic variation of these fluxes with temperature (Fig. 9.9, top). Both increase with temperature, until at some critical temperature they level out and peak. At temperatures warmer than the critical, the rates drop rather abruptly. This pattern has a straightforward explanation. In most chemical reactions, whether they occur in cells or in a beaker, rates of reaction roughly double with every 10°C increase of temperature. This is what drives the increase of both CO_2 consumption and CO_2 production over the cooler range of leaf temperatures. Most chemical reactions in cells, however, are mediated by protein catalysts called enzymes. Enzymes are rather delicate machines, and high temperature may disrupt their functioning or the ways they interact with other enzymes. A degradation of their function is the result.

In many leaves, the consumption and production rates for CO_2 respond differently to temperature: CO_2 production rates commonly peak at higher leaf temperatures than do the rates of gross photosynthesis. One interesting consequence of this mismatch is that the leaf's maximum *net* photosynthesis occurs at a cooler leaf temperature than the temperature of maximum *gross* photosynthesis. We shall call this the optimum temperature for photosynthesis, or $\hat{T}$.

Corporate decisions on costs and allocations are subject to the discipline of the market. If the corporation does not make wise decisions, its competitive strength will diminish, and it will be in danger of making little profit or going into bankruptcy. Natural selection provides a similar disciplinary environment; poor allocation "decisions" trees make with respect to their leaves are penalized with reduced fitness, perhaps even extinction. One would therefore expect to see a convergence between leaves' optimum temperatures for photosynthesis and the temperatures they commonly experience. In other words, natural selection should force plants toward the rule:

$$\hat{T} \approx T_{\text{leaf}} \qquad [9.1]$$

where T_{leaf} represents the average leaf temperature in a particular environment. The basic evolutionary prob-

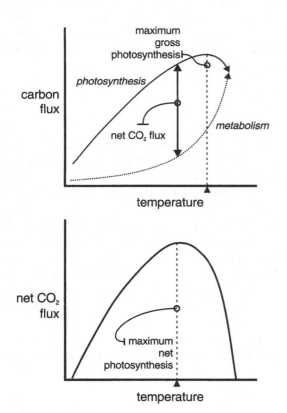

Figure 9.9 Optimizing photosynthesis of a leaf.

lem for plants, then, is to somehow match these two temperatures. There are essentially two ways plants can do this.

One way is for Mohammed to go to the mountain: push the leaf's $\hat{T}$ toward T_{leaf}. This involves modifying the complement of enzymes that control metabolism and photosynthesis in the leaf so that they function best at the temperature leaves are most likely to experience. In fact, many plants do just this: leaves of desert plants, for example, tend to have higher optimum temperatures for photosynthesis than do arctic plants. No surprises there, since deserts are commonly hot places. Plants also must cope with more acute variations of temperature: even deserts can be cold during some parts of the year. When faced with these occasional extremes, plants sometimes maintain optimum photosynthesis by keeping multiple sets of enzymes for photosynthesis and metabolism (called isoenzymes, or simply isozymes), each set to operate at different temperatures. If temperatures in the desert get very cold, for example, the plant could shut down its complement of "hot environment" isoenzymes (that is, those that result in a high $\hat{T}$) and activate its "cold environment" isoenzymes (the ones that result in a lower $\hat{T}$).

Such metabolic adaptations are not without cost. Having a strictly determined $\hat{T}$ leaves the plant at the mercy of a fickle and changeable environment. The benefits that accrue to extending $\hat{T}$ with isozymes also is limited, because each set of isozymes adds to the plant's genetic "overhead." Each set of isoenzymes requires the cell to maintain multiple sets of genes for the same reactions, which means more energy invested in overhead. Nevertheless, the fact that many plants do tune their biochemistry to prevailing temperature indicates this has been a good strategy for them.

Many plants do the opposite, though. Forcing T_{leaf} toward $\hat{T}$, they move the mountain to Mohammed. This seems odd, because it is difficult to see how a leaf's temperature could be anything but what the environment imposes on it. Nevertheless, the physical environment is a rich mosaic of various kinds of energy, and leaves can actually play these different sources of energy against one another to attain some degree of control over their temperatures. Leaves most commonly do this through variation in their shape. So, if an environment is too cold for its leaves to function efficiently, a tree could elevate its leaf temperatures to some extent simply by altering their shapes.

Physics of Leaf Temperature

To understand how plants do this, and how leaf galls might disrupt it, we must understand how heat flows between a leaf and its surroundings. The temperature of any object, leaves included, is a measure of the quantity of thermal energy, or heat, contained in it. Heat content is, in turn, determined by the balance between the heat coming in (q_{in}, expressed in watts per square meter) and the heat going out (q_{out}, also in W m^{-2}). Usually, there will be some temperature at which these flows of heat balance, or come into equilibrium, so that the net flux of heat is zero:

$$q_{in} + q_{out} = 0 \qquad [9.2]$$

We can estimate the temperature of a leaf at equilibrium if we can account for all the flows of heat into and out of the leaf and their dependence upon leaf temperature. For most leaves, these flows fall into three categories. First, solar radiation can warm a leaf that absorbs light. Second, evaporation can remove heat from the leaf. Third, wind blowing past the leaf can either warm the leaf or cool it, depending upon whether the air is warmer or cooler.

All these fluxes of heat combine into the leaf's heat balance equation. For a leaf at steady temperature, all the inflows and outflows of heat cancel and add to zero (equation 9.2). So, for a leaf exchanging heat with the environment by radiation (q_r), evaporation (q_e), and convection (q_c), the heat balance equation is:

$$q_r + q_e + q_c = 0 \qquad [9.3]$$

where each quantity q represents a flux rate of heat.

Each of these flux rates depends upon the leaf temperature. Convective heat exchange, for example, is a simple function of the difference in temperature between the leaf and air. Therefore, it is, in principle, possible to rewrite each of the heat flows as a function of leaf temperature. Doing so should, again in principle, make it possible to solve for the temperature at which the heat flows all add up to zero. This will be the temperature of the leaf in a particular environmental regime of sunlight, humidity, and air temperature.

The rub lies, as it so often does, in those deceptively simple words "in principle." In fact, the equations for each of the heat fluxes can be hellishly complicated. For example, radiative heat flux includes the leaf's temperature all right, but it must be the absolute temperature (in kelvins) raised to the fourth power. The equation also must include terms for orientation of the leaf with respect to the sun, position of the sun in the sky, scattering of sunlight by the atmosphere and particulate debris, and proximity and temperatures of other sources of radiation heat exchange, like reflection from nearby leaves or from the ground. The equation for evaporative heat flux is simpler, but scarcely so: evaporative heat flux is influenced by the vapor pressure of the leaf's water, which is dependent upon leaf temperature, and also by the humidity and the water's heat of vaporization, themselves complex functions of temperature. So one can write the heat balance equation all right, but it will very likely occupy the better part of a page.

So, as my daughter would have put it when she was ten, do you have to be "like, really smart" to be able to get a leaf temperature out of the heat balance equation? If you want an accurate temperature, the answer to this question, I am afraid, is "yes." Fortunately, the equation can be made much simpler and still yield reasonable approximations for a leaf's temperature. The first step is to realize that part of the solution is hard (radiation and evaporation), and part of it is simple (convection). We then try to do the simple bit as best as we can and fall back on approximating the hard bits.

Turning first to the relatively simple process of convection, q_c can be easily rewritten as a function of leaf temperature and air temperature (T_{air}):

$$q_c = h_c (T_{air} - T_{leaf}) \qquad [9.4]$$

where h_c is a quantity known as the convection coefficient (W $°C^{-1}$ m^{-2}). Equation 9.4 is easy to solve because it expresses convective heat loss as a simple linear function of the difference between leaf temperature and air temperature. We now substitute the quantity $h_c(T_{air} - T_{leaf})$ for q_c in the heat balance equation, so that it looks like this:

$$q_r + q_e + h_c(T_{air} - T_{leaf}) = 0 \qquad [9.5]$$

We next simplify the terms for radiation and evaporation by combining them into a single term, q_{net}, which is simply the sum of the radiative and evaporative heat fluxes. Expressing it this way enables us to put in reasonable numbers for net heat flux rather than solving for them directly. For example, if $q_{net} = 0$, this simply means that any radiative heating of the leaf is offset by an equal rate of evaporative cooling. Similarly, if $q_{net} > 0$, radiative heating exceeds evaporative cooling, and there is a net heating of the leaf. Typical net fluxes will vary between +300 and −300 W m^{-2}. A little simple algebra then lets us solve this equation for leaf temperature:

$$q_{net} + h_c(T_{air} - T_{leaf}) = 0 \qquad [9.6]$$

$$T_{leaf} = T_{air} + q_{net}/h_c \qquad [9.7]$$

Equation 9.7 allows us to estimate leaf temperature from three simple quantities: air temperature, the leaf's convection coefficient, and the net heat flux. It also expresses some convenient rules of thumb for how leaf temperature behaves and how a plant could control it. There are four:

1. Leaf temperature is a simple linear function of air temperature plus some temperature increment that is equivalent to the ratio of the leaf's

net heat flux and the leaf's convection coefficient.

2. Temperature increment is not determined by the absolute magnitude of any of the leaf's fluxes of heat, but by how these fluxes vary in proportion to one another. For example, if q_{net} = 600 W m^{-2} and h_c = 60 W °C^{-1} m^{-2}, the temperature increment would be 10°C. The same temperature increment would result if q_{net} = 200 W m^{-2} and h_c = 20 W °C^{-1} m^{-2}.

3. Temperature increment is magnified by increasing net flux with respect to the leaf's convection coefficient. Similarly, temperature increment is reduced by increasing the convection coefficient with respect to the leaf's net heat flux.

4. Temperature increment can be either positive, indicating an elevation of leaf temperature above air temperature, or negative, meaning that the leaf will be cooler than air temperature. Whether temperature increment is positive or negative depends upon which term in the net heat flux predominates. If evaporative cooling is more intense than radiative heating, the leaf will be cooled. If radiative heating is more intense, the leaf will be warmed.

Optimizing Leaf Temperature through Variations of Leaf Shape

The key to controlling leaf temperature, then, is manipulating the temperature increment. Suppose, for example, a plant lives in a hot sunny environment, where air temperature is high. The plant could keep its leaf from getting too hot if it could keep the leaf's temperature increment low. The terms of the temperature increment, q_{net}/h_c, suggest some strategies a plant might pursue. It could, for example, adjust the absorption of radiation to keep q_{net} as small as possible. Many desert plants indeed do just this: their leaves are covered with silvery hairs or whitish waxy coatings that reflect light away from the leaf. In a pinch, by increasing evaporation from the leaf, the plant might even drive q_{net} below zero (although this might be problematic in a desert, where water is scarce).

The temperature increment could also be lowered by elevating the convection coefficient, h_c. The convection coefficient is mainly a function of the boundary layers that limit heat exchange by convection. In general, thinner boundary layers make for steeper temperature gradients between the leaf surface and air. Increasing the temperature gradient boosts heat flow, irrespective of what the temperature difference might be. In other words, anything that thins a leaf's boundary layer will elevate the convection coefficient and reduce the temperature increment.

The boundary layer is the connection between leaf temperature and leaf shape. Consider what happens when wind encounters a flat surface like a leaf. At the leaf's leading edge, the boundary layer will be very thin, and its thickness will grow as the air moves along the leaf's surface. It follows that the boundary layer will be thinner over a narrow leaf than it would be over a broad leaf—narrow leaves give the boundary layer less distance along the leaf to grow. Plants in hot sunny environments, therefore, should have leaves that are smaller or narrower than plants in cooler or shadier environments. Indeed, that is generally the case. Desert trees like mesquite or acacia usually have tiny leaves, compared with those of related species that live in more temperate climes. The massive leaves of rhubarb, found only in shady conditions, further underscore this point.

Even within a single plant, though, leaves experience a wide range of environmental conditions. In the crown of a maple or oak tree, for example, leaves that are located at the crown's outer margins will experience sunnier and hotter conditions than leaves in the shadier interior of the crown. If a maple leaf has a standard size and shape, leaves that are inside the crown should therefore be cooler than leaves at the sunnier margins. If maple leaves have a particular optimum temperature for photosynthesis, the tree runs the risk that, at any particular time of the day, a considerable fraction of its leaves will not be functioning at their optimum temperatures. Some simple variations of leaf shape or size can ameliorate this problem. For example, leaves in the shady interior of the crown tend to be larger and smoother in outline, while leaves in the crown's sunnier margins tend to be smaller and "pointier," divided more along the margins. Consequently, "shade leaves" will have thicker boundary layers over them, reducing convection heat loss and increasing the leaves' temperature increment. "Sun leaves" have thinner boundary layers, and their "pointiness" acts further to disrupt the boundary layers. These features enhance convective heat loss, reducing temperature increment. The end result is a relative uniformity of leaf temperature throughout the crown, despite the substantial variation of radiative heat reaching the different leaves.

Leaf Galls and Convective Heat Loss from Leaves

We are now (finally!) ready to return to the matter of leaf galls. Let us review three things we have learned so far. First, affliction by leaf galls produces dramatic changes of leaf shape. Second, leaf temperature is an important factor in the efficiency of leaf function. Finally, leaf shape is an important component in a tree's ability to manipulate the temperature of its leaves. Putting these thoughts together, may we conclude that insects that induce galls are somehow manipulating leaf shape to alter the leaf's temperature, and so bias the leaf's metabolism, presumably to the inducer's benefit?

This is a pretty tall order, admittedly, but I'm going to explore it anyway. I warn you in advance that my argument is very speculative. You may, in the end, find yourself agreeing with Mark Twain when he famously disparaged scientists who make "such wholesale returns of conjecture out of such a trifling investment of fact." Nevertheless, the principle of Goldberg's lever suggests that my attempt may pay off, so here goes.

If galls affect leaf temperature at all, it is likely they will do so through an effect on the leaf's convective heat exchange. For example, galls that protrude above the leaf surface, like the hackberry gall or the needle galls of maple, could alter the boundary layers at the leaf surface. This in itself is not remarkable: many plants use protuberances like spines or hairs to alter the convective environment of their leaves. Just how plants use these spines is not always consistent, however. Many desert plants have spines or tufts of hair to promote turbulence in the boundary layers that gather at their leaves, and this turbulence thins boundary layers and enhances convective heat flux. On the other hand, some plants use these protruding structures to thicken boundary layers and produce an insulating layer of relatively still air that reduces convection fluxes. So it is not possible to say in advance what effect galls might have on an infested leaf's temperature. The presence of galls could either increase or reduce a leaf's temperature increment. And, of course, there is always the possibility that galls will have no effect whatsoever on the leaf's temperature.

Fortunately, this is a question that can be fairly easily settled by experiment. It is simple to construct a model leaf out of aluminum sheet and to heat it with a heating coil. The heat dissipated by the coil is calculated from the electrical current passing through it, and dividing this loss of heat by the model leaf's surface area gives the q_{net}. Leaf temperature and air temperature are also easily measured, which gives T_{leaf} and T_{air}. The convection coefficient is then calculated by rearranging equation 9.6:

$$h_c = q_c / (T_{air} - T_{leaf}) \qquad [9.8]$$

I have made some measurements in my laboratory of model leaves with two types of galls. I simulated "needle galls" by gluing small lengths of monofilament upright on the leaf surface, about the length and diameter of the real thing. For hackberry galls, which are more or less spherical, I glued small beads of expanded polystyrene onto the model leaf. I then placed the entire apparatus into a wind tunnel and made measurements of the model leaf's convection coefficient at various wind speeds. The result is clear: the presence of "galls" on a model leaf reduces the leaf's average

Table 9.1 Average convection coefficients, h_c, measured for heated model leaves. Each mean is significantly different from the others. Percent difference is calculated with respect to the model leaves with no galls.

Group	h_c (W °C^{-1} m^{-2})	% Difference
No galls	13.29	—
With 20 needle galls	12.00	−9.71%
With 20 spherical galls	11.62	−12.57%

convection coefficient, whether the galls are spherical or needle-shaped (Table 9.1). Spherical galls will have a greater effect in this regard than will needle galls, although the difference is slight. Galled leaves should therefore be warmer than ungalled leaves. If real galls have a similar effect on real leaves, then the presence of galls should increase the leaves' temperature increment.

Leaf Galls and the Energy Balance of Infested Leaves
What does this prove, exactly? Simply demonstrating that leaf galls can alter a leaf's convection coefficient is not enough to say that the change will have any adaptive value. To get *there*, we must ask two other ques-

tions. First, by how much will the actual temperatures of galled leaves change? Second, to what degree will this change of temperature affect the net photosynthesis of a galled leaf? Unfortunately, the answers to these questions are completely unknown, but let us not let a little thing like ignorance stand in our way—let us follow the thread and see where it leads.

To take the first question: what change of temperature should we expect for the 10–15 percent reduction in a galled leaf's convection coefficient? Equation 9.6 for T_{leaf} shows that the effect of any change in the leaf's convection coefficient will vary with the magnitude of the net heat exchange. I have made a few calculations of what temperature increments should be expected for values of q_{net} that more or less span the range that real leaves experience (Table 9.2). These calculations indicate that in some circumstances galls will have a small effect, but in others the effects will be pretty large. At low values of q_{net}—which might indicate conditions of low radiation, or high radiation with a considerable rate of evaporation—galled leaves should be less than one degree warmer than ungalled leaves. At higher values of q_{net}, though, which might occur under sunny, more humid conditions, galls on a leaf should warm leaf temperature by a few degrees. Under all circumstances, though, galled and ungalled leaves

Table 9.2 Temperature increments (ΔT, in °C) estimated for model leaves under various regimes for q_{net}. The first three rows tabulate temperature increments for model ungalled leaves, model leaves with simulated needle galls, and model leaves with simulated spherical galls. The bottom two rows indicate the difference in temperature increments (δT) between model leaves with galls and model leaves with no galls. These numbers indicate how much galls raise the temperature of the leaf above that of a leaf without galls.

Experimental condition	Increase in leaf temperature (°C) due to increases in q_{net} of 0–500 W m^{-2}					
	0	100	200	300	400	500
ΔT (no galls)	0.0	7.4	14.7	22.1	29.4	36.8
ΔT (needle galls)	0.0	8.2	16.3	24.5	32.7	40.8
ΔT (spherical galls)	0.0	8.4	16.8	25.2	33.6	42.0
δT (needle galls − no galls)	0.0	0.8	1.6	2.4	3.2	4.0
δT (spherical galls − no galls)	0.0	1.0	2.1	3.1	4.1	5.2

should never differ in temperature by more than about 5°C. So, in answer to our question whether galls significantly affect the temperatures of the leaves they afflict, about the best we can do is to say "sometimes yes, sometimes no."

We next must ask: would the temperature elevations expected for galled leaves have any effect on the leaf's net photosynthesis? To get there, we need some real values for how temperature affects net photosynthesis in a real plant. Fortunately, data like these are abundantly available for agriculturally important plants, so I have extracted from the literature a net photosynthesis curve for corn, *Zea mays* (Fig. 9.10). This curve shows a $\hat{T}$ of about 38°C, with net photosynthesis falling to zero at 12°C and 51°C.

To evaluate the likely effect galls might have, we simply estimate leaf temperatures as in the experiment described above, plot them on this curve, and compare where on the net photosynthesis curve the points lie. It does no good to estimate just any old leaf temperature, though: the estimates must be done systematically. A sensible approach might frame the estimates in the context of the leaf's optimum temperature. Let us then start with the simplest case: set conditions so that the temperature of a leaf unburdened with galls equals the optimum temperature (Fig. 9.10). Under these conditions, galled leaves are 2–3°C warmer than ungalled leaves. Because the net photosynthesis curve is somewhat flattened around $\hat{T}$, such small variations of temperature have little effect on net photosynthesis. Consequently, the small increase of temperature experienced by the galled leaves will have little effect. Leaves with needle galls, for example, have rates of net photosynthesis that are at about 99.3 percent of the optimum. Leaves with spherical galls photosynthesize at a net rate of about 98.7 percent of the optimum.

Leaves do not always function at their optimum temperatures, however—leaf temperature may vary through the day, or from day to day, or from place to place on the plant. To be thorough, our comparison must also consider what happens when leaf tempera-

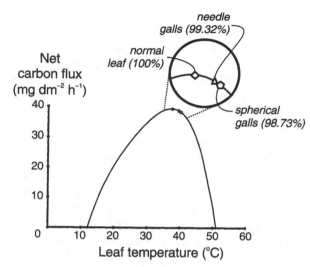

Figure 9.10 Estimated net carbon flux for normal leaves and leaves with simulated gall disease. Environmental conditions are set so that a normal leaf functions optimally; leaves with galls function somewhat less well.

ture differs from the optimum. Using the optimum temperature as a frame of reference, we set conditions so the ungalled leaf is photosynthesizing at 90 percent of its maximum net rate. Two comparisons must be made, one with leaf temperature cooler than the optimum (Fig. 9.11) and one with leaf temperature warmer (Fig. 9.12).

At these temperatures, we find that galls affect photosynthesis more strongly. Although the actual temperature differences between galled and ungalled leaves are small, their effect on net photosynthesis becomes larger as the change from the optimum increases. At cooler leaf temperatures, the small warming of the galled leaves is sufficient to boost net photosynthesis by a few percentage points: leaves with needle galls should function at about 93 percent efficiency while leaves with spherical galls work with about 94 percent efficiency (Fig. 9.11). At temperatures warmer than $\hat{T}$, the effect of galls is larger (Fig.

9.12), both because the temperature elevations are larger in the galled leaves and because the drop-off in net photosynthesis is steeper at temperatures above the optimum. When ungalled leaves photosynthesize at 90 percent of the optimum, galled leaves function at 76 percent efficiency for needle galls and 71 percent efficiency spherical galls.

The Parasite's Dilemma

In popular culture, parasites are bad things, "freeloaders" who suck their victims dry and throw them away. True, some parasites do just this (like the ichneumon wasps described in Chapter 8), but in general this is really not a very sensible strategy. The host, after all, is the conduit for the flow of energy and materials to the parasite. What serves a parasite's interests far better is to keep the host alive and as healthy as possible for as long as possible.[1] The parasite's dilemma, then, is this: how to keep the host healthy enough so that you can continue to steal as much from it as you can?

To illustrate the parasite's dilemma, consider a curi-

---

1. There are circumstances where using up the host and throwing it away might be a sensible strategy, however. The evolution of a parasite's life history is determined simply by what transmits the parasite's genes most readily into future generations, and not by any attention to the interests of the host. Whether the parasite comes to an accommodation with its host or becomes virulent, killing the host, depends upon how readily transmission from host to host can be accomplished. If transmission of the parasite between hosts is difficult, the best strategy for the parasite is accommodation. On the other hand, if transmission between hosts is easy, then the parasite's interests are less readily served by protecting the host, and virulence will likely be the result. This is one of the reasons why public health officials raise concerns about social policies and behaviors that make it easier to transmit disease organisms from person to person. For example, a hypodermic needle exchange program for drug abusers that does not simultaneously prevent needle sharing among abusers may transmit the more virulent strains of disease organisms more readily than a program that forces people to use the needles in the supervised environment of a hospital or clinic. The end result of a poorly supervised needle exchange program may be an increase in the virulence of otherwise manageable disease organisms.

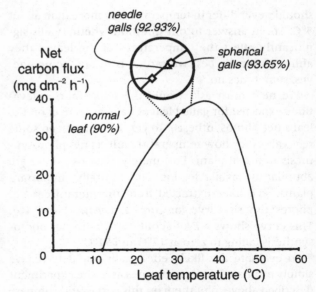

Figure 9.11 Estimated net carbon flux for normal leaves and leaves with simulated gall disease at cool temperatures (< 38°C). A normal leaf in this environment functions at 90 percent of optimum, and the warmer leaves with galls function at a slightly higher level.

---

ous political situation that has developed in the United States in the 1990s. During this time, our state and local governments took it upon themselves to impose substantial financial burdens on tobacco companies. The ostensible reason is to force tobacco companies to bear the social costs of their products. Arguably, the real purpose is to punish Big Tobacco for its sins, most notably the resolute indifference of cigarette makers toward those who—whether they be responsible adults or babes in their cribs—consume their products.

Here I am thinking of the government as the parasite and the tobacco industry as the host. I realize the casting is not politically correct, but the analogy doesn't really work the other way. Now government can, if it so chooses, extract so much money from tobacco companies that they go out of business. However righteous this course might be, though, it poses real problems for government: the sums of money being extracted from tobacco companies are already so enormous that many government programs could not be funded without this cash. If the tobacco industry goes belly-up, the loss in tax revenues would be painful, perhaps devastating. So government now finds itself in the rather ludicrous position of having to keep the demon tobacco industry alive and flourishing. This is the parasite's dilemma.

The usual way out of the parasite's dilemma is to live off margin, as bookies, mobsters, stockbrokers, and politicians do: you insert yourself into a stream of money and skim off a little as it passes by, but not too much. To make more money off the stream, you could increase your margin, but that increases the risk the stream of money will dry up. A more sophisticated way is to get people to give you more money voluntarily. State lotteries are a good example of this. Most governments already tax their citizens to the maxi-

mum tolerable amount. A government that wants still more revenue can easily get it by asking people to decide to divert their resources into a nearly hopeless get-rich scheme. And, remarkably, it works!

Something like this may be happening in leaf galls. Plants normally have vigorous defenses against animals that want to eat their leaves, and any herbivore that is too greedy will end up seeing its stream of energy dry up. Plants defend themselves through the so-called injury response. After the initial bite, the plant induces an initial proliferation of cells at the site of injury, called a hyperplasia. This is followed by lignification, essentially the formation of a scar that isolates the site of injury from the rest of the plant. The scar also separates the herbivore from the goodies left behind it. Generally, the more vigorous the attack, the more intense the injury response will be.

Gall inducers probably got their evolutionary start by figuring out, so to speak, how to intervene in this injury response. At its simplest, the intervention involves a simple prolongation of the hyperplastic phase. Among other things, encouraging hyperplasia provides the gall inducer with a steady supply of juicy, succulent plant cells to eat. This bonanza doesn't come free, however. Cell division and cell growth require energy. In a normal leaf, mobilizing this extra energy is an integral part of the injury response; the extra energy must come from uninjured parts of the leaf itself or from other, uninjured leaves on the tree.

By subtly altering the leaf's temperature, gall insects may be manipulating the signals plants use to manage their leaves' energy economy. The signals could be changed in a number of ways (Fig. 9.13). Consider, for example, an ungalled leaf that is operating at a temperature cooler than the optimum. Adding galls to that leaf increases its temperature slightly and, as we have seen, should increase the leaf's net photosynthesis. Net photosynthesis increases in this case because the increase in temperature accelerates photosynthesis proportionally more than it accelerates the leaf's metabolic rate. This results in an excess of sugar—an energy

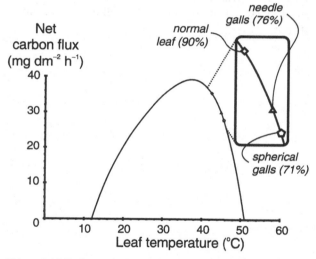

Figure 9.12 Estimated net carbon flux for normal leaves and leaves with simulated gall disease at warm temperatures (38°C). A normal leaf in this environment functions at 90 percent of optimum, and the warmer leaves with galls function at a slightly lower level.

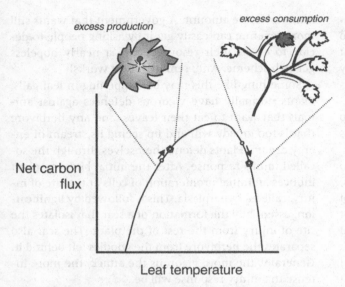

excess production   excess consumption

Figure 9.13 Playing the margins of leaf photosynthesis.

Net carbon
flux

Leaf temperature

margin—which can be diverted to fueling the hyperplasia the gall insect depends upon. The situation is reversed at temperatures higher than the optimum: net photosynthesis declines because metabolism starts to outstrip gross photosynthesis. Sugar may then be "im-

ported" from other leaves, but because galled leaves will be warmer, the diversion of energy to them may be greater. And in all cases, sitting there in the middle of these streams of energy is the gall insect, skimming its take.

First thou shalt arrive where the enchanter Sirens dwell, they who seduce men. The imprudent man who draws near them never returns, for the Sirens, lying in the flower-strewn fields, will charm him with sweet song; but around them the bodies of their victims lie in heaps.
—CIRCE'S WARNING TO ODYSSEUS,
  HOMER, *THE ODYSSEY,* BOOK XI

CHAPTER TEN

Twist and Shout!

Social beings that we are, we tend naturally to think of communication as an *inter*action: good communication is a two-way street, a clear *exchange* of thoughts, desires, or emotions between two or more parties. We want to be very sure we understand a contract before we sign it, we want to be very sure of a rival's intentions before we deal with him, we want to be very sure we will not be caught up in someone else's hidden agenda when we enter into a relationship with her. The interaction, spoken and unspoken, that humans use to ascertain both meanings and motives is prolonged, complex, and subtle. Despite the effort we put into it (or perhaps because of it), human communication is maddeningly prone to failure, to the enrichment of no one save lawyers and country music singers.

Communication between other animals is usually more straightforward. These messages are generally not intended to convey truth, understanding, or comity—they are intended to manipulate other animals' nervous systems to elicit from them particular responses that will, in the end, increase the likelihood that the sender's genes are passed to its offspring. Communication is judged solely by how reliably it accomplishes this goal. By this logic, Homer's Sirens were great communicators; using their songs to zero right in on and activate mate-seeking behavior in the sailors that heard them, they got the desired response without having to waste time on the protracted, subtle, and delicate negotiation that we call courtship.

For the physiologist, communication is an exchange of energy between the sender and recipient. Generally, the energy involved is deducted from the sender's metabolic energy budget. Premiums should accrue, therefore, to animals that communicate efficiently— that is, with minimum expenditure of energy but consistent with conveying the sender's message widely and clearly. Fireflies, for example, communicate with potential mates by converting ATP energy into light. The light signal should be sufficiently bright to attract the attention of potential mates. However, the signal should activate seeking behavior only in potential mates that are programmed to receive them. Consequently, signals between potential mates are usually modulated in some way, as by color or flash rate.

Communication presents an interesting problem to the physiologist, because it is very difficult to separate what goes on inside two communicating animals from the physical transmission of energy through the medium separating them. At what point in the chain of signal transmission does communication stop being physiology and start being physics? Let us put the question more concretely, staying with the example of fireflies. The eye of a firefly consists of two elements, sensory and optical. The sensory eye consists of cells that absorb photons and convert them into electrical signals suitable for transmission to the firefly's brain. Before the photons get to these cells, though, they pass through an array of refractive lenses and light guides that make up the optical eye. The functioning of the optical eye is governed by the same physical interaction of light with matter that governs inanimate optical systems. When the optical eye manipulates light, is it physics or is it physiology? In my opinion, the distinction is a false one: I would have difficulty arguing that anything that modulates, modifies, or transmits the energy in a message is not part of a physiological process of communication. And this leads us to an interesting quandary: if the built-in parts of sensory systems, like corneas, lenses, and irises, are properly organs of communication, why should structures that do the same thing, but are constructed outside the animal, not be considered the same?

In this chapter, we will turn to structures, built by crickets, that aid in transmitting auditory signals to prospective mates. We will first explore the question of why crickets should need such structures in the first place. As part of the answer, we must also familiarize ourselves with the acoustical principles underlying the production of cricket song, and how the structures they build make them more effective disseminators of sound. We will then turn to two remarkable examples of crickets that build external structures that modulate, amplify, or direct the sounds they produce. Along with the argument I wish to convey, I also have a hidden agenda, which, in the interests of good (human) communication, I now reveal: the physiology of communication, more than any physiological process of which I am aware, demolishes the artificial boundary that limits physiology to the organism.

Communication and Songs of Crickets

Crickets, like Sirens, are superb acoustic communicators. Males produce sounds and use them to broadcast their identity and location to prospective mates flying about looking for them. A female who intercepts this sound is guided by it to its source, just as we are drawn to the pleasant smells emanating from a kitchen. The messages are simple, direct, and clear.

Anyone who has sat outside on a warm summer's evening is aware of how well the cricket's sounds carry. This is a sensible thing for a mating call to do: a sound that travels far is more likely to be heard by a potential mate than a call that dies out quickly. Generally, making a sound travel far means making a *loud* sound. The curious thing is that by most principles of acoustical physics, crickets should only be able to produce relatively soft sounds, and inefficiently at that. Below, we will delve into the reasons why this is so, but first let us learn more about cricket songs and how they are produced.

A cricket song is a series of "chirps" (Fig. 10.1), each chirp consisting of a series of discrete tones, each lasting a few milliseconds, interspersed with short periods of silence. The number of tones in a chirp, which de-

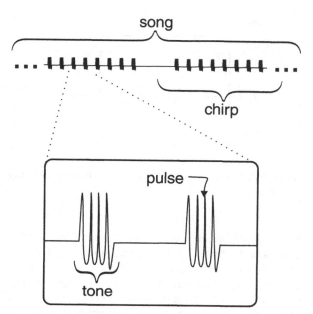

Figure 10.1 A cricket song song consists of repetitive chirps. A chirp, in turn, is composed of numerous tones, which consist of groups of sound pulses.

duced by setting the harp membrane into vibration, just as the skin of a kettle drum vibrates when it is struck by a mallet. The energy to power the harp's vibration is fed into the membrane via a complicated mechanism involving two specialized structures on the cricket's wings. On one wing is the file, which is decorated with a series of ridges, called the teeth. On the other wing is a stiff plectrum. As the wings are moved relative to one another, the plectrum slides across the file, alternately catching on and releasing from the file's teeth, in a kind of catch-and-pawl mechanism. While the plectrum is held at a tooth, the harp membrane is stretched slightly by the steady pressure of the wing muscles. When the plectrum and tooth disengage, the strain on the membrane is released abruptly, and the liberated energy sets the harp membrane into a damped vibration, which dies out after 20–30 cycles or so. The sound emitted by this vibration constitutes the tone. The chirp is produced by the plectrum being dragged across the many teeth on the file, with the number of tones in a chirp corresponding to the number of teeth.

Energetics of Sound Production

Despite the presumed benefits crickets might derive from producing loud sounds, they face serious physical obstacles to doing so. The most serious is how efficiently work done by the wing muscles is translated into energy in sound. Acoustical physics tells us that cricket harps should do this very inefficiently.

Crickets produce sound by translating energy from one form (mechanical work by muscle) to another (resonant vibration of the harp). The harp's vibration then must set air into the oscillatory motion that radiates away as sound. Each of these transformations of energy is governed by the constraints of the First and Second Laws: each must conserve energy, and each will proceed at a characteristic inefficiency. Sound communication cannot be fully understood without understanding these constraints.

We begin by considering a simple sound emitter, a

pends upon the species, ranges from roughly ten to at most a few dozen. Each chirp usually lasts about a half-second or less. Each tone in the chirp is a so-called pure tone.[1] The frequency of the tone, or the carrier frequency, also depends upon the species, ranging from 1,000 to 6,000 cycles per second, or 1–6 kilohertz (kHz). For those with a musical ear, these frequencies correspond to tones from sixth C (two octaves above middle C) to eighth C (four octaves above middle C).

Cricket song is produced by stridulation (derived from the Latin *stridulus*, "to creak"), a process in which the cricket's wings are rubbed against one another. There is more to it than that, however: the cricket wing is essentially a means of transmission, coupling the "engine" of sound production, the cricket's wing muscles, to the sound generator, a patch of flexible membrane on the wing called the harp. Sound is pro-

---

1. For a primer on terminology of acoustics, see Box 10A.

10A

Acoustic Terminology

Acoustics has a language all its own. Following is a short primer describing some of the terms that will be useful in this chapter.

Sound is a periodic oscillation of pressure about a mean that propagates through a medium in the form of a traveling wave. The medium can be fluid, like air or water, or it can be solid. The simplest type of sound manifests itself as a sinusoidal oscillation of pressure (Fig 10A.1), p (Pa), with respect to time, t (s), described by the equation:

$$p = A\sin(2\pi ft) \qquad [10A.1]$$

This equation is a periodic function. Two important numbers describe it. The amplitude, A, measures how far pressure is elevated above (or depressed below) the mean pressure. Variations of amplitude are perceived by us as variations in loudness, or intensity, of a sound: higher amplitudes correspond to louder sounds. The frequency, f, describes how frequently the pressure oscillation repeats itself in a second. Its units are cycles per second, sometimes abbreviated as s^{-1} ("per second") or hertz (Hz), after the German physicist, Heinrich Hertz. Variations of frequency are perceived by us as variations of pitch, or tone: high frequency sounds correspond to high-pitch tones. The range of frequencies perceptible by humans is from about 500 Hz to 5,000 Hz. Two other numbers are sometimes used to describe sound waves. One, the period, P, is simply the inverse of frequency: it describes the time required for a sound wave to complete one oscillation. The wavelength, λ, measures how far apart in space two similar points on a sound wave are from one another. (If the horizontal axis in Fig. 10A.1 were distance, not time, the length labeled P would be the wavelength, λ.) This distance will depend upon how rapidly the oscillation takes place and how rapidly sound waves propagate through the medium, also known as the speed of sound, c (m s^{-1}). Speed of sound varies with the medium: in air, $c \approx 330$ m s^{-1}, but sound travels much faster in water ($c \approx 1,500$ m s^{-1}). Sound travels fastest in solid media; in steel, for example, $c \approx 6,000$ m s^{-1}. For

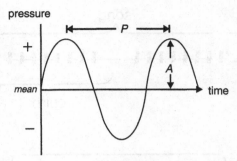

Figure 10A.1 A sine wave of pressure, indicating amplitude *(A)* and period *(P)*.

the range of frequencies perceptible to humans, wavelengths range from about 66 cm at the low end ($f = 500$ Hz) to about 6.6 cm at the high end ($f = 5,000$ Hz). It is important to keep these numbers in mind, because many aspects of the ability to generate and emit sounds are related to the wavelength of the sound emitted.

Sound intensity, I, refers to the energy being carried in sound. Sound pressure, p, described above, is a measure of the potential energy in a sound wave. As sound travels, this potential energy does work moving air molecules in a back-and-forth motion as the pressure wave passes by. These movements are very small, on the order of a hydrogen atom's diameter, but they still represent work. This work is the product of the movement and the force doing the moving. In sound, work rate, or power, is the product of the sound pressure and the velocity of the particles of air it moves. The power in sound is the sound intensity, which has units of watts per square meter (W m^{-2}). Sound intensity is related to pressure by the formula:

$$I = p^2/2\rho c \qquad [10A.2]$$

where ρ is the density of medium (in the case of air, roughly 1 kg m^{-3}).

Sound intensity can be measured directly. A pressure amplitude of zero in a sound therefore corresponds to no sound. It is frequently the case that sound intensities are measured in comparison to some standard, or in comparison to some other sound. The standard of comparison in this case is the decibel, abbreviated dB. The decibel level of a sound is calculated from a ratio of intensities in two sounds:

$$dB = 10\log_{10}(I_1/I_2) \qquad [10A.3]$$

Two sounds of equal intensity will differ from each other by 0 dB ($I_1/I_2 = 1$, and logarithm of 1 is 0). A sound ten times louder than the other ($I_1/I_2 = 10$) will be 10 dB louder ($\log_{10}10 = 1$). A sound a hundred times louder ($I_1/I_2 = 100$) will be 20 dB louder ($\log_{10}100 = 2$), and so forth. Conversely, a sound a hundred times softer than another ($I_1/I_2 = 0.01$) will differ from it by -20 dB ($\log_{10}0.01 = -2$).

The decibel comparison is used in a variety of ways. Commonly, the comparison is made against an agreed-upon standard sound. Unfortunately, the standard sound is anything but standard: engineers and audiophiles have, for their own good reasons, adopted different conventions for the standard sound. For physics, the standard is a pressure fluctuation of 200 mPa, which corresponds to an intensity of about 6×10^{-11} W m^{-2}. Audiophiles, because they are designing things to be heard by humans, build their standard around the acoustical behavior of the ear, which varies with frequency. More frequently, though, decibel comparisons are made between two non-standard sounds, say the peaks in a power spectrum. If a sound is composed of a dominant frequency and its harmonics, for example, the harmonics may be described as being so many decibels relative to the dominant. Alternatively, a sound may be described as being so many decibels louder than the noise that surrounds it.

---

disk being driven back and forth in an oscillatory motion. We assume that the disk is surrounded by a medium (let us assume it is air) that can carry sound. We need not concern ourselves for the moment with what the engine is, only that it is doing mechanical work on the disk at a rate Q_m. We know from the First Law that the work done by the engine must be conserved: energy transmitted to the disk from the engine must equal the energy transmitted from the disk to the medium surrounding it. Whether that energy emerges as sound depends upon what type of work the disk does on the medium. The type of work done, in turn, is influenced by the properties of the medium itself—its density, viscosity, and rigidity. The heart of the matter turns on two questions. First, what kinds of work does a vibrating disk do on the air surrounding it? Second, how do these different kinds of work translate into sound?

Regarding the first question, the answer depends in large measure on how fast the vibration is. Imagine first that the disk is oscillating back and forth very slowly. As the disk moves back and forth, it will drive synchronous motions of the air surrounding it. At the disk's leading edge—the one that is moving forward—air will be pushed ahead, while air behind the disk's trailing edge will be pulled along toward it. At the same time, air surrounding the edge of the disk will be moved in the opposite direction, flowing from the leading surface of the disk and around its edge to the trailing surface. This air has mass, and setting it in motion means accelerating it. Therefore, work must be done in overcoming the air's inertia, which we shall designate inertial work. The rate at which inertial work is done is I.

Now imagine that we speed up the disk's oscillation rate. At low vibration rates, nearly all the work done by the disk will be inertial work. As oscillation rates increase, though, appreciable pressures will appear around the disk, just as they do around a car speeding down the highway. Specifically, pressure will be elevated at the disk's leading surface, where the air is compressed, and will be reduced at the disk's trailing surface, where the air is temporarily rarefied. The faster the oscillation rate, the greater these pressures will be. We remind ourselves now that pressure is potential energy, and work must be done to impart it to the air. Consequently, the work done by the driver on the disk does not now go completely into doing inertial work: some portion of it goes into imparting potential energy to the air.

The secret of sound production lies in what happens to this potential energy. At the top or bottom of the disk's stroke, there will be a pressure difference between the disk's leading and trailing edges. Some of this pressure will help drive the disk back during its return stroke. In other words, some of the potential energy stored in air pressure is recovered to help drive the oscillation of the disk. This work we designate as capacitative work: the rate at which capacitative work is done is symbolized as C.

If the rate of oscillation is fast enough, not all the potential energy does capacitative work. Some of the compressed air at the disk's leading surface helps compress air ahead of *it*, setting in motion a wave of elastic recoil that propagates away as a wave of high pressure. As the disk draws back, it pulls the air back with it, again launching a wave of elastic recoil that propagates away, this time as a trough of low pressure. This propagating wave of up-and-down pressure, of course, is a sound wave. The portion of energy that powers this elastic recoil we designate as dissipative work, which is done at a rate D.

You are now in a position to grasp one of the essential facts of acoustics, be it physical or physiological: it is an energy balance problem (Fig. 10.2). The energy balance on the disk involves three components of work: inertial work, I; capacitative work, C, and dissipative work, D. The energy flow through the disk can be expressed as a straightforward energy balance equation, where the work done by the disk on the air must equal the work done on the disk by the engine, Q_m:

$$I + C + D = -Q_m \qquad [10.1]$$

(The minus sign in front of Q_m is needed to differentiate between work done on the disk from work done by the disk.)

No Loud Crickets
Equation 10.1 is at the heart of the reason cricket songs should not be loud. Sound is produced by only one type of work, dissipative work. Efficient sound production involves manipulating the disk's energy

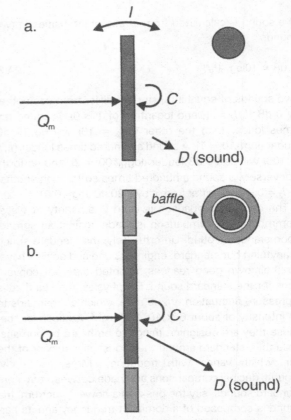

Figure 10.2 The energy budget at a sound emitter, in this case a disk shown from the side (at left) and from above (at right). *a:* Work done on the emitter by a driver, Q_m, must emerge from the emitter as inertial work, I, capacitative work, C, or dissipative work, D. *b:* Placing a baffle around the emitter enables a larger proportion of energy to emerge as sound.

budget so that energy fed into it by the motor does mostly dissipative work, because capacitative and inertial work do not produce sound.[2] Unfortunately, most sound-producing insects are small, and small sound

2. This is not strictly true for insects that transmit sounds over very short distances, on the order of fractions of the wavelength of the sound.

emitters generally are poor sound emitters—too much of the energy fed into them goes to capacitative and inertial work. A few calculations illustrate the difficulty.

It is an acoustic rule of thumb that a sound emitter should be large with respect to the wavelength of the sound it produces. Wavelength is the distance (not the time) between the high points of pressure in adjacent sound waves. Specifically, the diameter, d, r of the emitter should exceed the ratio of the sound's wavelength, λ, and π:

$$d > \lambda/\pi \qquad\qquad [10.2]$$

This is bad news for crickets. Consider the harp of a common cricket, *Oecanthus*, as illustration. These crickets produce chirps with a carrier frequency of about 2 kHz. The wavelength λ (in meters) of any sound wave is the ratio of the speed of sound, c, which in air is about 330 m s^{-1}, and its frequency f (Hz):

$$\lambda = c/f \qquad\qquad [10.3]$$

A sound wave with $f = 2{,}000$ Hz therefore has a wavelength in air of about 165 mm. The diameter of the *Oecanthus* harp is only about 3.2 mm, though, fifty times smaller. The rule of thumb indicates the harp could emit sound pretty well at wavelengths of (3.2 mm $\times$ π) $\approx$ 10 mm, which correspond to frequencies in air of roughly 33 kilohertz (kHz) or higher. But it does not—it emits sound at 2 kHz. Therefore, the cricket harp should be a very poor emitter of sound. Let me hasten to add that this is not necessarily a bad thing—even a very inefficient emitter of sound can, with some clever manipulation of various properties of sound, be useful in communication. (For some examples, see Box 10B.)

The Baffle Leaf of Oecanthus burmeisteri

Even with an inefficient harp, however, crickets must still compete with other crickets for the attention of mates. All other things being equal, benefits should accrue to crickets that manage even modest increases in efficiency of sound production. One simple solution is to mount the sound emitter in a baffle—a wall with a hole cut in it to accommodate the oscillating disk. Baffles work by presenting a physical obstacle to the movement of air around the disk's margins (Fig. 10.2). Energy that would have powered this movement is now available to increase pressure across the face of the disk, some of which may now produce sound. This property of baffles is familiar to anyone who has ever worked with loudspeakers or high-fidelity sound reproduction: you simply must baffle a loudspeaker, or the sound quality is dismal.

Some insects have learned this trick: a remarkable example is found in a South African cricket, *Oecanthus burmeisteri*. The *Oecanthus* harp oscillates at a frequency of about 2 kHz and, as we have seen, its production of sound should be very inefficient. Baffling the harp seems an obvious thing to do. All crickets baffle their harps to some extent by situating the harp membrane in the center of the wings, where the surrounding membranes will act as baffles. The wing's effectiveness as a baffle is itself limited, however, by the wing's small size. A baffle is most effective when its diameter is one-third of the sound's wavelength or larger: for a 2 kHz tone with a wavelength of 165 mm, the baffle should have a diameter of at least 56 mm. Baffles smaller than this will help, but not as much as a 56 mm baffle would. The cricket's own wings measure about 10 mm by 5 mm, which means they should not be very effective baffles.

Oecanthus burmeisteri gets around this problem by building itself a larger baffle (Fig. 10.3). These crickets normally feed on leaves, chewing out small holes or portions from the edge of the leaf. Prior to calling, however, the male cricket chews out of the center of a leaf a pear-shaped hole, measuring roughly 8 mm by 14 mm. The leaves they call from usually measure from about 70 by 80 mm at the smallest to as large as 170 by 300 mm. During calling, the cricket positions itself so that its wings are located in the center of the pear-shaped hole. The dimensions of the leaf and hole and the positioning of the cricket are consistent with

Playing with Sound

The inefficient production of sound by cricket harps is superficially troubling: why should natural selection allow the development of an inefficient system? The use of energy for capacitative and inertial work offers no advantages in attracting mates, and it takes energy that otherwise could be used to produce offspring. It makes no sense, but there it is. In this regard, however, the constraints of acoustics can illuminate the subtle interplay of costs and benefits that shape the selective *milieu* in which crickets operate.

The *Oecanthus* harp should, in the interests of efficiency, produce sounds at 33 kHz or higher. Frequencies as high as this are well above those most animals can hear, but that in itself is no reason why insects could not use them to communicate. Some, in fact, do: certain species of stridulating insects generate sounds at frequencies as high as 55 kHz. One of the big problems with high-frequency sound, though, is that it does not propagate well in air. A 33 kHz tone, for example, loses about 1 decibel of sound intensity for every meter it propagates: 90 percent of the sound's power will have been lost by the time it radiates 10 m from the insect. At 2 kHz, however, the loss of power is much less, only 0.01 decibel to 0.1 decibel per meter. At this rate, the insect can broadcast sounds that travel from 100 m to 1,000 m before they lose 90 percent of their power.

This problem throws a rather large monkey wrench into the presumption that natural selection should always favor efficient production of sound. In fact, the physical constraints of acoustics sometimes make inefficient sound production work quite well. Suppose, for example, a cricket has X joules of energy to invest in sound production. It can use this energy to produce sounds at high efficiency (that is, at high frequency) but with limited range, or it can use it to produce sounds that will project further (that is, at lower frequencies) but only very inefficiently. In fact, there is great potential in pursuing the latter strategy: energy converted to 33 kHz sound at a particular rate of efficiency will project to about the same volume as will 2 kHz sound produced at 0.1 percent of that efficiency.

In laying out this argument, I am not trying to make the point that efficiency in the conversion of energy into sound is not important in the *milieu* of selective forces that have influenced the evolution of cricket song. Rather, I wish to lead you to two conclusions. First, and most obvious, absolute efficiency of conversion of energy into sound is not in itself a sufficient criterion for determining the success of a sound-producing organ. Second, and more indirect, even very modest increases of efficiency in an inefficient sound-producing system can yield big payoffs. For example, if X joules of energy were used to produce a 2 kHz sound at 0.2 percent of the efficiency of the 33 kHz sound, the sound projects to eight times the volume. This gives the cricket's song quite a boost.

the leaf acting as a baffle. The proof lies in whether or not it improves the emission of sound. Indeed, it does: songs from crickets sitting in leaf baffles are 2.5 to 3.5 times louder than songs from crickets standing alone. Thus, a cricket singing in a leaf baffle can project its song to a volume 15 to 47 times larger than the volume produced by a cricket standing alone.

Boosting the Performance of Small Sound Emitters

Baffles around a sound emitter obviously help, but there are better ways to increase the efficiency of sound production. For one thing, baffles do not completely eliminate the inertial component of work. To some extent, inertial energy is spent because of imperfect baffling. The leaf baffle of *Oecanthus,* for example, does not fit tightly around the wing. Although it reduces the diversion of energy into inertial work, it does not eliminate it, because there still will be some back-and-forth movement of air between the outer margins of the wing and the inside margin of the hole in the leaf. Even if the leaf baffle were solidly joined to the wing, however, it still would not eliminate the inertial work. As the harp membrane bulges alternately in and out, it will also accelerate air away from the membrane

Figure 10.3 A male *Oecanthus burmeisteri* singing from its leaf baffle. *[From Prozesky-Schulze et al. (1975)]*

parallel to the baffle's surface. This also saps energy from production of sound.

People who work with sound, such as musical instrument makers, have known for centuries of a clever trick that substantially boosts the acoustic performance of a small sound emitter. Indeed, an understanding of the physics of this trick points to the possibility of a perfect sound emitter, one which eliminates completely the diversion of energy into capacitative and inertial work. The trick involves enclosing the sound emitter in an infinite horn.

Horns are familiar elements of sound amplification systems and musical instruments. Popularly, horns are thought to direct sound as the sound waves are re-flected off the horn's internal walls, similar to the way the curved mirror in a flashlight reflects light from a bulb. In this conception, sounds seem louder because the horn concentrates sound. For the most part, though, acoustical horns simply don't work this way: the long wavelengths of most audible sounds limit the degree of internal reflection that is possible. Internal reflection of sound will likely be important only at very high frequencies (very short wavelengths).

An infinite horn works by making it impossible for the sound emitter to do either inertial or capacitative work.[3] By making the horn infinitely long, for example, you also make the mass of the air enclosed in it infinitely large. No matter how hard you push on an infinitely large mass, you will not accelerate it: this eliminates the emitter's ability to do any inertial work along the axis of the horn. Enclosing the sound emitter at its sides also eliminates that component of inertial work that acts parallel to the sound emitter's surface. The result: the sound emitter does no inertial work at all. Finally, enclosing air in an infinite horn makes it infinitely capacious, and an infinitely capacious volume cannot store energy that could do capacitative work. This leaves only one avenue through which energy can escape from the sound emitter: through the waves of elastic recoil that comprise dissipative work, that is, sound.

There is no such thing as an infinite horn, of course. Any horn that is finite in length will enclose some finite mass and volume of air, and a sound emitter will end up doing at least some inertial and capacitative work on that air. But there are ways to nudge a finite horn closer to being a perfect emitter of sound. One simple trick is make the horn a particular shape. Horns come in a variety of shapes, commonly three. Conical horns, as the name implies, are simple cones with straight sides, like the megaphones once used by

---

3. In the jargon of acoustical science, an infinite horn imposes a completely resistive load on the emitter, so called because the rate of dissipation of sound energy is limited by the elastic properties of the air, just as electrical current is limited by a resistor.

cheerleaders. Parabolic horns flare widely at the throat, where the sound emitter is, and less steeply toward the mouth, where the sound leaves the horn. Exponential horns flare more widely at the mouth than at the throat, like the graceful flare of a trumpet. Of the three, the exponential horn most closely approximates the performance of an infinite horn.

How Horns Modulate Frequency

Finite horns also offer some interesting ways to turn acoustical shortcomings to advantage. The air enclosed in a finite horn has inertia, because it has mass, and it has elasticity, because it springs back after it is compressed. The air in the horn can therefore resonate. The resonant frequency of an exponential horn is related to its length: in general shorter horns resonate at higher frequencies than do longer horns. When the air in a horn is set into resonant motion, however, it turns the tables a bit on what is emitter and what is driven. So far we have been assuming that the emitter drives vibration of the air, but when air in a horn is set into resonant motion *it* can now drive the vibration of the *emitter*. This helps boost sound production, but it also helps purify the sound. In other words, the sound emerges as a purer tone. As illustration, consider the vibration of a common sound emitter in many musical instruments, a vibrating reed.

When air is blown past the edge of a reed, the reed vibrates back and forth at some resonant frequency determined by its own mass and stiffness. The reed's vibration then sets the air around it into vibration, generating a sound. The sound that emerges from the reed will have a primary frequency equivalent to the reed's vibration frequency. However, the sound also emerges at lots of other frequencies. These arise in a variety of ways: the reed may vibrate imperfectly; air may flow turbulently around the reed and so forth. To the ear, these secondary frequencies give the reed a "buzzy" sound. The "noise" may be visualized in a plot of the energy in the sound versus its frequency—the frequency spectrum (Fig. 10.4c). Most of the sound energy is concentrated at the reed's resonant fre-

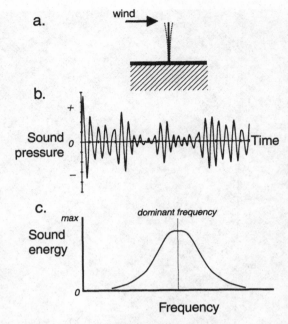

Figure 10.4 Vibration and sound emission at a vibrating reed standing alone. *a:* Air blowing past the edge of a reed generates turbulent eddies that set the reed into vibration. *b:* The chaotic variation of sound pressure resulting from secondary components of the reed's vibration. *c:* The power spectrum of an undamped vibrating reed.

quency, but the "buzziness" shows up as a "smearing" of the energy at frequencies on either side of the resonant frequency. The sound energy coming out of the reed is distributed among all these frequencies. Consequently, the loudness of the sound *at the resonant frequency* will be limited by the sound energy expended in all those ancillary frequencies.

If the relatively massive body of air in a horn is set into resonant motion, the air will now drive the vibration of the reed (Fig. 10.5). The energy in those messy secondary vibrations of the reed is now entrained to the resonant vibration of the air in the horn, with two benefits. First, it makes the sound emerging from the reed purer—that is, dominated more strongly by the resonant frequency. Second, the sound emerging at the resonant frequency is absolutely louder. Energy in those secondary frequencies has not disappeared—it has simply been diverted to the resonant frequency.

The Klipsch Horn

Musical instrument makers know another trick for turning the acoustical shortcomings of a finite horn to advantage. Horn and wind instruments do not consist of just the sound emitter and the horn—the mouth cavity of the musician is also part of the instrument. Acoustically, the entire system is a series arrange-

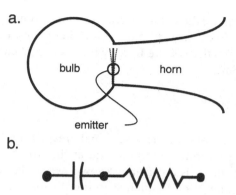

Figure 10.6 The Klipsch horn. *a:* Relative placement of the sound emitter, the horn, and the bulb. *b:* Equivalent circuit for sound energy in a Klipsch horn. The bulb acts as a capacitor in series with the emitter and the horn, which acts as a resistor.

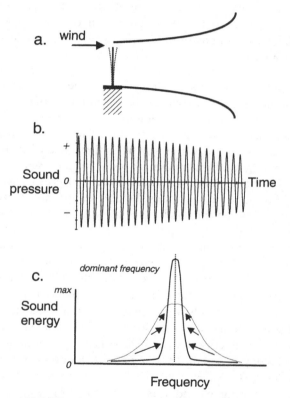

Figure 10.5 Vibration and sound emission at a reed contained in an exponential horn. *a:* The reed enclosed in the horn. *b:* More uniform vibration of an enclosed reed. *c:* The power spectrum shows energy concentrated at the dominant frequency.

ment of the air space of the horn, the vibrating emitter of sound—either the lips of the musician or a reed—and a capacious bulb corresponding to the musician's mouth cavity (Fig. 10.6). This configuration is known as a Klipsch horn.

The sound emitter in a Klipsch horn does both capacitative and inertial work on the air that surrounds it. Because the horn is open at one end, however, the emitter will do mostly inertial work on the air in it. Air in the bulb, being enclosed, will have mostly capacitative work done on it. Acoustically, the Klipsch horn is like a resistor and capacitor in series. This arrangement makes the air in the horn resonate more strongly and leads to a very efficient conversion of inertial and capacitative work to sound. Just as the frequency of a resonant system of springs and masses can be "tuned" by adjusting mass and stiffness, the resonant frequency of a Klipsch horn can be tuned by adjusting the volume of the bulb and the length of the horn.

Klipsch horns also expand the range of frequencies that horns can transmit. Despite all the wonderful things horns do, they do labor under some limitations. Horns generally will not perform well if the sound's frequency falls below a critical value known as the cutoff frequency, f^*. The cutoff frequency depends upon the flare of the horn, which is described by the equation:

$$A_x = A_0\, e^{\mu x} \qquad\qquad [10.4]$$

A_x and A_0 are the cross-sectional areas of the horn at distance x from the throat and at the throat, respectively, and μ is a constant known as the flaring constant. The cutoff frequency, f^*, is estimated reasonably well by the equation:

$$f^* = \mu c/4\pi \qquad\qquad [10.5]$$

To transmit low frequencies well, therefore, the horn should have a small flaring constant. A well-designed horn is flared so that the cutoff frequency is about an octave below the lowest frequency to be transmitted. Performance is also improved by making the cutoff as sharp as possible. Ordinary horns do badly in this respect, because acoustical performance degrades appreciably even at frequencies much higher than the cutoff. In the jargon of acoustical engineering, the "skirts" of the frequency response curve are wide. Configuring the horn as a Klipsch horn, though, helps make the cutoff frequency "sharper"—that is, the horn maintains high performance even at frequencies very close to f^*.

The Singing Burrows of Mole Crickets

Most of these design principles for horns were known to musical instrument makers centuries before acoustical engineers came along to tell them they got it right. I don't mean to be flip—acoustical physics has put the design principles for musical instruments on a solid theoretical footing, and this is unequivocally a Good Thing. But the original designers of musical instruments did not bring this type of knowledge to their work. Their knowledge of design was based on a collection of cultural norms, passed down from master to apprentice, continually modified and improved upon in the process.

If this description sounds like an analogy to natural selection, it is a deliberate one. Enormous rewards flowed to those who introduced new and successful innovations to instrument design and fabrication: Antonio Stradivari died a rich man. Similarly high rewards (not monetary, of course, but reproductive) might accrue to animals that refine the ability to communicate with sound, either by transmitting it well or by receiving it well. Indeed, natural selection has refined the acoustical devices used by insects to a remarkable degree. Insect ears, for example, are on a par with vertebrate eyes as organs of remarkable quality in design and function. But there is one outstanding example of refinement in an acoustical structure crafted by an insect.

These acoustic devices are constructed by a group of relatively large crickets, members of the insect family Gryllidae, known collectively as the mole crickets, so called because of their habit of constructing extensive networks of underground tunnels. These tunnels may be so dense in pastures and fields that the crickets become significant agricultural pests. But it is their song, not their economic importance, that concerns us here.

Although mole crickets themselves are rather large (up to six centimeters length), they do not have especially large harps. The tones in mole cricket songs range from about 1.5 kHz up to about 3.6 kHz, which means their harps radiate sound at comparatively long wavelengths. As we have seen, this should make mole crickets poor emitters of sound: they should, like most crickets, emit only a soft chirp. Yet, mole crickets actually produce one of the loudest sounds made by animals. Mole cricket songs can be heard by humans as far away as 600 m. The song of one, a European mole cricket, *Gryllotalpa vinae*, sets the ground around it visibly vibrating to a radius of 20 centimeters or so.

Mole crickets accomplish this remarkable feat by building singing burrows, which happen to incorporate many of the features that make horns such marvelous acoustic devices. The singing burrows constructed by an American mole cricket, *Scapteriscus acletus*, nicely illustrate these design features (Fig. 10.7). One of the cricket's tunnels is expanded to a capacious bulb. The singing burrow itself extends from the bulb first horizontally and then turns upward, expanding in cross-section as it goes. At the opening to the surface it is quite wide. The tunnel extending from the bulb is called the horn, which in fact flares like an exponential horn. Separating the bulb from the horn is a narrow constriction. When the male cricket sings, it positions itself at this constriction, facing inward, so that the wings are placed right at the constriction. This configuration should look familiar to you—it is a Klipsch horn. It would appear that these insects have hit upon the same solutions for boosting acoustic performance that musical instrument makers have.

Before getting carried away by the similarity, though, we need to ask some critical questions. As intriguing as its shape might be, the burrow might only be *shaped* like a Klipsch horn. Does it really *work* like a Klipsch horn? Here is where the physical foundation laid down by acoustical engineers comes in handy, because we can use it to predict how the burrow, given its shape, *should* perform. If its expected performance as a horn matches what we know about the song produced by the cricket, then it may indeed be working as a Klipsch horn. Three questions immediately come to mind. First, will the burrow's cutoff frequency fall comfortably below the frequency of the sound it needs to amplify? Second, will the air enclosed in a burrow resonate at a frequency that corresponds to the resonant frequency of the cricket's harp? Third, could the bulb in the tunnel act like the bulb of a Klipsch horn?

We have already seen (equation 10.5) that the cutoff frequency of an exponential horn is determined by its rate of flare. The measured flaring constant for the *Scapteriscus* burrow is about 49.5 m^{-1}, and equation 10.5 tells us the burrow's cutoff frequency should be about 1.3 kHz. The harp membrane of the *Scapteriscus* wing resonates at frequencies ranging between 2.5 kHz to 3.0 kHz. The resonant frequency of the harp, therefore, is about an octave higher than the burrow's

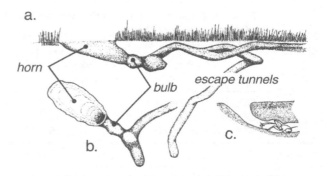

Figure 10.7 The singing burrow of *Scapteriscus acletus*. *a:* In side view, the horn and bulb are seen to be arranged in the configuration of a Klipsch horn. *b:* Top view. Shaded lines inside the horn are contour lines added by the artist. *c:* Cutaway showing the location and orientation of a singing cricket. [*From Nickerson, Snyder, and Oliver (1979)*]

cutoff frequency. You will recall this to be the "rule of thumb" that instrument makers use for flaring the horns they make. So far, so good.

The horn should also have a resonant frequency that comes close to the resonant frequency of the harp. Acoustical theory helps in answering this question too. Unfortunately, the formula is rather complicated, only approximate, and loaded with seemingly arbitrary terms. Its single virtue is that it works well. The horn's resonant frequency, f_0, should be:

$$f_0 \approx c/2L \times \sqrt{[1 + (L / \pi h)^2]} \qquad [10.6]$$

where L is the horn's effective length (which is its actual length plus 0.6 times the radius of the horn's opening), and h is the length of horn required for its diameter to increase by roughly 2.72 times (actually, by the base of the natural logarithm, *e*). All these quantities are easily estimated from plaster casts of the burrows. For the *Scapteriscus* burrow, the value of h is about 40 mm. Lengths range from 65 mm to 85 mm and average radii of the mouth range from 20 mm to 25 mm. Plugging these values into equation 10.6 gives

Table 10.1 Estimated and actual volumes of the capacitative bulbs in singing burrows of three species of mole cricket.

Species	Dominant frequency of song (kHz)	Estimated bulb volume (mm³)	Measured bulb volume (mm³)
Scapteriscus acletus	2.75	4,100 to 6,500	7,000
Gryllotalpa vinae	3.40	12,700 to 19,700	10,250 to 15,500
Gryllotalpa gryllotalpa	1.60	15,600 to 21,000	12,500 to 16,000

an estimated resonant frequency of the burrow of 2.6 to 2.8 kHz. This nicely overlaps the harp's resonant frequency of 2.5 kHz to 3.0 kHz: again, the features of the burrow are consistent with the idea that the burrow is acting as an exponential horn.

The bulb also is about the size one would expect for a Klipsch horn. Here, the formulas acoustical engineers give us are even more approximate, but the best one (from Klipsch himself) indicates the volume V of the bulb of a Klipsch horn should be:

$$V = 2.9AR \qquad [10.7]$$

where A is the cross-sectional area of the narrow opening between the bulb and horn and R is the distance along the horn required for its cross-sectional area to double (R is related to the flaring constant, μ [equation 10.4]). For the *Scapteriscus* burrow, A ranges from 110 to 150 mm², and R ranges from 13 to 15 mm. The formula yields a range of bulb volumes from about 4,150 mm³ to as large as 6,530 mm³. The actual volume of the turnaround gallery is about 7,000 mm³: pretty good correspondence, given that the measurement of volume itself is somewhat inexact. In fact, this correspondence holds for the singing burrows constructed by other mole crickets as well (Table 10.1).

So the shape of the mole cricket burrow is consistent with its acting as a Klipsch horn. But does the burrow actually amplify the sound coming from the cricket's harp, as a Klipsch horn should? Again, the evidence suggests strongly that it does. Amplification

by the burrow can be estimated by measuring the intensity of sound emanating from a sound generator—essentially a simulated cricket harp—placed inside the burrow. Comparing this sound with that of the simulated harp standing alone gives a direct estimate of amplification. Indeed, the *Scapteriscus* burrow amplifies sound, not only at the harp's resonant frequency but over frequencies ranging from 2.5 kHz to 3.0 kHz. The amplification is very large and is greatest—250 times—at the resonant frequency of the *Scapteriscus* harp (2.7 kHz). This strongly suggests the burrow is tuned to resonate at the harp's natural frequency. Further evidence comes from alterations made to the burrow's structure. Caving in the bulb, for example, should reduce amplification, because doing so compromises the diversion of capacitative work to the bulb that is characteristic of the Klipsch horn. In fact, caving in the bulb drops sound emission from the burrow by 6 to 12 decibels.

So the evidence seems to confirm that the singing burrow, at least that of *Scapteriscus*, amplifies the sound emerging from the harp. The benefits a mole cricket may accrue from this are manifold, but one obvious advantage lies in the more efficient conversion of muscle work into sound, the *raison d'être* for horns in the first place. Conversion of muscle work to sound has been estimated for two species of European mole cricket, *Gryllotalpa vinae* and *Gryllotalpa gryllotalpa*, which also construct singing burrows (which we will examine in detail momentarily). For comparison, this estimate has also been made for a field cricket that

Table 10.2 Muscle power used to generate sound compared with emitted sound intensity for three species of European cricket.

Species	Muscle power (mW)	Mean sound output (mW)	Efficiency (%)	Song frequency (kHz)
Gryllotalpa vinae	3.5	1.2	34	3.6
Gryllotalpa gryllotalpa	1.0	0.025	1.5	1.6
Gryllus campestris	1.2	0.06	5	4.4

does not build a singing burrow, *Gryllus campestris*. The numbers are compiled in Table 10.2.

The results are mixed, in an instructive way. First of all, *Gryllotalpa vinae* in its burrow is clearly a superstar: a whopping 34 percent of its muscle power is converted to sound. This is impressive performance, better than anything we humans are capable of: commercial loudspeaker systems commonly operate at conversion efficiencies of about 2 percent. *Scapteriscus* burrows probably would yield conversion efficiencies similar to the burrows of *Gryllotalpa vinae*. Not all "singing burrows" are good acoustic amplifiers, however. Conversion efficiency for *Gryllotalpa gryllotalpa* is only about 1.5 percent. This performance is poorer than that of a garden-variety field cricket, *Gryllus campestris*, which uses no external acoustical contrivances and still manages a conversion efficiency up around 5 percent.

This variation in performance may reflect imperfections of burrow construction, or it may simply reflect different approaches to attracting mates that each work well in their own ways. For example, the singing burrow of *Gryllotalpa gryllotalpa* opens to the surface through numerous small openings rather than through a single flaring horn as characterizes *Scapteriscus* burrows. The perforated opening of the *Gryllotalpa gryllotalpa* burrow may muffle the emission of sound, rather as a pillow placed over a loudspeaker would (Fig. 10.8). The relatively low-frequency song (1.6 kHz), even though it emerges with less power, should nevertheless travel further, because lower-frequency sounds generally do. *Gryllus campestris* may have a relatively higher conversion efficiency because

it "tunes" its harp to resonate at a higher frequency, which should improve the baffling provided by the wings and hence the efficiency of sound production. These crickets' high-frequency song may also reflect off surfaces the cricket sits on more effectively than would the lower-frequency songs of mole crickets. Whatever the explanation, exceptions to the theory remind us of two maxims that biologists forget at their peril: (1) perfection isn't everything and (2) there are many ways to crack a nut.

Broadcast Beacons and Guidance Beacons

Mole crickets' singing burrows fall into two functional types, and the shapes of the burrows likewise fall into two categories, as illustrated in Figure 10.9. The first, exemplified by the *Scapteriscus* burrow, opens to the surface through a single flared horn, with the long axis parallel to the axis of the burrow. In the second type, common to the various burrows constructed by *Gryllotalpa*, the opening is more oblong and is oriented with its long axis perpendicular to the burrow's axis. In some species, like *Gryllotalpa major*, the horn opens through a single oblong opening, while in others, like *Gryllotalpa vinae*, the horn has two openings whose centers are arrayed perpendicular to the burrow's long axis. In some, like *Gryllotalpa gryllotalpa*, the openings are screened by overlying soil.

The two types of burrows emit sound in different ways. The single opening of the *Scapteriscus* burrow broadcasts sound as a point source, so that sound radiates from the opening as a fairly uniform hemispheri-

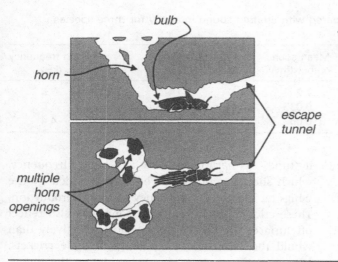

bulb

horn

multiple horn openings

escape tunnel

Figure 10.8 Cross-section *(above)* and top view *(below)* of the singing burrow of *Gryllotalpa gryllotalpa*. Doublet horns lead to several small openings at the surface. [*After Bennet-Clark (1970)*]

cal wave. The *Gryllotalpa* burrows, on the other hand, broadcast as a so-called line source, in which the sound radiates as a semicircular disk oriented at right angles to the line.[4]

These two types of "sound beams" each solve the problem of getting the male's message to potential mates, but in different ways. In both, the male must project sound at sufficient power so that a female can hear it when she intercepts the sound wave. Female crickets, for their part, are surprisingly hard of hearing. For a male cricket's song to get the attention of a female, it has to be loud, about 60 decibels or higher—about the intensity of loud chatter at a cocktail party. The male cricket sings louder than this: at one meter from the hole, the song is about 90 decibels—about the sound intensity of an automobile horn blaring one meter from your ear.

The hemispherical bubble of sound emerging from a *Scapteriscus* burrow radiates to about 30 meters before it becomes too soft to attract the female's atten-

tion. What would happen, though, if the same power was used to drive sound from a line source like the *Gryllotalpa* burrow? Flattening the sound bubble will project sound further in one dimension, just as squashing a spherical wad of pie crust dough makes it wider (precisely, $\sqrt{2}$ times wider than the hemispherical bubble, or out to 42 m), although it would have a narrower breadth. This might pay off for a male cricket in three ways.

First, flattening the sound envelope also broadens it, increasing the probability that a female flying about at random will encounter the 60 dB edge of the sound bubble. A few simple calculations show that the "sweep area" of a disk-shaped "sound net" is about 50 percent larger than that of a hemispherical sound bubble emitted at the same intensity. Flattening the sound bubble may also enable a male cricket to distinguish itself more clearly from other males. *Gryllotalpa* males often sing in leks, large assemblages of males concentrated in a limited area, all competing at once for the attention of females. Leks derive their name from the remarkable communal displays of galliform birds, such as prairie chickens. Several dozen males may line up to display for a single female. The female comes to the lek and sits back and enjoys the show until she chooses

---

4. The fact that the *Gryllotalpa vinae* burrow has a double opening does not matter acoustically, since two point sources emit sound as a line source does if they are positioned closely enough.

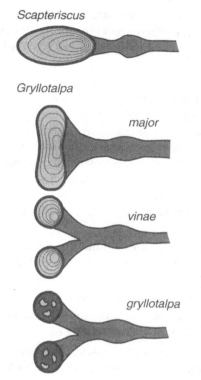

Scapteriscus

Gryllotalpa

major

vinae

gryllotalpa

Figure 10.9 Various shapes of the singing burrows of four species of mole crickets.

the male with which she wishes to mate.[5] In the visually spectacular display of prairie chicken leks, it is simple to distinguish one male from another. However, distinguishing one male from another in an acoustical lek, like a mole cricket lek, is more difficult. First, one male must ensure that his sound envelope does not overlap another's. Hemispherical sound envelopes present obvious difficulties here, since intense calls will tend to merge into a mega-envelope of sound. This may work fine for attracting females *to* the lek, but overlapping may not be much help in guiding her to the burrow of any particular male. Flattening

5. The word *lek* probably has a Swedish origin, from the word for "playground" or "sporting ground."

the sound envelope may reduce the chance that one male's sound envelopes will merge imperceptibly into another's. A flattened sound envelope also acts more effectively as a guidance beacon for the female.

Singing Burrows as "Organs of Extreme Perfection"

The singing burrows of mole crickets are certainly wonderful structures. Now, wonder is a good nursemaid of science, because it is the fount of curiosity. It makes a poor partner, though. Science works best when it asks critical questions, and it is our nature as human beings that we do not ask probing questions of that which holds us in thrall. So, when confronted with a truly wonderful structure like the mole cricket burrow, it is important to splash cold water on your face and ask: how does the singing burrow come to be such an apparently well-designed acoustic structure?

An easy answer to this question would invoke the perfecting power of natural selection. Natural selection, so the story goes, rewards "good" adaptations and penalizes "bad" ones. If singing a loud, pure song is important to a male cricket's reproductive success, we could expect natural selection to force the gradual perfection of the singing burrow into the wonderful structure it is. This may very well be so, but it is not a wholly satisfactory explanation. All it really does is transfer our sense of wonder from the singing burrow itself to the supposed power of natural selection. This mode of reasoning has led biologists and others to ascribe almost magical powers to natural selection and to the genes it works on. As many others have pointed out, this is a problem that has plagued evolutionary biology ever since Darwin.

Darwin himself fretted about the implications for his theory of natural selection of what he called "organs of extreme perfection," of which the vertebrate eye is the exemplar. In a nutshell, the problem with organs of extreme perfection is this: how can one credibly assert the spontaneous origin of a complex and perfectly formed structure like the eye? To do so, one must posit a constellation of simultaneous developmental events, all of which are required to make the structure work properly: the retina must be arranged just so; the cornea must be shaped to refract the light just so, the diameter of the eyeball must correspond just so to the focal length of the cornea and lens; and so forth. If any one of these numerous requirements is not met, you do not have a properly functioning structure. It strains credulity to propose that all these things could be made to happen all at once and properly in the helter-skelter *milieu* of natural selection.

Evolutionary biologists have since come to grips with the problem of organs of extreme perfection, although the problem survives in some fringe elements of biology, like creation science. For one thing, most biologists now have no problem in believing that even a poorly functioning eye is an improvement over an eye that functions even worse, or over no eye at all. So approaching perfection through the gradual improvement in function implied by natural selection does not now seem as high a barrier as it seemed to Darwin and his contemporaries. We also know that the development of eyes is not as complicated as it was once thought—the arthropod homeotic gene *eyeless* has related genes scattered throughout the animal kingdom. These closely related genes nevertheless promote the development of a wide variety of eye types, including eyes as seemingly disparate as the compound eyes of insects and the camera eyes of vertebrates and some molluscs.

Nevertheless, this "solution" to the matter of organs of extreme perfection still doesn't banish wonder from biology: a feeling of awe still pops up in the view of genes as wondrous arbiters of evolutionary destiny. I grant that this feeling may be natural when we are considering structures, like eyes, that must arise through a process of organic development. But is it so credible when we are talking about *external* organs of extreme perfection, like the singing burrow of mole crickets? If natural selection explains the construction of such "amplifiers," how do these genes work their wonder? Does the cricket carry around in its DNA a blueprint, a plan, of the perfect acoustic bur-

row, which it uses to construct the actual burrow? The principle of Goldberg's lever means that we cannot dismiss the possibility out of hand, but it is still hard to believe.

Feedback Tuning of the Singing Burrow

Some insight into this question may be gleaned by observing how *Scapteriscus* constructs its burrow. The cricket begins by excavating the bulb, initially to give itself room to maneuver while it is tunneling. The cricket then pushes its way, head first, out to the surface, opening up the burrow to the outside. Then it goes through a repeating series of three behaviors:

- The cricket pushes dirt out the opening, using its mouth parts and spreading movements of its forelimbs. This action itself will help to flare the mouth of the tunnel as it expands.
- The cricket backs into its tunnel, turns around at the bulb, and emits a short chirp.
- Following the chirp, the cricket moves deeper into the burrow and begins to push soil backwards, moving it out of the burrow. The reworking takes place mostly at the bottom of the tunnel, but the cricket also scoops dirt from the top and sides. Both the bulb and horn of the singing burrow are worked during this stage.

The cricket repeats this sequence several times for about an hour, at which time he settles down and begins to sing for several hours.

During burrow building, the characteristics of the song change dramatically (Fig. 10.10). At the beginning of the job, the harp resonates at about 2.7 kHz, but there is considerable sound energy in both the second and third harmonics (5.5 kHz and 8.2 kHz, respectively). The energy in these secondary frequencies is wasted energy for the cricket, because they are relatively high-frequency sounds and so will not radiate very far. Nor is the female programmed to hear sounds at these frequencies. Furthermore, they represent en-

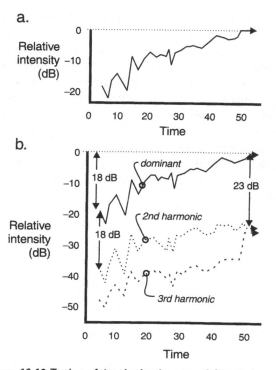

a.

b.

Figure 10.10 Tuning of the singing burrow of *Scapteriscus acletus. a:* Intensity of sound relative to normal singing volume. *b:* Frequency distribution of sound during construction of the burrow. [*After Bennet-Clark (1970)*]

and the sound becomes progressively more powerful and increasingly pure.

The excavation of the burrow has been described as a "tuning" process, which clearly is the case. Tuning a guitar string, for example, involves an alternating sequence of tightening and loosening of the string. At each step, the sound emanating from the string is compared to a standard—say, the sound from a pitch pipe—until the correspondence between the two is satisfactory. You will recognize this process as a negative feedback loop. The cricket modifies its burrow, emits a chirp (which it can hear), makes a further modification, and then repeats the process (Fig. 10.11). The operation of the feedback loop is evident

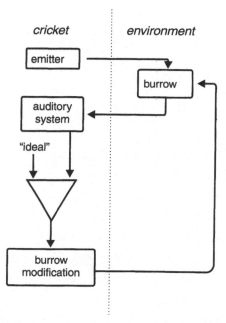

Figure 10.11 Simple model for the construction of the tuned singing burrow of the mole cricket. The cricket monitors burrow performance by how it perceives a test chirp it emits. If the perceived test chirp does not meet an ideal criterion, a round of burrow modification is initiated, which alters the energetic interaction between the sound emitter and the burrow. Note that the feedback loops extend outside the organism.

ergy from muscle that is not available to power the radiation of sound at the resonant frequency, which she can hear.

As the burrow is excavated, the energy emitted in the song undergoes two progressive changes. First, the sound energy levels in all harmonics rises, because of the increasingly horn-like, and therefore amplifying, properties of the singing burrow. Second, energy is shifted away from the second and third harmonics and toward the resonant frequency. We have already encountered this phenomenon in our discussion of horns driving the vibration of reeds. The relative damping of the harmonics and enhancement of the carrier frequency in the mole cricket harp arise in the same way, in the progressive, but not steady, improvement in function: after modifying the burrow, its acoustical performance sometimes improves and sometimes worsens. But by making certain modifications after performance worsens, and other modifications after performance improves, the cricket eventually creates a high-performance acoustical system. The only thing the cricket need carry around in its genes is the fairly simple behavioral program for burrow building and a sensory system capable of assessing the burrow's acoustical properties and correcting its structure as needed.

As physiology, there is nothing really very remarkable about this process. Feedback loops like this operate at all organizational levels within organisms to control their shapes and functions. What is remarkable about the construction of the singing burrow is that the feedback loop extends outside the cricket's body, involving a structure created by the animal and the cricket's use of energy to impose orderliness on its physical environment. The burrow, in short, is as much a part of the physiological process of communication as is the cricket's muscles, nervous system, and body.

And truly it is strange that, with such a government, the ter-mitary should have endured through the centuries. In our own history republics that are truly democratic are in a very few years overwhelmed in defeat or submerged by tyranny; for in matters politic our multitudes affect the dog's habit of preferring unpleas-ant smells, and will even, with a flair that hardly ever fails, single out the most offensive of them all.

—MAURICE MAETERLINCK, *THE LIFE OF THE WHITE ANT* (1930)

CHAPTER ELEVEN

The Soul of the Superorganism

The social insects, which include the bees, wasps, ants, and termites,[1] seem to attract a lot of pesky meta-phor-makers. At various times, the social insects have been pressed into service as exemplars of well-ordered societies, as arguments for the necessity of monarchs, as lessons on the conflict between free will and respon-sibility, as linguists, as practitioners of slavery and of warfare, or, as Maeterlinck suggests, as metaphors for the failures of democracy. I could extend the list indefinitely, for the tendency to metaphor is pan-demic. I might seem to be picking on philosophers, but scientists are just as prone to the disease.

In this chapter and the next, I examine one of the most powerful of the scientific metaphors that have built up around the social insects, the concept of the superorganism. I have used the word before in its adjectival form—superorganismal physiology—to de-scribe physiological processes that extend beyond the conventionally defined boundaries of the organism. The noun has a much broader meaning: a super-organism is any association of living things that, through the coordinated actions of its members, be-haves with all the attributes of an organism. Thus de-fined, the term has been used to describe things as varied as ecosystems, symbioses, human societies (par-

---

1. The bees, ants, and wasps are closely related members of the insect family Hymenoptera ("membrane wing"). The ter-mites are only distantly related to the Hymenoptera, being members of the family Isoptera ("similar wing"), descended from the cockroaches. Estimates put the origin of termites at 75 to 150 million years ago, while the ancestors of wasps are thought to be at least 100 million years older than this.

ticularly transient and unruly associations like mobs), as well as the colonies of social insects.

In taking up the idea of the superorganism, I will be treading onto controversial ground, because it challenges strongly held notions that the living world is composed of discrete organisms. The superorganism idea has had many ups and downs in its history, for a variety of reasons. Sometimes there are just more interesting and powerful ways to think about sociality (Box 11A), and sometimes the superorganism as a model has been associated with thoughts, particularly about humans and their societies, deemed to be impure or politically distasteful. Nevertheless, no matter how assiduously the scholars of the day try to smother it, the superorganism seems, *a la* Mr. Micawber, to keep turning up. Indeed, the 1990s has seen a renaissance of the idea, most notably in the earth sciences, where it has reappeared in the form of the Gaia hypothesis, James Lovelock's and Lynn Margulis's remarkable conception of the Earth as a single living entity, a superorganism.

In this chapter, we will explore the idea of the superorganism in the relatively prosaic context of the social insects. In particular, I wish to explore a phenomenon known as social homeostasis. Referring to the supposed regulation of the physical environment of the nest,[2] social homeostasis is the superorganismal analogue of homeostasis (or, as I shall call it from here on, organismal homeostasis). Like the regulated internal environment of an organism, the environments of many social insect nests are very stable. This stability is not evidenced by individual members of the colony; rather, it emerges only when the many individuals organize into the colony. This emergence of homeostasis in the context of the colony is what we will call social homeostasis.

My aim in this chapter is to present social homeostasis in the context of the "breathing" of social insect colonies, that is, the exchange of respiratory gases between an insect colony and its environment. In particular, I will focus on how structures built by social insects aid in the regulation of their nest environments.

What Homeostasis Is

In any discussion of social homeostasis, it is important at the outset to distinguish between the *outcome* and the *process* of homeostasis. The stability of some property, like an animal's body temperature or the concentration of some substance in its blood, is one obvious outcome of homeostasis. Indeed, that is the literal meaning of the word: *homeo* (steady)–*stasis* (state). Stability in and of itself, though, is not homeostasis. A fish living in the abyssal oceans will have a stable body temperature and a stable composition of salts in its blood in large part because it lives in an environment that is itself very steady. Nor is it necessarily homeostasis if an animal's internal environment remains steady in an *unsteady* environment. In contrast to your average small-bodied lizard, for example, a large monitor lizard (which can weigh in at a few dozen kilograms) has a stable body temperature throughout the day, even when faced with large daily fluctuations of environmental temperature. However, the stability of the monitor lizard's body temperature reflects more its thermal inertia than homeostasis: a large rock could be considered as "homeostatic" as a large lizard, by this logic. So simple steadiness of the internal environment—the outcome—is not sufficient evidence for homeostasis.

Homeostasis results, rather, from a regulatory *process*. To qualify as homeostasis, a system should display the signs of that process in operation—ideally, signs that are independent of the context within which it is working. That way, we are not limiting the phenom-

2. Some clarification of terminology is in order here, because each group of social insects has a jargon associated with it. I shall use the term *colony* to describe the assemblage of individual organisms that make up a familial unit. For example, a termite colony represents the descendants of a single queen, as well as the symbionts associated with them. The *nest* is the structure in which a colony is housed. Among honeybees, the nest is sometimes referred to as a *hive*. Among termites, the nest often has associated with it ancillary structures, the most spectacular being a mound.

Why Are Social Insects Social?

The social insects have long posed a challenge to biologists, mostly because they exhibit altruism: individuals engage in seemingly selfless sacrifice for the benefit of others. At its most extreme, altruism can involve the actual suicide of an individual member of the colony—a worker bee dying, for example, after stinging an attacker threatening the colony. Altruism is usually more mundane: most workers in a social insect colony are sterile and exhibit their altruism by forgoing the option to reproduce, a type of "reproductive suicide."

This behavior posed great difficulty to the early Darwinian evolutionists, who could not explain why natural selection would allow such a scheme to evolve. Imagine that there is an "altruism gene," which causes individuals that carry it to give up their own efforts to reproduce and to help other individuals reproduce. By not reproducing, those organisms with the altruism gene would not pass it on; that is, the gene would be selected against. One would expect any altruism gene that appears soon to be driven from the population. So great was this difficulty that Charles Darwin considered the social insects a dagger pointed at the heart of his theory of natural selection. If some reasonable explanation could not be found for them, then his whole theory would have to be dismissed.

The explanation Darwin was looking for finally emerged with the advent of the synthetic theory of evolution. One of the fundamental insights of the synthetic theory was to treat organisms as mere vessels for genes in a population. From that insight, it is a short leap to realize that there is no significant difference between an individual passing on a gene and an entirely different individual passing on an identical copy of the gene. Therefore, one's own evolutionary fitness (the likelihood you will reproduce and pass your genes on) is not just a matter of your own likelihood of reproducing; it must also include the likelihood that every individual that carries a copy of your genes will reproduce.

This concept of *inclusive fitness* provides the explanation for sociality that Darwin so avidly wished for. The likelihood that another individual carries a copy of your genes follows some well-behaved rules of inheritance. For example, you and your siblings have, on average, about a 50 percent chance of sharing identical copies of a gene. The more distantly related you are, the smaller this likelihood would be. Half-siblings share your genes only about a fourth of the time, and first cousins would share a gene with you only about an eighth of the time. Evolutionarily, your fitness would be the same if you reproduced yourself or if you altruistically decided not to reproduce and at least doubled the chance that your siblings would reproduce. Or if you quadrupled the chance your half-siblings would reproduce or octupled the chance your first cousins would reproduce. Thus, altruism should be reasonably common in family units, and it should become rarer among more distantly related individuals.

Among the Hymenoptera (the bees, ants, and wasps), the rules of inheritance are somewhat peculiar, and this peculiar mode of inheritance seems to predispose these insects to altruistic sociality. The mode of inheritance is called haplodiploidy. The queen of the colony (the sole reproductive member) produces haploid eggs, that is, eggs that contain one copy of her diploid genome. In this sense, she is like any other sexually reproducing animal. In the normal course of events, the haploid egg would unite with a haploid sperm, carrying one copy of a male's genome, and at fertilization the two sets of genes would form a normal diploid individual, carrying genes of both the father and mother. This also occurs in hymenopterans, and when it does, the offspring is inevitably a diploid female. The queen also produces eggs that are never fertilized by a sperm, however, and these go on to develop into a haploid male. This unusual means of reproduction skews the patterns of inheritance that determine inclusive fitness. I don't wish to bother with the details of the calculation here, but to put it simply, haplodiploid reproduction makes a female bee more closely related to her sisters than she would be to her own daughters. For a worker bee faced with the "choice" either of reproducing herself or forcing her mother to produce more sisters, she will tend to go with whatever increases her inclusive fitness the most. This means forcing her mother to produce sisters. From this reproductive fact follows the peculiar aspects of social insects, like the legions of sterile workers, the single reproductive individual, the high degree of altruistic behavior, and so forth.

Like most beautiful explanations, the haplodiploid theory of sociality only goes so far. Altruism and social behavior are found among animals that do not reproduce by haplodiploidy, and it obviously cannot explain sociality among them. Human beings are an obvious example, but since our concern here is with insects, the most dramatic exception among them has to be the termites. These insects are only distantly related to the Hymenoptera, the termites being descended relatively recently from the cockroaches and the Hymenoptera descended from an as yet unknown "protowasp." Termites reproduce in the conventional way, that is, male and female produce haploid sperm and egg, which combine to form diploid offspring that are either male or female. Yet these insects, too, form large social colonies, with one or a few reproductive females tended by an enormous number of sterile workers, an organization remarkably similar to that of the bees and ants. In the case of termites, the sterile workers are both male and female.

The probable explanation for sociality among termites rests with their unusual digestive physiology. As everyone knows, termites eat wood (cellulose, actually). This is a peculiar thing for an animal to do because animals generally do not produce cellulase, an enzyme required to cleave sugar off the cellulose molecule. If an animal is to digest cellulose, it must enter into an alliance with an organism that possesses cellulase, which can be found among many bacteria, some flagellate and ciliate protozoa, and fungi. For most termites, these organisms are cultivated in the termites' guts to form a rich intestinal flora. Among others, including the macrotermitines described in the text, cellulose digestion has been "outsourced" to fungi cultivated outside the gut. The termite provides its symbionts with a rich source of cellulose for food, and the termites get in return glucose cleaved off the cellulose, as well as protein, vitamins, and essential amino acids produced by the symbionts.

The problem for termites is that they hatch from their eggs without this essential intestinal flora. To acquire it, they must be inoculated with it, which the colony's inhabitants do by feeding the newly hatched termites with feces and drops of regurgitate that contain the symbionts. Thus, the digestive physiology of termites forces them into a social interaction right from the get-go; it is the foundation of their very strongly developed social lives.

enon of homeostasis to organisms—we could objectively look for *social* homeostasis in *assemblages* of organisms.

What might these signs be? Let us look for them in the context of temperature regulation. Suppose there is an organism with a stable body temperature in an environment with a fluctuating temperature. This disparity means that body and environmental temperatures will frequently differ. Any difference in temperature between the body (T_b) and the environment (T_e) establishes a potential energy difference that can drive a thermodynamically favored flux of heat (a *TFF*, in the jargon introduced in Chapter 3) across the boundary separating the animal from its environment:

$$Q_{TFF} \propto (T_e - T_b) \qquad [11.1]$$

where Q_{TFF} is the *TFF* of heat. Temperature, being a measure of heat content, would be affected by this flux: as the *TFF* drives heat into or out of the body, the body's temperature must change—the antithesis of homeostasis. Homeostasis arises only if the organism has some way of actively heating or cooling the body,[3] which means driving an opposing flux of heat, designated in Chapter 3 as a physiological flux *(PF)*. The *PF* must match the *TFF* in magnitude, but not, obviously, in sign:

$$M_{PF} = -Q_{TFF} \qquad [11.2a]$$
$$M_{PF} + Q_{TFF} = 0 \qquad [11.2b]$$

3. Among animals, active heating involves the generation of heat directly from the hydrolysis of ATP, the so-called metabolic heat production, or thermogenesis. Active cooling of the body can only be accomplished by evaporation. See Box 11B.

Thermal Energy Balance in Vertebrate Thermoregulation

The metabolic response to cold described in the text is a limited subset of the manipulations of heat flow that mammals and birds actually engage in. The complete suite of responses involves a complicated mix of variations of heat production by metabolism, adjustments of the body's thermal conductance, and heat dissipation by evaporation.

The foundation of these responses is the relationship between thermodynamic heat flux and environmental temperature expressed as Fourier's law. Where environmental temperature (T_e) is less than body temperature (T_b), the thermodynamically favored flux *(TFF)* of heat is outward, and the physiological flux *(PF)* must be inward. Where $T_e > T_b$, the *PF* must be outward. Of course, at $T_e = T_b$, no *PF* is needed. The trouble is this: most mammals and birds cannot lower their metabolic heat production below a minimum called the basal metabolic rate *(BMR)*. Therefore, the Fourier's law relationship between metabolic heat production and environmental temperature applies only at those temperatures where metabolic heat production (M_h) exceeds *BMR*. The upper limit of environmental temperature for Fourier's law is known as the lower critical temperature *(LCT)*.

At environmental temperatures greater than the LCT, metabolic heat production is steady at the BMR (Fig. 11B.1). The animal still cannot escape the constraints of Fourier's law, however: at steady heat production and steady conductance, the temperature difference between the body and environment likewise must be steady. This difference obviously cannot persist if body temperature is to be regulated. Instead, the body's conductance is adjusted over a *range* of environmental temperatures, roughly between $T_e = LCT$ and $T_e = T_b$, through variations of peripheral blood flow, postural adjustments, flattening of the fur or pelage, and so forth.

This strategy has its limitations, however, because the body's thermal conductance can be increased only by so much. Above this maximum conductance, K_{max}, the minimum BMR must be offset by some active cooling of the

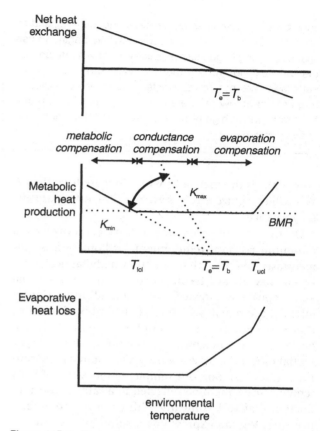

Figure 11B.1 The energetics of temperature regulation in a homeothermic endothermic animal.

body, which generally involves evaporation of water from the body surface (Fig. 11B.1). Specifically, the body's net heat production, M_n, consisting of the sum of the *BMR* and the evaporative heat loss, Q_e, must equal the *TFF* for heat demanded by Fourier's law:

$$M_n = BMR + Q_e = K_{max}(T_e - T_b)$$

Within a few degrees of $T_e = T_b$, moderate increases of evaporation are sufficient to keep the body in heat balance. However, above a certain environmental temperature, called the upper critical temperature *(UCL)*, increasing evaporation rate involves an increasing metabolic cost, because mechanisms like sweating, panting, and so forth cost energy to run. Consequently, M_h begins to rise again, requiring large increases in evaporation rate so that the net heat flux meets the demands imposed by Fourier's law. The limits on tolerable environmental temperature are set at low temperatures by the maximum metabolic rate, and at high temperatures by the maximum rate of evaporation.

where M_{PF} is the metabolic work done to drive the *PF*. You will recognize this by now as a straightforward energy balance.

On the basis of equations 11.2, we can now state a minimum physical requirement for homeostasis: homeostasis requires that the thermodynamically favored flux always be matched exactly by an equal and oppositely directed physiological flux. Consequently, the rate at which physiological work is done, that is, the power requirement for homeostasis, is proportional to the magnitude of the thermodynamic potential difference between the animal and its environment (Fig. 3.3). So, for example, homeostasis of body temperature should be reflected in a rate of heat production that varies linearly with environmental temperature. We can express this relation in an equation, Fourier's law, that explicitly represents the power requirement for temperature homeostasis:

$$-Q_{TFF} = M_{PF} = K_h(T_b - T_e) \qquad [11.3]$$

where K_h is the thermal conductance (W K^{-1}).

The outward signs of thermal homeostasis are evident when we plot equation 11.3 as a graph of metabolic heat production versus environmental temperature (Fig. 11.1). A straightforward pattern emerges: a linear decrease of heat production with increasing environmental temperature, with a slope of $-K_h$, falling to a heat production of null at an environmental temperature equal to body temperature (this oversimplifies the case for vertebrate thermoregulators—see Box 11B for the rest of the story). This graph represents the distinguishing mark of homeostasis—any living system, whether it be cellular, organismal, or superorganismal, that exhibits this *insigne* can fairly be said to be homeostatic. I will ask you now to tuck this tidbit away for a moment, while we look more deeply into the process of organismal homeostasis, again in the context of the regulation of body temperature of animals, mammals specifically.

Functional Elements of Organismal Homeostasis

The process of homeostasis implies an infrastructure to make it work. Among mammals, the "machinery" for temperature homeostasis involves the entire body, but a substantial part of it resides in the anterior hypothalamus of the brain (in us, located just behind and above the backs of the eyeballs). The anterior hypothalamus receives information from temperature-sensitive cells all over the body. These encode and transmit information on the distribution of temperature throughout the body. In addition, there is a group of temperature-sensitive cells in the anterior hypothalamus itself, and these encode information on brain temperature. Some of these cells respond to elevations of hypothalamic temperature while others respond to drops in hypothalamic temperature. Finally, there are clusters of hypothalamic cells that are spontaneously active and whose activity does not vary with temperature.

Here we see the elements for negative feedback control of body temperature (Fig. 5.18). The sensor signals arise from the various temperature-sensitive cells throughout the body and brain. The setpoint signal arises from the temperature-insensitive cells in the anterior hypothalamus. There, the signals are processed and compared in a neural circuit that acts as a comparator. Emerging from this circuit are outputs (error signals) that activate various effectors that alter the generation of heat within the body or modulate the flow of heat between the body and environment. Together, these parts operate as a negative feedback controller for body temperature.

This feedback loop was demonstrated in elegant

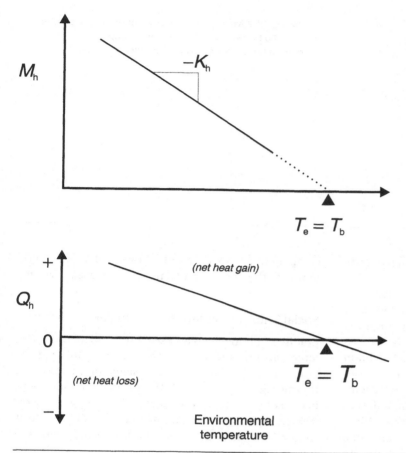

experiments that tampered with the hypothalamic "thermostat" of dogs, sheep, and pigeons. The rationale for these experiments requires some explaining, and it will help if I share a personal experience with you. I was a graduate student during the boycott of petroleum sales to the United States by oil-producing countries in the 1970s. In response, President Jimmy Carter issued a series of stringent energy conservation decrees, among them a dictate that all thermostats in government offices be turned down to 65°F during the winter. Because I was a student at a state university, this decree applied to the small office I shared with some of my compatriots. Unfortunately, the thermo-

stat in my office was defective, making the room considerably cooler than the decree allowed and keeping my fellow inmates and me in Dickensian misery as we shivered over our spreadsheets. We could not simply turn the thermostat up, though, because it was locked, and we had been warned of Dire Consequences should we attempt to change the setting. So, clever fellows that we were, we draped wet paper towels over the thermostat. The evaporation from the paper towels cooled the thermostat below the room's actual temperature, tricking it into thinking the room was cooler than it was. In response, the thermostat dutifully turned on the heat, which brought our actual room

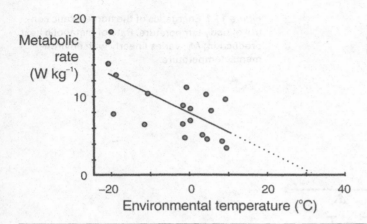

Figure 11.2 Metabolic heat production rate of a colony of 2,000 honeybees, *Apis mellifera,* depends on environmental temperature. [*After Southwick (1983)*]

temperature up to a reasonably comfortable 65°F or so.

One can pull a similar trick on the anterior hypothalamic "thermostat." Instead of a wet paper towel, of course, a small probe, called a thermode, is surgically inserted into the animal's brain. The thermode can raise or lower the temperature of a small patch of brain independent of the surrounding tissue. The consequences of localized tampering of hypothalamic temperature are remarkably similar to those that followed my tampering with the thermostat in my office. When the anterior hypothalamus of a dog, say, is warmed a degree or two, the dog will begin to pant, blood flow to the peripheral regions of the body will increase, and it will adopt postures, like sprawling on a cool floor, all of which increase the loss of heat from the body. All this happens even though the body temperature itself (that is, the rest of the body not including the anterior hypothalamus) is normal. Conversely, cooling the dog's anterior hypothalamus elicits shivering, a reduction of blood flow to the extremities, and postural adjustments, like curling up, that increase heat production and reduce the thermodynamic losses of heat from the body. These elegant experiments show that there is a negative feedback system operating in the thermoregulation of a dog, that its components can be localized and discretely manipulated, and that these component parts will respond in ways that are consistent with the operation of a negative feedback system.

Social Thermoregulation in Honeybee Colonies

Colonies of honeybees also seem to regulate their hive temperatures, and the outward signs of this regulation are very similar to the signs of mammalian thermoregulation (Fig. 11.2). For example, a honeybee colony's total rate of metabolic heat production (which represents the summed contributions of heat from all the individual bees) varies linearly with the outside temperature, and with the same result: the hive temperature is stabilized within a range of 4–5°C.

Keep in mind that we are not talking here about the stability of a bee's body temperature but of the *hive* temperature, which, from the bee's point of view, represents its external environment. Keep in mind also that the stability of hive temperature is not simply a consequence of large numbers of bees all individually regulating their own body temperatures. Suppose, for example, all the bees in a hive were individually homeothermic, each regulating its own body temperature. Each bee would exhibit the linear increase of individual heat production as external temperatures cool, similar to that seen in Figure 11.1 (top graph). That the colony's production of heat also behaves this way might simply reflect the summed contributions of the thousands of individual bees all behaving the same way. However, organismal homeostasis *en masse* cannot explain the homeostasis of hive temperature: the effectors for colony thermoregulation have a social dimension that is simply lacking in the individual bees. For example, on cool mornings bees tend to huddle into a compact cluster: the cooler the morning, the more tightly they huddle (Fig. 11.3). Huddling confines heat within the cluster, and huddling more closely restricts heat loss more. Furthermore, not all the bees in a cluster are doing the same things. Some of the bees occupy a circular shell spanning roughly the middle third of the cluster. These bees shiver their

a.

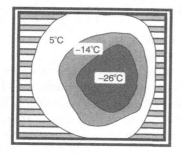

b.

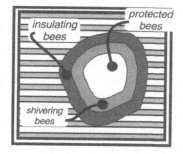

Figure 11.3 Operation of social effectors for social homeostasis of honeybee hives. *a:* Distributions of bees on a comb in three different environmental temperatures. Temperatures indicate the environmental temperature. [*After Wilson (1971)*] *b:* Functional distribution of bees in a typical huddle, showing the different locales within the huddle of bees engaging in the indicated behaviors.

flight muscles, generating the bulk of the heat produced by the cluster. At the same time, another group of bees, located at the outer margin of the cluster, adopt postures that interweave the chitinous hairs of each bee with its neighbors, forming a kind of downy coat that helps insulate the cluster. This social differentiation into shivering and down-weaving bees cannot be explained as organismal homeostasis *en masse*. Social homeostasis, then, exists—objective and sensible criteria can be used to identify it, and it seems clearly to operate at a social level of organization, at least among honey bees.

The crux of the problem for social homeostasis then becomes the deeper question: does similarity in the outward signs of homeostasis imply similarity of function? Individual bees undoubtedly carry around within them the neural machinery for the negative feedback control of their own body temperatures. There must exist within each bee sensors for body temperature, some neural machinery to process this information, and effectors that will alter the bee's heat budget. Social homeostasis, however, implies that there are feedback loops that extend beyond the boundaries of the individual bees—meta-loops, if you will, that control and coordinate the individuals in the colony. While it is easy to accept the conventional view of neurally based negative feedback controllers operating within each bee, the idea of meta-loops governing social homeostasis is more difficult to comprehend.

In fact, social homeostasis often operates through novel mechanisms of feedback control, some of which do not involve negative feedback at all. Interestingly, many of these mechanisms involve structures built by the colony inhabitants, structures which harness and transform both metabolic energy and physical energy in the environment to power the external physiology that social homeostasis requires.

The following discussion of social homeostasis will focus on regulation of the nest atmosphere. The air inside most social insect nests is more humid, richer in carbon dioxide, and poorer in oxygen than the outside air. These concentration differences drive fluxes of oxygen into the nest and fluxes of carbon dioxide and water vapor out. The nest atmosphere appears to be under the colony's adaptive control, because its composition is frequently steady, despite substantial variation in the flux rates for these gases.

The Little Big Bang: Positive Feedback Meta-Loops

Let us turn first to an example of social homeostasis that does not involve a negative feedback operation in the organisms. The mechanism I shall describe is, in fact, common among social insects. It involves an explosive (figuratively, of course) mobilization of workers in response to a perturbation of the colony—what I call the "Little Big Bang."

Anyone foolhardy or unfortunate enough to have broken into a nest of bees or wasps has experienced the Little Big Bang in one of its more dramatic forms: the defensive swarm that explodes in a furious frenzy from the disturbed nest. In a bee hive, the frenzy is set off when one or a few bees experience conditions that elicit an alarm reflex, which includes the emission of an alarm signal. Often, the alarm signal is a volatile chemical, an alarm pheromone, released from specialized glands in the worker's body.[4] When another bee senses the alarm pheromone, this triggers an alarm reflex in it, which initiates the reflex in yet others, and so forth. Thus, an alarm signal from a single perturbed worker bee can spread rapidly through all the members of the colony, eliciting the frantic and aggressive behavior that soon turns into a defensive swarm. You will recognize this as positive feedback, which we have already encountered in the context of diffusion-limited accretion growth of sponges and corals.

---

4. In some cases, the alarm pheromone is a single chemical. Among bees, for example, the alarm pheromone is isoamyl acetate, released from the worker bee's sting pouch. In other cases, the alarm pheromone is a complicated chemical cocktail. Among ants, alarm pheromone is released from glands around the head, the mandibular glands, which emit a mixture of terpenes, hydrocarbons, and ketones, some or all of which serve as alarm pheromones.

Positive feedbacks turn up frequently in physiological systems, because they are very handy switching mechanisms, and organisms often need to switch their physiology from one function to another. Physiology dominated by negative feedback, which resists change, would be a stable but boring place. For example, at some point in an organism's life cycle, it must switch its use of energy from maintenance to reproduction. If the use of energy for maintenance is governed by negative feedback, something must either negate, oppose, or switch off these systems so that the energy can go elsewhere. Frequently, that something is activation of a positive feedback loop.

The Little Big Bang, being a social response, involves positive feedback meta-loops. The curious thing about the Little Big Bang is that it is often employed for social homeostasis, as exemplified by the defensive response of termites to a breach in the structure that houses their colony. In Africa and Asia, termites often build above-ground structures, or mounds, to house and protect their colonies. These termites must often deal with holes made in their mounds, either by predators, such as aardvark and aardwolf, or by inclement weather, such as torrential summer thundershowers. Breaching the mound wall is roughly the termite's equivalent of a broken window in an air-conditioned building. In the rooms near the broken window, the building environment is perturbed by the outside air, and people who sit closest to the broken window feel the greatest change. People down the hall or on another floor may not even be aware that a window has been broken somewhere in their building. The air in a termite colony is similarly affected by a hole in the mound wall. In the immediate vicinity of the breach, the air is perturbed by locally steep gradients in the partial pressures of oxygen, carbon dioxide, and water vapor.

One way to deal with a broken window in an air-conditioned building is to isolate the break somehow—close the door to the office containing the broken window, for example. Most likely, the person to do this would be someone sitting close to the broken window, since that person would feel the perturbation most strongly. If you add to this scenario an interaction between office workers, such that the sight of one person getting up to close the door to the office with the broken window elicits other office workers to begin closing off other nearby doors, or the halls, or the stairwell doors, then you have something akin to the response of worker termites to a breach of their mound wall. If a termite encounters the locally perturbed atmosphere around a breach—such as anomalously high pO_2, low pCO_2, low humidity, or breezy conditions—it responds with an alarm reflex. First, it will grab a handy grain of sand and cement it into place with a drop of gluey secretions from its mouth—this is the daubing reflex. Second, it releases a chemical alarm pheromone that wafts through the rest of the nest. Finally, disturbed workers tap their heads rapidly against the mound's walls, sending a sound vibration spreading through the mound like a telegraph.

For about the first ten minutes following the breach, the activity in the vicinity of the opening is fairly haphazard. Termites stumble into the perturbed environment around the breach almost by chance, and when they do, their alarm reflexes are triggered. There is little evidence of coordinated activity at this point: daubs of cement are placed seemingly at random around the breach. After a few minutes, the number of workers at the breach begins to increase slowly, and in about ten minutes the first termites recruited by the initial alarm signal start to arrive. These too deposit their grains of sand and emit their own alarm signals. Then the number of workers at the breach increases rapidly. Eventually, so many recruits appear that the initially haphazard building activity merges into a marvelous construction project, with teams of termites erecting pillars, walls, and galleries. After about an hour or two the breach is sealed off.

What I have described here is social homeostasis—a perturbation of the colony atmosphere elicits a response that returns the colony atmosphere to its state prior to the perturbation. It sounds like classical negative feedback, but it is not. Rather, it involves two positive feedback meta-loops: a "fast" loop that mediates a rapid response between individual workers; and a "slow" loop that involves the colony as a whole. Let us take each in turn.

The fast meta-loop involves the daubing reflex. When another worker termite comes along and detects the residue of a previous termite's daubing reflex, a daubing reflex is elicited in it. This has been dubbed stigmergy (from the Latin stigma, "sign" or "mark," + ergon, "work," literally, "driven by the mark"). It is a form of positive feedback, and it results after time and repetition in the construction of pillars and walls (Fig. 11.4). Stigmergy alone is not sufficient for the Little Big Bang, though. To set it off, the fast response— stigmergy—must be coupled to a slower positive feedback meta-loop that drives the recruitment of new workers to come to the site of the breach. The speed of the recruitment response is limited by how rapidly the alarm signal can be transmitted from the termites initially encountering the breach to new termites elsewhere in the nest and how quickly these new termites can come to the site. In termite colonies, this is a relatively slow process, because the spread of any chemical signal may require minutes to spread throughout the nest. Termites are not very fast runners, so they require some time to make their way to the site of the breach. Once set in motion, though, every termite that is recruited to the breach itself emits alarm signals, and the recruitment increases rapidly in the manner characteristic of the Little Big Bang. The result is a large mobilization of energy (in the form of the work done by the workers) that keeps on increasing until the perturbing influence is literally overwhelmed. Only then does the intensity of the "slow" loop begin to diminish and recruitment decline.

Pushmi-Pullyu Ventilation: Manipulating Colony-Scale Energy Gradients

Honeybees sometimes meet their colony's gas exchange needs by actively ventilating the hive. Workers positioned at the hive's entrance beat their wings to create a fanning effect. The mechanism, so seemingly

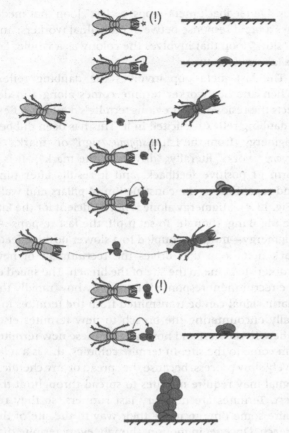

Figure 11.4 Stigmergy and positive feedback. When a termite encounters either a hole or a fresh daubing, it sends out an alarm signal *(!)* and contributes to the plug. At right, the fresh daubings, shown from the side, pile up to form a new column or wall.

simple, in fact involves a complicated interplay between multiple gradients of potential energy in the hive, some of which are metabolic and some of which are physical. The architecture of the hive plays an important role in this interplay.

This mechanism of social homeostasis takes its name from the famous pushmi-pullyu of Dr. Dolittle. The pushmi-pullyu, to remind you, was an extraordinary camel-like creature with two front halves, each facing in opposite directions, but with no hind quar-

ters. Normally, the pushmi-pullyu coordinated its two front halves very nicely: one watched while the other slept, one talked while the other ate, and so forth. But the pushmi-pullyu could get confused about which way was forward. If both heads were equally determined to go in the direction each thought was forward, the animal would go nowhere. Only if one head was more determined than the other could the pushmi-pullyu move.

Social homeostasis of the atmosphere of honeybee hives involves a process I call pushmi-pullyu ventilation. In short, pushmi-pullyu ventilation works like an on-off switch for gas exchange in the hive. Just as a heater can regulate the temperature of a room by being switched alternately on and off, so too can pushmi-pullyu ventilation of a hive regulate the hive's atmosphere. The phenomenon itself will require some explaining, but before I do so, let us first lay some background about bees and their hives.

Before beekeepers came along, honeybees commonly housed their colonies in hollow cavities in trees. These cavities are naturally isolated from the outside world, but if a bee colony takes up residence in one, the worker bees typically isolate it even more, sealing off all openings to the cavity, save one, with "bee-glue" or propolis, a very hard and resinous wax. Commonly, the one remaining opening is located below the colony, and it serves as a port of egress and entry for the worker bees shuttling back and forth between the hive and their sources of food. It also serves as an exchange port for the hive's respiratory gases: oxygen, carbon dioxide, and water vapor.

Oxygen and carbon dioxide must move through the hive's entrance hole at a rate adequate for the colony's needs. Each gas's rate of movement and rate of consumption must be balanced, as an organism's physiological state is balanced, according to equation 11.2. The flow of oxygen, for example, through the hive involves two processes: the consumption of oxygen by the bees (M_{O_2}, with units like grams per second or milliliters per minute) and an exchange of oxygen (Q_{O_2}, with similar units) across the hive entrance. Note that we are treating the hive environment now as functionally equivalent to the internal environment of an organism. In the terminology used in equation 11.2, M_{O_2} is a physiological flux *(PF)* while Q_{O_2} is a thermodynamically favored flux *(TFF)*. Homeostasis requires the two to match:

$$Q_{O_2} + M_{O_2} = 0 \qquad [11.4]$$

To some degree, oxygen will flow passively into the hive simply as a consequence of the colony's consumption of oxygen. As the colony consumes oxygen, the pO_2 in the hive atmosphere will drop, and the partial pressure difference thereby established drives oxygen passively across the hive entrance. Thus, the *TFF* of oxygen can be represented by an equation similar to Fick's law:[5]

$$Q_{O_2 TFF} = K_p(pO_{2o} - pO_{2h}) = -M_{O_2} \qquad [11.5]$$

where pO_{2o}, and pO_{2h} are the partial pressures of oxygen outside and inside the hive, respectively, and K_p is the passive conductance for oxygen. For our purposes, we shall assume that passive conductance is determined by the size and shape of the hive's entrance hole. This is intuitively easy to grasp: oxygen moves through larger holes faster than through smaller holes.

A bee colony that relies on passive movements of gas alone cannot develop social homeostasis. Consider, for example, what must happen if the colony's demand for oxygen increases, as it might if the outside temperature cooled. Equation 11.5 tells us there are two ways this increased demand can be met.

First, increased oxygen consumption could be supported by an increase in the partial pressure difference driving oxygen across the hive entrance (pO_{2o} − pO_{2h}). Because the partial pressure of oxygen in the atmosphere is essentially fixed, an increased partial pressure *difference* can only come about through a drop of the hive's pO_2. That result would be the antithesis of homeostasis. Alternatively, the bees could increase the hive's passive conductance (K_p), which could increase flux without any change of the partial pressure difference. This also is problematic. Any alteration of the hive's conductance would require structural modification of the hive entrance: widening it, or shortening it, or some combination of the two. Bees do this, obviously, but it is a lot of work, and it is suitable only for long-term or chronic alterations of metabolism, not for the minute-by-minute modulation of flux that social homeostasis would presumably require.

Social homeostasis of the hive atmosphere must therefore circumvent the constraints of equation 11.5. Bee colonies do so by a social behavior: stationing workers at the hive entrance where they fan their wings in place. The fanning adds a component of forced convection to the oxygen's flux, supplementing the passive exchange rate with an active ventilatory component, Q_v. Because the ventilatory flux depends upon the number of bees fanning (n_b) and how energetically each individual fans (q_v), the flux of oxygen can be modulated minute-by-minute simply by altering the number and activity of bees recruited to do the fanning:

$$Q_{O_2} = Q_{O_2 TFF} + Q_v = K_p(pO_{2o} - pO_{2h}) + n_b q_v(pO_{2o} - pO_{2h})$$
$$= (K_p + n_b q_v)(pO_{2o} - pO_{2h}) \qquad [11.6]$$

If fanning could be modulated by a meta-loop coupled to oxygen demand, then variation of the hive's oxygen consumption could be supported with no disruption in the gas composition of the hive atmosphere: in short, social homeostasis.

At least in honeybee colonies, we know that such a meta-loop exists. For bees, the signal for recruitment

---

5. This equation is *similar* to Fick's law because Fick's law technically refers to a flux driven by the mechanism of diffusion. In the example, I am being deliberately agnostic about just what the mechanism of exchange is; my point is just that it occurs by some passive mechanism, which may partially involve diffusion but may be supplemented by other mechanisms, like convection induced by winds blowing past the hole. If such processes behave similarly to diffusion, we say the flux is governed by the Fick principle.

to fanning is the hive's pCO_2. This can be demonstrated by the colony's response to imposed perturbations of the hive atmosphere. The hive's pCO_2 can be increased artificially by adding CO_2 to the hive from a gas cylinder, independently of any variation of the colony's oxygen consumption or metabolic rate. This is akin to the wet paper towel trick I described earlier: altering the local environment independently of the effectors intended to control it. As CO_2 is added to the hive, workers begin to show up at the hive entrance and fan in place until the hive pCO_2 declines, even though the "production rate" of carbon dioxide is artificially elevated. The greater the perturbation of hive pCO_2, the more workers will be recruited to fan. The hive pCO_2 will then return to its pre-disturbed value, even though the CO_2 flux has been increased.

The Metabolic Meta-Loop

While the meta-loop linking fanning behavior to hive pCO_2 is one of the components of pushmi-pullyu homeostasis, it is not the whole mechanism. The colony's metabolism, in addition to its effects on the hive atmosphere, also elicits other potential energy gradients within the hive. These gradients interact with the hive's architecture to promote still other feedback meta-loops. It is these that are at the heart of pushmi-pullyu ventilation, so let us understand what they are and how they arise.

Metabolism of glucose liberates a considerable amount of heat, which warms the hive air. In addition, the consumption of oxygen and its replacement with carbon dioxide and water vapor reduces the average molecular weight of the air in the hive.[6] Both reduce the density of air in the hive, and gravity therefore imparts buoyant forces to it. As the colony metabolizes, therefore, spent air will be lofted upward to the top of the hive (Fig. 11.5a).

What happens next depends upon the configuration and number of holes that open into the hive. Let us look first at the arrangement of holes common to many commercial bee hives: two openings, one at the bottom, which serves as the normal entry and egress port for the hive, and one at the top, which can be opened or closed by the beekeeper. In a hive with this configuration, the buoyant energy imparted to the air will do work: it will drive a natural convection of air through the hive, lofting the hive air out through the top hole and drawing a steady flow of fresh air in through the hive entrance (Fig. 11.5b). This mechanism is known as thermosiphon ventilation.

A hive so constructed will be self-regulatory to a degree, because the physiological flux (the rate of oxygen consumption) is physically coupled to the flux rate of oxygen into the hive. The physical coupling arises, of course, because of the link between colony metabolism and the hive's air density. Increased levels of oxygen consumption will dissipate more heat from the bees' bodies to the hive air, magnifying the metabolism-induced change of air density, which will, in turn, drive faster rates of ventilation, ultimately increasing oxygen flux into the hive. Indeed, this is why commercial hives have two holes. The self-regulatory tendency of a two-hole hive, supplemented by the beekeeper's control over the upper hole, spares bees from the need to drive regulatory ventilation themselves. And if bees are relieved of the chore of fanning, they can spend their energy gathering pollen, making honey, or maintaining the hive—increasing the colony's productivity, in other words.

The picture is very different in a wild hive with the more common natural configuration of a single entrance hole near the bottom. Because there is no hole at the top of the hive, the buoyant forces imparted to the spent air cannot do ventilatory work, and the energy in the buoyant forces accumulates at the top. To ventilate the nest, fanning bees must do work against

---

6. Oxygen has a molecular weight of about 32 g mol^{-1}, and it is replaced by equal quantities of carbon dioxide and water vapor. Carbon dioxide, with a molecular weight of 44 g mol^{-1} is somewhat heavier than the oxygen it replaces, but its addition to the hive's atmosphere is more than offset by the addition of the much lighter water vapor (molecular weight 18 g mol^{-1}). The overall change is a reduction in the average "molecular weight" of air inside the hive.

a.

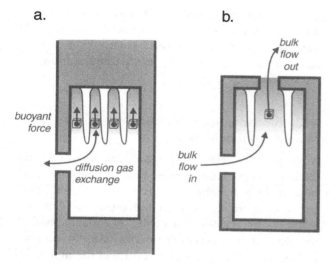

b.

Figure 11.5 Effects of hive architecture on respiratory gas exchange in a metabolizing honeybee colony. *a:* In a closed hive, the metabolism-induced buoyant forces (indicated by upward-pointing vectors) drive spent air up to the upper portions of the cavity, resulting in gradients of pCO_2, pO_2, and pH_2O from the top to the bottom of the cavity (indicated by shading). *b:* In a hive with an opening at the top, the metabolism-induced buoyant forces drive a bulk flow of air through the hive.

this tendency of hive air to rise upward. This they do by orienting their bodies with their heads pointing inward so that they drive hive air *outward* through the hive entrance. What emerges from this balance of forces is a dynamic movement of air that is at the heart of pushmi-pullyu ventilation.

Pushmi-pullyu ventilation arises because the bees recruited to fanning duty do not fan continuously. Rather, the activity of the fanning bees is synchronized somehow so that they all fan for a time, then all stop fanning for a while. During the period of active fanning, the bees power a bulk flow of air outward. In so doing, they draw down the stratum of spent air to a level below where its buoyant energy alone would take it (Fig. 11.6). This is akin to storing energy in a spring—the spent air moves downward because the

fanning bees do work against the buoyant forces driving it upward. When the fanning stops, this stored energy is now available to drive a bulk flow of air upward, which draws a bolus of fresh air inward through the hive entrance. The alternating phases of synchronized fanning (pushmi) and relaxation (pullyu) result in a tidal flow of air across the hive entrance, similar to the alternating cycle of exhalation and inhalation that characterizes our own breathing (Fig. 11.7). The combination of cycle time between fanning and not fanning and recruitment of workers to fanning results in very sensitive control of the hive atmosphere.

Social homeostasis in this case arises because of two meta-loops that govern the behavior of the bees in the colony. One we have already discussed—the meta-loop responsible for recruitment of worker bees to fanning duty. An additional meta-loop operates just among the bees that are actually on fanning duty, the interaction that controls the synchronization of fan-

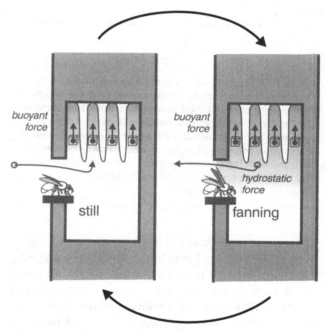

Figure 11.6 "Pushmi-pullyu" ventilation of a honeybee hive.

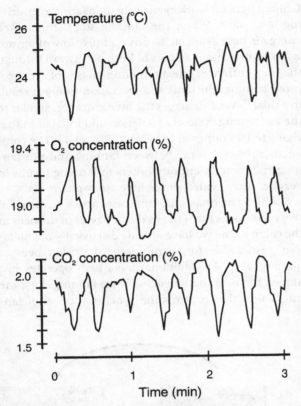

Figure 11.7 Properties of air at the hive entrance during pushmi-pullyu ventilation of a honeybee hive. Periods of high oxygen concentration, low carbon dioxide concentration, and low temperature coincide with periods of no fanning and the movement of fresh air into the hive. Periods of low oxygen concentration, high carbon dioxide concentration, and high temperature coincide with fanning periods and the movement of spent air out of the hive. [*After Southwick and Moritz (1987)*]

ning. A bee at the hive entrance can assess the magnitude of the buoyant forces in the hive's air by assessing how hard it needs to work to maintain the air's outward flow. Specifically, faster rates of metabolism in the hive will impart larger buoyant forces to the hive's air, which the fanning bees will have to work harder to oppose. The ability to make this assessment depends crucially upon the number and placement of holes at the hive entrance. Consequently, social homeostasis in this case is tied not only to the existence of meta-loops but also to the architecture of the structure housing the colony.

Adaptive Structures: Gas Exchange in Macrotermitine Mounds

A bee hive is a relatively passive component in its inhabitants' social homeostasis. Bees may seal off holes here or there, but the structure of the hive itself does not alter dramatically with the metabolic demands being placed on it—most of the adaptation involves recruitment of workers to particular tasks. In some social insects, though, the architecture of the nest is dynamic, varying according to the homeostatic demands placed on it by its resident colony.

Such "adaptive structures" pose particularly acute challenges to conventional views of homeostasis. Adaptability of structure implies there is some way to assess the structure's performance and then to convey this information to effectors, which will modify the structure as necessary. We saw in the last chapter how this system operates for adaptive structures built by individual animals: the singing burrow of mole crickets is tuned by a feedback loop that allows the cricket to assess the burrow's acoustical performance as it is being built. Once the structure achieves the "right" architecture, it is ready to work. Adaptive structures built by social insects are not so straightforward. Suppose, for example, that the structure of a social insect nest is an important component in the colony's capabilities for social homeostasis. It follows that the nest must have the "right" architecture to carry out this task. Yet, how do the many thousands, or even millions, of individuals "know" they have built the "right" structure? If the architecture is not right, how do these many individuals collectively assess how the actual structure, which may be many orders of magnitude larger than the builders themselves, deviates from its "correct" structure? And how does the colony make the large-scale modifications that help bring the structure into conformity with its "proper" architecture? Is there a "collective consciousness" of the colony, what some of the more, shall we say, philosophical students of social insects have called a soul of the colony? Or are there other, more prosaic explanations for how these structures work?

One of the most interesting examples of adaptive structures is built by the macrotermitine termites, a relatively advanced group of several termite genera, most commonly the genus *Macrotermes*. The macrotermitines build large above-ground structures (Fig. 11.8), called mounds, which are prominent features of the landscapes they occupy. In my opinion, they are the most spectacular animal-built structures on the planet.

As it happens, macrotermitine mounds played an interesting role in the history of the superorganism concept. Our understanding of how these mounds function owes a great debt to a Swiss entomologist, Martin Lüscher, who in 1961 published a remarkable paper claiming that the mounds of *Macrotermes natalensis*, a southern African termite, functioned as colossal heart-lung machines for the colony. He fur-

ther claimed these mounds were constructed to regulate the nest environment automatically, requiring little in the way of physiological feedback. The only role for the termites (aside from building the structure in the first place) was as a source of heat energy to drive a circulation of air in the mound.

Because the mound's architecture is so important to Lüscher's conception of social homeostasis, let us look at it in some detail (Fig. 11.9). The *M. natalensis* nest is a compact globular structure, roughly 1.5 to 2 meters in diameter, placed so that about two-thirds of the nest is located below the plane of the ground surface. Above the nest, a large conical mound rises 2–3 meters above ground level. The mound is permeated by air spaces that surround the nest. Below the nest is a large space, the cellar, so open that the nest actually appears to be supported on pillars. To the sides of the nest is a series of radial tunnels that extend downward to the cellar and upward toward the ground surface. Just above the nest, the mound is penetrated by a large-diameter vertical chimney that extends upward toward the top: it does not open to the outside, though, but into a large, enclosed collecting chamber. Just below the mound's outer surface is another series of vertical surface con-

Figure 11.8 A mound of *Macrotermes michaelseni* from northern Namibia.

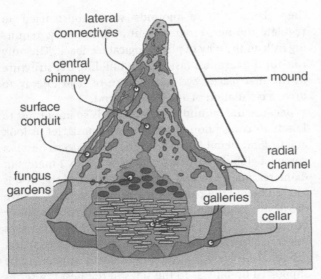

Figure 11.9 Cross-section of the mound of *Macrotermes michaelseni.*

labels: lateral connectives, central chimney, surface conduit, fungus gardens, mound, radial channel, galleries, cellar

The colony maintains these high flux rates while maintaining a remarkable constancy of temperature, humidity, pO_2, and pCO_2 in the nest. Martin Lüscher thought that the high flux rates for respiratory gases and their steadiness in the colony resulted from a thermosiphon similar to that described for commercial honeybee hives (Fig. 11.10). Heating and humidification of nest air imparts buoyant energy to it and lofts it upward out of the nest. Because the *M. natalensis* mound is closed at the top, the buoyant energy drives a circulatory movement of air, reminiscent of the circulation of blood to the lungs. As spent colony air is forced downward through the surface conduits, it is refreshed, exchanging heat, water vapor, and respiratory gases through the surface conduits' thin, porous walls. Once it is refreshed, air in the surface conduits becomes cooler, dryer, and more dense. Gravity then pulls this dense air downward to the cellar, aiding, through a siphon effect, the upward movement of air in the chimney. The refreshed air is then ready to be powered on another circuit through the mound. Because metabolism is directly coupled to ventilation, the

duits that extend upward to the collecting chamber and downward to the radial tunnels. The rest of the mound is permeated by an extensive network of lateral connective tunnels that connect the chimney with the surface conduits.

A mature *Macrotermes* colony has a very high rate of metabolism, which must be supported by a considerable flux of oxygen and carbon dioxide and which dissipates a considerable quantity of heat. The 1–2 million individual termites in a typical *M. natalensis* colony consume roughly 1.5 liters of oxygen per hour. In addition, these termites cultivate extensive fungus gardens, which consume oxygen at a rate about 5.5 times the rate of the termites themselves. This brings the colony's total oxygen flux up to about 9.5 liters per hour. The dissipation of heat is proportional to this rate of oxygen consumption. The termites themselves generate heat at a rate of about 8 W, and adding the fungal metabolism boosts this to about 55 W. This is roughly equivalent to the metabolic rate of a 42 kg mammal, about the size of a goat.

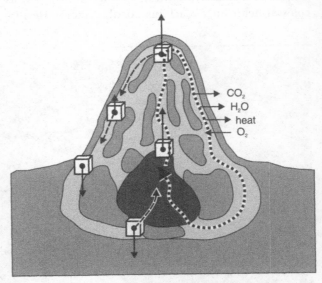

CO_2
H_2O
heat
O_2

Figure 11.10 Martin Lüscher's model for thermosiphon ventilation of the *Macrotermes natalensis* mound.

exchange of gas is autoregulatory, as it is in the ventilation of an open honeybee colony.

Although Lüscher's model for the *Macrotermes* mound is very elegant, it has not been immune from criticism. For example, Lüscher's model presupposes that the walls of the surface conduits will necessarily be cooler than the mound interior. This is not the case if the sun shines on the walls of the mound, for sunlight can warm the surface conduits several degrees above the temperature inside the colony. If this happens, the thermosiphon no longer operates, because air in both the surface conduits and the chimney will tend to rise. For me, though, one of the more interesting defects of Lüscher's model rests in one of its predicates: that the purpose of a social insect's nest is to insulate the colony from a harsh external environment. Embedded within this assumption is the necessary preclusion of any energetic interaction between the colony and the external environment. Any force that weakens the link between the circulation rate of air and the colony's metabolism necessarily disrupts the mechanism that ensures the homeostasis of the colony in the first place. Placing the colony into splendid isolation in the mound also limits the possibilities for the feedback meta-loops that would seem necessary for the mound to be an adaptive structure. If the mound cannot interact energetically with the outside environment, how can the performance of the mound be assessed and adjusted?

Nest-Environment Interaction

How reasonably can we assume that the mound insulates the colony from the outside world? On the one hand, the assumption seems quite plausible: termites are fairly fragile creatures that seem to require a high degree of protection. Also, nearly all termite colonies are located inside nests of some sort, and these damp out the wild fluctuations of temperature and humidity that characterize the outside world. On the other hand, many species of termites build nests that are relatively open and exposed to the environment. Small arboreal nests, for example, experience substan-

tial variations in internal temperature and humidity. One could, of course, view these "exceptions" as relatively inferior steps on the grand progress to the evolutionary pinnacle represented by *Macrotermes*. But then, these supposedly inferior species seem to be doing quite well, despite not being very well insulated from the environment. Even among the supposedly advanced macrotermitines, the degree of isolation of the colony varies both between and within species. Many macrotermitines build mounds with large openings that seem contrived to capture wind energy in various ways (Fig. 11.11). The mound of the Kenyan *Macrotermes subhyalinus*, for example, is permeated with large holes, which seem contrived to act as wind scoops or vent holes. Some races of these termites even top their mounds with high chimneys. Success of these species does not seem to depend upon isolating the colony from the outside world. Indeed, they are doing just the opposite, building structures that draw in kinetic energy from the outside world in the form of wind.

Nevertheless, these relatively open macrotermitine nests still could be considered relatively imperfect steps on the way to the closed mounds of *M. natalensis*, with their virtually clockwork mechanisms of ventilation and ensuing homeostasis. The key to the matter, then, is this: does a closed mound like that of *M. natalensis* interact significantly with energy in the outside world? If so, then these interactions could be the basis of feedback meta-loops that turn the mound from a structure built by "air-conditioning engineers," termites that carry around with them a blueprint of the correct structure of the mounds they are destined to build, into a rather mundane adaptive structure, something like the tuned singing burrow of the mole cricket.

I have been exploring this problem in colonies of *Macrotermes michaelseni*, a macrotermitine that is closely related to *M. natalensis* and very similar in its biology. Like *M. natalensis*, *M. michaelseni* build massive, completely enclosed mounds. The principal difference between the two is a tall cylindrical spire that tops the

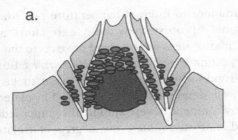

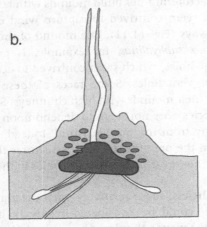

Figure 11.11 Some examples of "open" mounds built by *Macrotermes subhyalinus* nests. *a:* Cross-section of the "Bissel" type mound. This mound is a low dome, roughly 1.5 to 2 m high, permeated by numerous vent holes. *b:* Cross-section of the "Marigat" type mound. These mounds are topped with a tall chimney that can extend several meters high. Fungus gardens are indicated by the small dark ovals. The colony is the large, dark-shaded mass in the center. [*After Darlington (1984)*]

conical mound of *M. michaelseni*. This spire, which has a pronounced northward tilt, can extend the mound to substantial heights, as high as 9 meters. It turns out that the movements of gases in these mounds, unlike those in *M. natalensis* mounds, are strongly influenced by wind. So these colonies, despite having completely enclosed mounds, do not sit in splendid isolation from the environment.

The interaction of the *M. michaelseni* mound with

wind is straightforward. You will recall (from Chapter 8) that any fluid, including air, has mass, and when a mass is set in motion it has kinetic energy. When wind interacts with an object in its path, so that it is accelerated or decelerated, the air's kinetic energy is converted to potential energy in the form of pressure. How this conversion occurs is determined by the Bernoulli principle: if the air is slowed down, the kinetic energy is converted to a positive pressure that pushes on the object. If the air is accelerated around the object, a negative or suction pressure results.

The *M. michaelseni* mound, extending upward through the wind boundary layer, develops around it a complicated pressure field. Positive pressures develop at the mound's upwind or leading face, and negative pressures develop at the mound's trailing and lateral faces. The pressures are small close to the ground but become larger higher up on the mound's surface, reflecting the variation of wind speed in the boundary layer. These pressures can be quite substantial, reaching as high as a few hundred pascals. At the porous walls separating the surface conduits from the outside, these pressures can drive air flow into or out of the surface conduits. If a surface conduit is at the mound's leading face, fresh air is driven into it. If the surface conduit is at the lateral or trailing face of the mound, air is sucked out of it.

This complicated distribution of pressure drives a flow of air within the *M. michaelseni* mound that is quite different from the circulatory motion of air Lüscher posited for the *M. natalensis* mound.[7] Air movements in the *M. michaelseni* mound occur differently in three different zones within the mound (Fig. 11.12). The forced convection zone encompasses the surface conduits which form a network of tunnels in which air is well-mixed by the force of the wind

7. Movements of air in a structure like the *M. michaelseni* mound can be followed by use of tracer gases and arrays of sensors that detect them. In this case, the tracer gas is a dilute mixture of propane and air that is injected into the mound at various points. The movement of the injected gas is detected by arrays of combustible gas sensors.

outside. Because wind is turbulent, the flow of air in the forced convection zone is highly variable, moving fast as the wind speed picks up or slowing down as the wind ebbs. As the wind changes direction, air sloshes through the surface conduits from one side of the mound to the other. A second zone, the natural convection zone, encompasses the nest itself and the lower parts of the chimney. Here, air movements are governed mostly by buoyant forces imparted to the air by colony metabolism, similar to those in the *M. natalensis* mound. Air movements here, in contrast to those in the surface conduits, are slow and steady. Finally, sandwiched between the two and encompassing the lateral connective tunnels, is a mixing zone, where air in the chimney mixes with air in the surface conduits.

The rate of colony gas exchange depends upon the degree of mixing of surface conduit air and chimney air. The degree of mixing in turn depends upon the interaction of two sources of energy: metabolic energy generated by the colony and wind energy admitted to the mound across its surface conduits. Gas exchange in the *M. michaelseni* mound is therefore reminiscent of

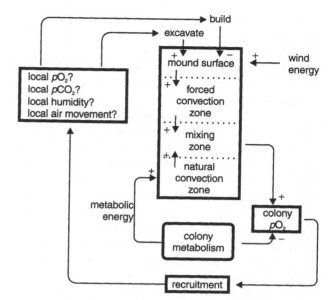

Figure 11.13 Schematic diagram of the construction and maintenance of the adaptive structure of a *Macrotermes michaelseni* mound.

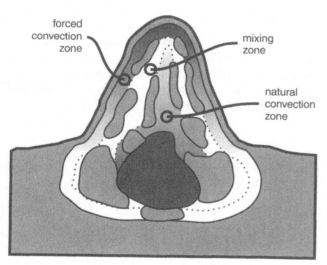

Figure 11.12 Zones of gas exchange in mounds of *Macrotermes michaelseni*.

the exchange of respiratory gases in the alveolus of the mammalian lung. Variation in wind speed and direction induces a tidal ventilation in the surface conduits. The admitted energy is damped by the lateral connectives, where it helps to mix chimney air with surface conduit air. Variation of gas exchange is driven partly by variation of wind speed and partly by variation of the colony's metabolism.

Together, these two energy sources form the basis of a feedback meta-loop that conveys information to the colony about the mound's performance as a gas exchanger (Fig. 11.13). Coupling this information to the stigmergic reflex described above probably completes the feedback, and in this way the mound can act as a truly adaptive structure. It probably works like this. The gas composition of the air inside the nest is determined by the balance between the colony's rate of production of carbon dioxide and its consumption of oxygen and the rate at which these gases are exchanged between the nest air and the environment. As always, homeostasis of the nest atmosphere requires that these fluxes be matched (equation 11.5).

Any departure of the nest atmosphere from that preferred by the termites indicates a mismatch between the colony's *PF* and the *TFF* through the mound. If, for example, the colony pO_2 is low, the *TFF* is comparatively low. This indicates that insufficient wind energy is available to power the mixing of colony air and surface conduit air. Conversely, if the colony pO_2 is too high (or the pCO_2 is too low), the *TFF* is relatively high, which means that too much wind energy is driving the mixing. Indeed, the rate of exchange of gases inside the mound is strongly influenced by wind (Fig. 11.14): high wind speeds increase flux, and the increase in flux alters the gas composition inside the colony. If these colony-level variations of atmospheric composition are coupled to recruitment of workers and the activation of the stigmergic reflexes described above, then the mound can be considered an adaptive structure.

For example, consider what must happen as the colony grows. From its inception (the queen and a few hundred workers in her nuptial brood) to its maturity, the colony's collective metabolic rate increases by about six orders of magnitude. Despite this large increase in demand for oxygen, the composition of the nest atmosphere stays pretty constant, with CO_2 concentrations hovering between 2–5 percent at all stages of colony growth. Such large increases of respiratory flux are supported mostly by the upward extension of the mound into sufficiently energetic winds to power

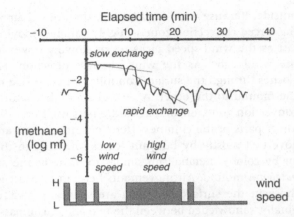

Figure 11.14 Variation of gas exchange rates in colonies of *Macrotermes michaelseni*. Gas exchange is assessed by the clearance rate of a propane tracer gas injected into the colony. A shallow slope of the plot of the tracer's log mole fraction *(log mf)* against time indicates slow exchange of gas, while a steep slope indicates rapid exchange. During periods of low wind speed, the tracer is cleared from the colony slowly. During periods of high wind speed, the clearance of tracer is approximately doubled.

a sufficiently high rate of ventilation. Far from shielding the colony from the disruptive influence of the wind, the *M. michaelseni* colony uses its mound to seek out and capture the energy in wind, appropriating it to power its ventilation. In short, the *M. michaelseni* mound is a fairly straightforward adaptive structure, no different in principle from the adaptive structures of the tuned singing burrow of mole crickets.

Similarly, the earth is affected by harmony and quiet music. Therefore, there is in the earth not only dumb, unintelligent humidity, but also an intelligent soul which begins to dance when the aspects pipe for it. —JOHANNES KEPLER

CHAPTER TWELVE

Love Your Mother

In this, the final chapter of this book, we delve more deeply into the idea of the superorganism. We will explore a radical question: is the Earth a global superorganism that, like a colony of social insects, exhibits a coordinated global physiology, perhaps even a global homeostasis? One of the more remarkable ideas to come out of biology in the late twentieth century is that it does. In this view the Earth is portrayed as Gaia, a single, self-regulating living entity. What has come to be known as the "Gaia hypothesis" is the brainchild of James Lovelock, nominally an atmospheric chemist but, like all novel thinkers, one not so easily characterized. His idea was prompted originally by a practical question put to him by NASA: how do we decide whether or not a planet supports life? Lovelock's answer to this question was simple and obvious: look for evidence that the chemistry of a planet is held persistently out of thermodynamic equilibrium.

This answer did not endear Lovelock to NASA, because it meant that the answer could be determined from observations that could be made from Earth. This did not help NASA justify sending expensive spacecraft to Mars, and so at that point the mind of the bureaucratic beast lost interest. Lovelock's idea did not die, though, and Gaia has grown into what its proponents say is a comprehensive new theory of Earth biology, one which unifies previously disparate fields of scientific inquiry, like geology, ecology, and the life sciences. It has also resurrected a holistic view of nature that has largely been eclipsed by the triumphant reductionism that has dominated science throughout the late twentieth century.

In essence, Gaia is the superorganism writ large, and

for reasons both good and bad it has been a controversial idea. I don't intend to use this chapter as a venue to air the various arguments pro and con. Rather, I wish to consider Gaia in light of the theme of this book: is there a global physiology that is mediated by structural modification of the environment, of which animal-built structures are an example? If you have been persuaded up to now that there is indeed physiology that extends beyond the organism, then you should be willing to entertain the idea that this external physiology could extend to large-scale systems, like communities, ecosystems, perhaps even the biosphere itself. Of course, it is a long reach from a termite mound to Gaia, but I will try to bridge the gap anyway.

The "Physiology" of Global Climate

To start, let us examine one of Gaia's claims—that the biota of the Earth regulates the Earth's climate. One obvious objection would be that the biota—all the

flora and fauna collectively—is fairly weak in comparison with other forces that drive climate. Recall the discussion in Chapter 2: only a tiny fraction of the energy streaming into the Earth's atmosphere is captured by green plants for powering the ATP energy economy that all life depends upon. Far more energy is available in the physical energy stream to drive climate. How can the comparatively minuscule engine of the Earth's biota regulate such a powerful force?

It may not have to. Climate may in fact be rather delicately poised and susceptible to small biasing forces, just as a finely balanced beam might be tipped by the movement of a flea. For the past million years or so, climate has shifted between cold periods, called ice ages, and warmer periods, called interglacials, once every 100,000 years or so, almost as if it were being switched between two relatively stable states. This cycle is nicely illustrated by climate records for the past 35,000 years or so (Fig. 12.1), which encompass the latter part of the last ice age and the climb into the

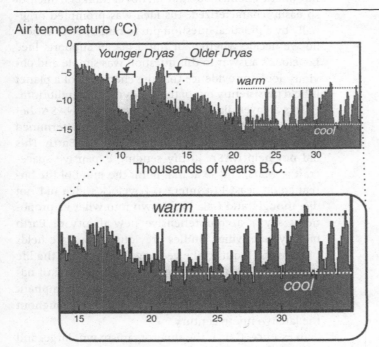

Figure 12.1 Temperatures over the Greenland ice sheets for the past 35,000 years. [*After Kerr (1993)*]

interglacial climate we now enjoy. Obviously, the record is rather "noisy," but look in particular at the period between 20,000 and 35,000 years before the present (B.P.). During this time, the climate seems to switch periodically between a warm phase and a cool phase roughly seven or eight degrees cooler. This switching also occurs during the climb out of the ice age, beginning around 17,000 years B.P. The warming trend is clear, but twice, during the Older and Younger Dryas periods, the Earth plunged back into extended periods of cold. Only at about 8,000 years B.P. or so does the climate—represented in Figure 12.1 by the air temperature over the Greenland ice sheets—finally climb into the relatively steady and equable interglacial range.

This behavior is very reminiscent of a type of control known as ON-OFF regulation, actually the most common method for thermostatic control. ON-OFF control is most easily understood by comparing it with proportional control, which is exemplified by mammalian thermoregulation (Chapter 11). In proportional control, the rate of heat production by the organism varies proportionally with the magnitude of the error in temperature—that is, the difference between the set temperature and the actual temperature of the system. ON-OFF control operates by switching a heater either fully on or fully off: heat production rate is modulated by the duration of the ON-phase relative to the OFF-phase, the so-called duty cycle.

If the oscillation of climate reflects the operation of a "climate switch," what then controls the switch? There is, in fact, a strong physical component in the switching mechanism. The actual energy required to throw the switch may be small, however, perhaps small enough that the biota might help trip it—and thereby regulate climate to a limited degree.

The Oceanic Heat Conveyor

The identity of the switch controlling the climate is not clear at present, but it is probably safe to conclude that changing patterns of ocean circulation are important.

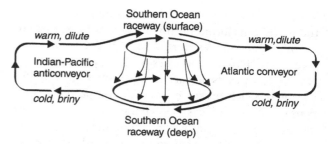

Figure 12.2 The major circulation patterns in the world's oceans. [After Broecker (1997)]

Water in the oceans is a major conveyor of heat absorbed from the Sun. Because the distribution and movement of heat are what control weather, these patterns of oceanic circulation are probably the major determinants of the Earth's climate. The current thinking, so to speak, is that the climate switch is driven by the ocean switching between two very different patterns of circulation.

Presently, water circulates through the world's oceans along three major pathways (Fig. 12.2). One is in the Southern Ocean that surrounds Antarctica, the only ocean where the east-to-west movement of water is not blocked by continents. The Earth's rotation imparts a strong east-to-west movement to these waters, forming a current known as the Southern Ocean raceway. The Atlantic conveyor circulates water between the Southern Ocean and the north Atlantic, as far north as Greenland. Finally, the Indian-Pacific anticonveyor circulates water between the Indian and Pacific oceans and the Southern Ocean. In the Pacific, the anticonveyor transports water up to the Bering Straits.

All three currents have strong vertical components to them, driven by gravity acting on variations in water density. In the Southern Ocean raceway, water along the ice shelves freezes, producing a dense, cold brine, which then sinks to the abyssal depths as a skirt-like vertical current surrounding Antarctica. The Atlantic conveyor is fed by the comparatively warm and buoyant surface waters of the Southern Ocean, and these are transported northward as a surface current. As this water moves north through the tropics and across the Equator, it loses water by evaporation, which increases its density. At the same time, it is also warmed, which makes it buoyant and partially compensates for the effects of evaporation. Once the conveyor has passed the tropics, though, it loses heat to the now cooler atmosphere, giving northern Europe the relatively benign climate it now enjoys. As these surface waters cool, the combined effects of evaporation and cold make them dense enough to sink. At the lower depth they feed the return leg of the Atlantic conveyor as a cold, salty current.

The Indian-Pacific anticonveyor also sends water north, but here it is the northward current that is cold and salty, fed from the deep waters of the Southern Ocean raceway. At its northern extreme near the Bering Straits, these waters are mixed with relatively brackish waters from the Arctic Ocean, which receive considerable quantities of freshwater runoff from Siberia and North America. Now the water in the anticonveyor is lighter, so it rises and then feeds a return surface current of relatively warm and dilute water. When this water joins the Southern Ocean raceway, heat may be transferred between the Indian and Pacific oceans and the Atlantic.

These patterns of circulation probably are common to interglacial periods. The connection is not hard to see: the more widely heat can be distributed across latitudes, the more temperate climates will be across the globe. The tropics and equatorial regions will be cooler than they otherwise would be, because heat is transported away from these regions to the poles. Similarly, the temperate and polar regions will be warmer than they otherwise would be, because their climates receive a subsidy of heat from the tropics.

Ice ages, on the other hand, probably correlated with different patterns of ocean circulation. Most likely, ice ages come about when the conveyor systems shut down. If that happens, heat absorbed by waters at tropical latitudes would tend to stay there, rather than being distributed to higher latitudes as it now is. Tropi-

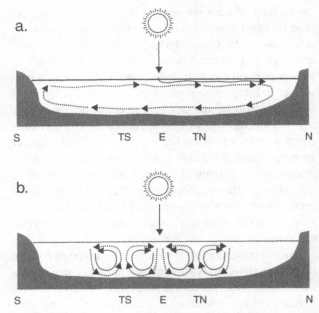

Figure 12.3 Hypothetical patterns of ocean circulation and heat transport during (a) an interglacial period and (a) an ice age. This schematic view represents a cross-section from the South Pole (S) to the North (N), cutting through a temperate southern region (TS), the Equator (E), and a temperate northern region (TN).

cal waters would be warmed more, evaporate more water, and might even become dense enough to sink of their own accord. Consequently, the mixing of the oceans would be dominated by locally delimited cells of convection, with a strong vertical component of heat transport but only a weak latitudinal component (Fig. 12.3). Because the temperate latitudes would no longer receive a heat subsidy, they would be colder, snowfalls would be heavier, and snow packs would be retained longer. Extensive glaciation and an ice age could follow.

What might cause the oceans to switch between these markedly different patterns of circulation? Clearly, powerful physical forces are at work, and it may be that the oceans would periodically make the switch irrespective of whether or not there was a biota. Consider, for example, the following scenario for shutting down the conveyor systems and plunging the climate into an ice age. Gravity acting on the markedly increased densities of water in the north Atlantic is a significant force driving the Atlantic conveyor. Presently, the waters of the north Atlantic are among the saltiest and coldest on Earth. Suppose, though, that the northern hemisphere experiences a period of warmer temperatures and increased rainfall. If these provide a sufficiently large influx of fresh water, the waters of the north Atlantic would not be so salty as they now are, they would not be so dense, and gravity could no longer act forcefully enough to make them sink. The result: no more Atlantic conveyor.

This switching between warm and cool climates may result if each of the climate patterns is self-limiting in its behavior (Fig. 12.4). If an interglacial period gets

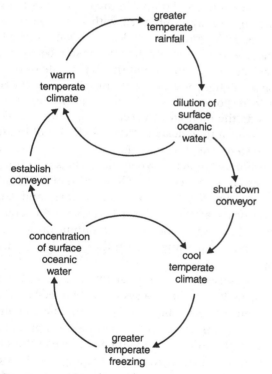

Figure 12.4 Possible mechanisms for self-limitation of climate patterns and switching between interglacial periods *(top circle)* and ice ages *(bottom circle).*

too warm, as we have just seen, the ensuing changes could shut down the oceanic conveyors, forcing the climate back to an ice age. Conversely, if glaciers become very widespread, the consequences of long-term cold could re-establish the coldness and saltiness of waters in the north Atlantic and give the conveyor systems a push. If the push was forceful enough, it could start the conveyors moving again, ushering in an interglacial period.

It should be clear now how the biota could bring about significant changes in Earth climate, even without controlling most of the power that drives climate. If the biota could somehow bias the movement of water and energy that drives the ocean circulation, they could exert a regulatory influence by altering the duty cycle of the climate's ON-OFF controller. I confess right now that I haven't a clue how such biasing could work. But just to illustrate the point, allow me to offer one possible scenario.

Ice-nucleating bacteria (INB) are a class of microorganisms, mostly of the genus *Pseudomonas*, that are commonly found on the surfaces of vegetation. They have the interesting capability of altering the temperature at which ice forms from water vapor, and they play an important economic role in protecting crop plants from frost damage. When a plant dies and decomposes, some of its INB are lofted into the atmosphere, where they can serve as nuclei for the formation of cloud droplets. When they do so, they bias the tendency of cloud droplets to stay liquid or freeze. Raindrops that coalesce around INB would tend to stay liquid, but raindrops that coalesce around ordinary dust particles might be more likely to freeze. Thus, these bacteria may play a role in determining whether precipitation in a cold climate comes down as rain or snow.

Imagine now that we are in an emerging interglacial climate. As the climate warms, the extent of green vegetation on the Earth expands toward the poles. With expansion comes an increased aggregate leaf surface, which can, in turn, support larger populations of INB. If the airborne burden of INB also increases as a result, we would expect greater rainfall in the northern latitudes. Increased rainfall, in turn, might shut down the oceanic conveyor system. In this scenario, a biological entity (INB) has biased the switch so that it may be thrown sooner, or with a smaller perturbation from the physical energy stream, than if it had not been so biased. The biasing would be physiological, in that it involves the biological manipulation of a flow of mass and energy. It also would involve positive feedback: a warming climate encourages the spread of vegetation, which promotes further climatic warming, and so forth.

Homeostasis and Symbiosis

That the biota might exert wide-ranging and subtle influences on climate is still a far cry from the central claim of Gaia: namely, that the biota *regulates* the climate of the Earth. To get *there*, two conditions must be met. First, the biota itself should have strong tendencies to self-regulation—homeostasis—and this tendency should be universal or nearly so. This conditions implies that substantial benefits accrue to homeostasis per se. Second, this tendency must involve flows of matter and energy outside the organism; the physical environment must be drawn into a physiological conspiracy, so to speak, that will confer homeostasis not only upon the organisms *in* the environment but upon the environment itself.

Feedback and Homeostasis

What might the benefits of homeostasis be? So far, we have been content to assume simply that homeostasis is a good thing. But why should that be? What are the real benefits organisms derive from doing all that work of homeostasis? The answer to this question lies, I believe, in an understanding of negative feedback control that goes deeper than previous chapters have gone. Up to now, I have simply declared that homeostasis is associated with negative feedback. While this often is true, it is not always true. In what follows I discuss precisely how homeostasis and negative feedback correlate, for only when this point is understood is it clear

how homeostasis might be beneficial and how it might result in Gaia.

Let us start with a simple example of a negative feedback process. A driver steering an automobile keeps a point on the car's hood aligned with the stripe painted down the middle of the road. If the car deviates to one side or the other of the stripe, the driver takes corrective action and brings the car back to its proper trajectory. Good control is evident as a trajectory that deviates little from the "planned" trajectory that is indicated by the stripe.

Engineers have long been concerned with analyzing and designing controlled systems, and they have developed a set of tools, control systems theory, to help them do so. The first step is to divide the system into a series of "black boxes," conceptual devices that receive an input and produce an output. The relationship between the input and output is the black box's transfer function, usually symbolized with a Φ. The transfer function is determined by at least two values. The first value, the *gain*, is simply the relationship between the magnitude of the input and the magnitude of the response. High gains produce a large change in the output in response to a small change of the input. Low gains do the opposite. The second quantity, the *phase*, represents a time delay between a change of input to the system and the initiation of the response. The black box is "in phase" when the delay is zero (that is, the response immediately follows a change of input). The transfer function takes on a larger and larger phase delay as the time lag between input and response increases. Whenever the output of one black box serves as the input of another, this is a feedback loop.

We can use these tools to analyze steering of an automobile. We construct a system with a feedback loop between two black boxes (Fig. 12.5). One contains a transfer function which translates a deviation of the car from the stripe into a particular angle of the steering wheel. The other translates the angle of the steering wheel into a deviation from the planned trajectory. The gain is the relationship between the deviation of

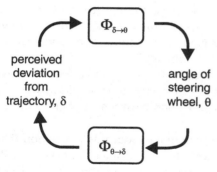

Figure 12.5 The transfer functions that govern the steering of an automobile.

the car from the stripe and the degree to which the steering wheel is turned. Low gain corresponds to a slight turning of the steering wheel in response to a large deviation of the car from its intended trajectory. High gain corresponds to a large rotation of the steering wheel in response to a small deviation in the car's position.

It is easy to simulate the behavior of this simple negative feedback controller for steering, and I have reproduced the results of some of these simulations in Figure 12.6. The obvious lesson to be drawn from them is this: "successful" control of the car (that is, keeping the deviation of the car from the stripe small) results only from a particular combination of gain and phase. By tweaking the gain and phase of the model, it is actually quite simple to send it into erratic and wild behavior. It is not hard to think of real-life examples of this. A new and inexperienced driver, for example, frequently over-corrects the steering in response to a slight deviation of the car from its intended trajectory. We can simulate this by increasing the gain of the system, as is shown in Figure 12.6. The failure of control is evident. To take another example, a driver's reaction time can be impaired by various legal and illegal drugs. The result is an increased phase delay, with the driver taking corrective action only some time after the deviation is registered. Clearly, the car behaves erratically.

Feedback and Symbiosis

Let us now use our new conceptual tools to explore the matter of homeostasis in a simple cooperative assemblage of organisms, a symbiosis. A nice example is the lichen, a symbiotic association between a fungus and an alga. The fungal partner produces a substratum for the alga to live on, and some fungi form elaborate structures, called thalli, as housing. The photosynthetic alga, in turn, provides energy that supports the fungus. The interchange of matter between the symbionts can be represented with a systems diagram—that is, with a collection of black boxes linked into a feedback loop—and we could analyze this association using control systems theory (Fig. 12.7).

Our simple model system consists of a photoautotroph, P (the alga), and a heterotroph, H (the fun-

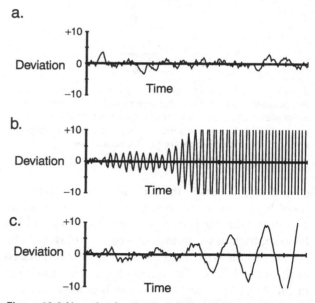

Figure 12.6 Negative feedback, stability and instability. *a:* Gain and phase of the transfer functions are matched properly: deviations from the "desired" value are small. *b:* Gain of the transfer function is too high (driver over-corrects for deviations): the system goes into an unstable oscillation. *c:* Phase response of the transfer function is too slow (driver's reaction time is delayed): the system also is unstable.

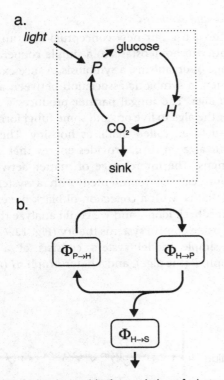

a.

light

glucose

P

H

CO_2

sink

b.

$\Phi_{P \to H}$

$\Phi_{H \to P}$

$\Phi_{H \to S}$

Figure 12.7 A simple symbiosis consisting of a heterotroph (H), photoautotroph (P), and a carbon sink (S). *a:* Possible movements of carbon between the heterotroph, the photoautotroph, and the sink. *b:* The transfer functions governing the symbiosis operate in a feedback loop.

gus). Energy enters in the form of light, which is captured by P and stored in chemical bonds in glucose. The energy in the glucose is transferred to H, which breaks it back down to carbon dioxide and water. The carbon dioxide is then fed back to P to be incorporated again into glucose. Carbon in this system can exist in three energy states. The highest is glucose, boosted there by the capture of light energy. Lower is carbon dioxide, produced by the heterotroph. Lowest of all is a carbon sink, which could be some stable mineral salt of carbon, like calcite. Let us say that neither P nor H can mobilize carbon from this sink—once it is there it is lost to both forever. The cycling of carbon between

P and H therefore form a feedback loop, governed by two transfer functions: $\Phi_{H \to P}$ for the flow of carbon through the heterotroph and back to the photoautotroph, and $\Phi_{P \to H}$ for the flow of carbon through the photoautotroph to the heterotroph. There is also a third transfer function, $\Phi_{H \to S}$, which governs the flow of carbon between the heterotroph and the sink *(s)*.

Is there anything about this association that would make it tend toward homeostasis? Let us follow the flow of energy and matter through this system. The symbiosis is successful only when carbon cycles perpetually between the photoautotroph and the heterotroph. If, for some reason, P or H "drop the ball" and allow the carbon to leave the loop that joins them, the carbon falls into the carbon sink and is lost to both. The symbiosis can survive only if it keeps the carbon "in play." That is to say, it will survive only if there is homeostasis. We can state this more explicitly: successful symbiosis results only if the transfer functions $\Phi_{H \to P}$ and $\Phi_{P \to H}$ are matched. If they are not, carbon, and the capacity for work it carries with it, will be lost. Thus, the continued success of the symbiosis depends crucially on the control of one symbiont's transfer function with respect to the other's. This type of operation implies feedback, cooperation, and control. What is truly interesting, though, is that benefits arise only in the context of the assemblage and not in the context of its individual members. There is no intrinsic value to *either P* or *H* being individually homeostatic—regulating their internal pH, temperature, water balance, whatever. Homeostasis only has value if there is physiological coordination between the members of the assemblage.

Homeostasis, Symbiosis, and Fitness
When homeostasis is viewed in this light, Gaia begins to make just a little more sense. It does not yet make complete sense, though, because the real world is not composed of nice cooperative symbioses. Rather, it is full of organisms that have a vital interest in forcing existing assemblages to "drop the ball"—in short, there

is competition in the real world. If Gaia is to be credible, it must somehow explain how dog-eat-dog competition between organisms nevertheless results in the cooperation and coordination of living entities that global homeostasis seems to require.

In fact, it may not be as great a leap as it seems. Let us admit competition into our model symbiosis. Assume again a single photoautotroph, P, but now there are two heterotrophs, H_A and H_B (Fig. 12.8). The cycling of carbon through this system is now governed by six transfer functions, two each for the carbon flowing between P and each H, and two more for carbon lost to the sink. There now is competition for carbon, but it is only superficially between the two heterotrophs: the real competition is between the two

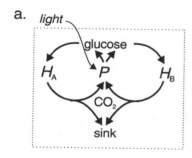

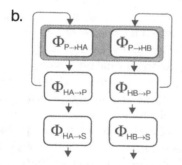

Figure 12.8 Competition between two heterotrophs, H_A and H_B, for carbon cycling through a photoautotroph, P. *a:* The movement of carbon forms two feedback loops. *b:* The competition is actually between the loops and is only secondarily between the heterotrophs.

loops through which carbon can flow. The presence of competition does not negate or otherwise minimize the necessity for coordination between the component members of the loops. There are certainly many possible outcomes to the competition: the extinction of either H_A or H_B or some stable association of the two. But no matter who the survivors are, all the members of the assemblage still must keep that carbon moving, and the only way they can do that is to coordinate their respective transfer functions.

This line of argument can be extended to a physiological definition of evolutionary fitness, which is crucial if Gaia is to be credible. The conventional theory of natural selection looks upon fitness as the likelihood of passing on a gene. To the physiologist, though, fitness is a matter of energetics. Reproduction is, at root, making copies of yourself, which in turn requires the imposition of orderliness on otherwise inanimate matter. The genes might provide the template, but without the energy, the genes are useless. The rate of reproduction depends directly upon the rate at which this energy can be mobilized—in other words, on the reproductive *power*. In any collection of self-reproducing or autocatalytic entities, those that work at high power will produce more copies of themselves and be more fit than those that work at lower power.

Now we are in a position to see the real value of homeostasis. Let us return to our simple assemblage of one photoautotroph and two competing heterotrophs (Fig. 12.8). Focusing on carbon as I did is actually a bit of legerdemain. What is really powering the assemblage is a flow of electrons: the atoms are important only as electron carriers. For work to be done, the electrons must make a controlled migration from bonds in glucose to lower-energy bonds in carbon dioxide and water. More powerful electron currents will perform more order-producing work than will less powerful currents, but the current will never flow unless something provides electrons that can be boosted into the high-energy bonds in glucose. What provides them are the lower-energy bonds of carbon dioxide and water. Thus, the individual organisms in an assemblage are secondary: an organism has a high level of fitness only if its metabolism is coordinated with that of other members of the symbiosis.

This energetic perspective on fitness offers a natural check to unrestrained competition by organisms and ties fitness to coordination and cooperation. Suppose, for example, H_A attempted to maximize its fitness with respect to P, which it could easily do by diverting its carbon dioxide directly to an insoluble mineral like calcite. This move would make selfish sense for H_A, but it would break the loop for electron flow and soon H_A would no longer be provided with the electron carriers it needs to power its own physiology. Its only hope for survival, in fact, is to cooperate metabolically with P.

Homeostasis and Telesymbiosis

Perhaps you are persuaded by now that a successful symbiosis is a "well-tuned" symbiosis—one in which matter and energy flow between the symbionts with a sort of harmony. But does this model get us to the global physiology of the type posited by Gaia? I would suggest it does. The requirement for "tuning" a symbiosis follows from elementary principles of conservation of mass and energy. Even organisms far removed from one another are subject to these constraints, and it is conceivable that they could engage in a kind of telesymbiosis (literally "symbiosis at a distance"). The challenge for Gaia is to explain how telesymbiosis can work. If telesymbiosis is credible, perhaps a global physiology, one that involves the entire Earth—the oceans, crust, and atmosphere—and not simply the organisms in it, is not so far-fetched an idea.

In fact, many natural ecosystems do exhibit a degree of physiological coordination that might qualify as telesymbiosis. The possibility may be assessed by comparing how readily matter cycles between organisms—is kept "in play"—against how readily it drops into a sink and is re-mobilized. The two can be combined into what Tyler Volk has called the cycling ratio. For carbon, for example, the cycling ratio is about 200 to 1, which means that carbon atoms on average cycle between heterotrophs and photoautotrophs about 199

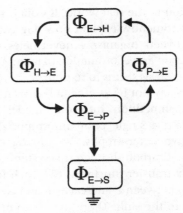

Figure 12.9 A telesymbiosis between a photoautotroph and a heterotroph separated from one another by an environment through which matter and energy flow.

times before being lost to the sink. Fortunately, the carbon can be re-mobilized fairly easily. Other atoms, like nitrogen, seem to be even more avidly kept in play: the global cycling ratio for nitrogen is estimated to be between 500 and 1,200 to 1. These ratios are maintained by guilds of organisms each with complementary metabolisms: nitrogen, for example, must be cycled through complicated communities of bacteria before it can be used by plants or animals for proteins, and the nitrogen in proteins must in turn be converted by other bacteria back into nitrogen gas. The high fidelity with which nitrogen is passed from one "metabolic guild" to another speaks strongly for a highly coordinated superorganismal physiology.

"Tuning" a symbiosis presumably is easier if matter and energy can be transferred directly between the symbionts. This may be one reason why conventional symbioses are so intimate. The problem with a telesymbiosis is that matter and energy must pass through the environment that separates the putative telesymbionts. If this environment is fickle or unstable, tuning the telesymbiosis might be difficult. We can express this in a systems diagram (Fig. 12.9) representing the flows between organisms by environmental transfer functions. For example, a transfer function $\Phi_{P \to E}$ governs flow from the photoautotroph to the environment, and another, $\Phi_{E \to H}$, governs it from the environment to the heterotroph. The transfers obviously are more complicated, but the requirement for matching them does not change. A successful telesymbiosis, like a conventional symbiosis, still must keep matter "in play." Again, this can be done only if *all* the transfer functions of the loop are matched. Telesymbioses in which the matching is better will work at greater power and will have greater fitness than those whose loops are less well tuned.

In telesymbiosis, adaptation is not simply the response of organisms to the environment: it also involves the environment adapting to organism. Let us illustrate this concept with another model world, one that includes two heterotrophs, H_A and H_B, and one photoautotroph, P (Fig. 12.10). Assume that H_A and P have adapted well to the environmental transfers that separate them and that they form a well-tuned telesymbiosis; H_B and P do not. Carbon should cycle preferentially between H_A and P and do more work for them than it will between H_B and P. How can H_B successfully compete? One option would be for H_B to modify its internal transfer functions, probably through conventional genetic selection of its internal physiology. The other option, though, and one per-

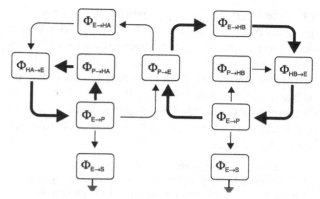

Figure 12.11 Adaptation of the environment in telesymbiosis. H_B can tune its loop more closely to that of P by modifying the transfer functions for flow of matter through the environment separating them. H_B can now compete against H_A.

haps quicker to implement, involves altering the transfer function of the *environment* separating it from P (Fig. 12.11). If H_B can improve the match by tuning the environment to *it*, it is now in a position to compete with H_A for the carbon put into play by the photoautotroph.

Gaia and the Extended Phenotype

It seems a natural step from the extended phenotype to the global physiology posited by Gaia. As a rule, evolutionary biologists have been unwilling to take the step, though. This is why I said in Chapter 1 that there would come a point where the evolutionary and physiological perspectives on the extended phenotype might part company. We are now at that point.

The bone of contention seems to be the belief that the evolution of Gaia requires natural selection to operate in a manner that nearly all evolutionary biologists have ruled out of bounds. Group selection is the name given to any kind of selective process that operates at levels above the organism or gene. Classically, group selection has been invoked to explain altruism

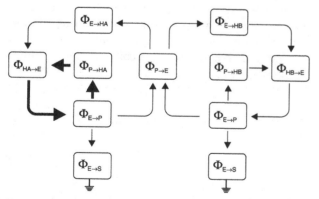

Figure 12.10 A competition between two telesymbioses. H_B competes poorly against H_A because its loop is out of tune.

of some sort: an example would be a gene that prompts certain members of a herd to sacrifice themselves to predators "for the good of the species." Such genes would not last very long, of course, and this is why most Darwinian models for altruism posit genetic subterfuges, like kin selection, that actually promote the interests of the "altruist's" genes. With respect to Gaia and group selection, the point is well made by Richard Dawkins:

> . . . if plants are supposed to make oxygen for the good of the biosphere, imagine a mutant plant which saved itself the cost of oxygen manufacture. Obviously, it would outreproduce its more public-spirited colleagues and genes for public-spiritedness would disappear. It is no use protesting that oxygen manufacture need not have costs: if it did not have costs, the most parsimonious explanation . . . would be the one the scientific world accepts anyway, that oxygen is a by-product of something the plants do for their own good.   (Dawkins 1982, p. 236)

I have no disagreement with this statement, except that it is incomplete—it fails to ask *why* producing oxygen should be good for plants. After all, oxygen holds its electrons very avidly—remember its high oxidation potential—and it takes a lot of energy to strip electrons away from water and force them to take up residence in glucose. Why should the plant work so hard to get electrons when it could get them more easily elsewhere?

The real benefit of oxygen manufacture for plants does not derive from how hard it is to produce oxygen, but from how powerfully oxygen attracts its electrons back. Oxygen's large oxidation potential increases the voltage difference that draws electrons away from glucose, and these "liberated" electrons can be made to do more work. In short, oxygen as an electron donor is sensible only if it can also be an electron acceptor. What confers fitness in this case is that rates of electron donation and acceptance are matched: the genetic identity of the donors or acceptors is of secondary importance. A self-contained loop, such as one which retains oxygen within a plant, may actually generate less power than one which cycles the oxygen to a heterotroph. The plant's fitness could indeed be increased by increasing the fitness of its predators.

Here, then, is the crux of the argument. The evolutionary biologist looks at Gaia and is stopped short by the apparently insurmountable barrier of group selection. The physiologist looks at Gaia and is drawn to it as perfectly reasonable physiology, even if it does require us to define physiology a bit broadly. Who is right, if either? I confess that I am not enough of an evolutionary biologist to say that Gaia is indeed fatally incompatible with current thinking on evolution. But I am a physiologist, and to me, Gaia is where a physiological analysis of the extended phenotype leads.

Animal-built structures come into the picture because these are the agents whereby organisms adaptively modify flows of matter and energy through the environment. I have offered several examples of this kind of adaptation through this book—the expansion of fractal dimensions of coastlines by coral reefs, the control of energy flow through tidal flats by tube-dwelling worms, the transformation of wind energy by termite mounds, and so forth. In such structures, organisms co-opt the environment into a physiology that extends well beyond their conventionally defined boundaries. And that, in summary, is the point of this book.

Epilogue

I began with a question: where do we fix the boundary between an organism and its environment? I will now return to it.

In the early years of the twentieth century, biologists seemed to be more willing than they presently are to ask radical questions about the nature of the organism and its relation to the inanimate world in which it lives. Indeed, the idea of a superorganismal physiology, of the type I have outlined and tried to defend in this book, was mainstream biology back then. Ecologists will recognize throughout this book the strong influence of thinkers like Frederic Clements, Arthur Tansley, Alfred North Whitehead, Raymond Lindeman, G. Evelyn Hutchinson, Howard Odum, and other biologists who brought a deliberately holistic perspective to the matter of life and the environment, in which the boundary between the organism and its environment was not so strongly drawn as it is today. Such a point of view is no longer in the mainstream, though. Presently, biology that is not strictly materialist or reductionist is commonly regarded as somehow suspect or deficient in intellectual rigor. Even among ecologists, I think it fair to say, "good" ecology is defined by the distance one can put between it and the holistic philosophical leanings of the early ecologists like Clements and Whitehead.

Have we lost something we once had in those years? I think we have, but if the loss is to be regarded with any sentiment other than nostalgia, we need to ask why biology developed the way it did. Put another way, was the holistic perspective in biology lost because it really was inadequate, or did it simply go out of fashion?

Most likely, it was a little of both. The twentieth century has been biology's first Golden Age: the merging of chemistry, physics, and biology into the modern science of molecular biology has resulted in magnificent insights into the very nature of life itself. Similarly, neo-Darwinism put powerful conceptual tools in the hands of those interested in a scientific theory of evolution. The combination of the two has been one of science's great tag teams. In comparison, the early holistic biology simply offered less in the way of intellectual rigor and rewards: it was difficult to pin down exactly what a superorganism was, never mind how to recognize one when you saw one, or even what critical questions to ask of it when you did. It is no surprise, then, that biologists flocked to the disciplines that offered them the greater prospects for intellectual reward and professional advancement. Evolutionary biology and molecular biology offered these in abundance: "holistic" biology simply did not.

I usually am not one to argue with success, but I think holistic biology has also suffered for reasons quite apart from its intellectual merit. Looking back from our present vantage, it is easy to forget just what a revolutionary period the twentieth century has been, not just in science, but in politics, art, economics—the entire realm of human existence. Just as people were caught up seemingly at random in the turmoil of the twentieth century, so too was science, and the course of biology and ecology was altered in ways that we are only now beginning to comprehend. For example, holistic ecology flowered during the 1920s, in Germany, where the German *Naturphilosophen* had prepared fertile ground for it, and most fruitfully in the newly fledged Soviet Union, in the hands of thinkers like V. V. Dokuchaev and Vladimir Vernadsky (whom we have to thank for the term *biosphere*). Unfortunately, holistic concepts in biology and ecology shared a provenance with some of the philosophical underpinnings of National Socialism in Germany and of fascism generally. In the bloody turmoil that followed, it would seem inevitable that the entire stock, and not just its odious branches, would come under attack. Despite its promising early start among the Soviet ecologists, holistic biology would be strangled in its crib by Stalin and his minions. To the west, holistic biology was left gasping in the smoldering ruins of post-Nazi Germany, shunned as politically suspect. England was left too poor and exhausted to take up where Tansley and his students had left off. And in America, holistic biology was smothered under massive mountains of cash devoted to "big ecology" programs like the International Biological Program, which, in its brisk way, set about to "solve the ecosystem problem" with organizational charts, five-year plans, and the peculiarly American myth of the can-do everyman with little use for dreamy speculations about superorganisms. So perhaps there is something to the claim that the holistic perspective in biology faded because it became unfashionable (and for some of its proponents, literally deadly).

Every revolution carries within it the seeds of its own destruction, though, and the intellectual revolutions in twentieth-century biology are no exception. Molecular biology, which still has a very rich future in developing new pharmaceuticals, treatments for disease, enhancement of crop productivity, and so forth, has become transformed essentially into a branch of industrial biology. There is nothing wrong with that, mind you, but, I would suggest, there is also nothing to come from it that will make us think about the world in a fundamentally different way, as did the discovery of the structure of DNA. Neo-Darwinism, meanwhile, having explained the world to its own satisfaction, is now looking a bit frayed and dowdy, its proponents insisting to anyone who will listen that they really are the keepers of the keys to understanding the history of life on Earth. In short, evolutionary biology has become scholastic, with all its best insights behind it, and its adherents engaged in endless rancorous debate over ever more arcane and abstract subjects.

So, now that the revolution has become *bourgeois*, we at the end of the twentieth century are in the privileged position of being able to give serious consideration to the question: how much farther can these approaches to biology go? Put another way, can they lead us into biology's next Golden Age? Perhaps, but I think not. Maybe somewhere, someone will stumble across a gene for consciousness—now *that* would make us think differently about the world, wouldn't it?—but I doubt that such a thing exists. Similarly, it is hard to see what, short of a direct glimpse at the very origin of life, could penetrate the internally consistent and cozy world of modern Darwinism and make its adherents ask really critical questions of themselves. Rather, I think the path to biology's next Golden Age will involve breaching the essentially arbitrary boundary between organisms and the environment, to create a biology that unifies the living and the inanimate worlds.

Readings / Credits / Index

Readings

1. The Organism's Fuzzy Boundary

Benson, K. R. (1989). Biology's "Phoenix": Historical perspectives on the importance of the organism. *American Zoologist* 29: 1067–1074.

Bowler, P. J. (1992). *The Norton History of the Environmental Sciences*. New York: Norton & Company.

Collias, N. E., and E. C. Collias, eds. (1976). *External Construction by Animals*. Benchmark Papers in Animal Behavior. Stroudsburg, PA: Dowden, Hutchinson & Ross.

Donovan, S. K., ed. (1994). *The Paleobiology of Trace Fossils*. Baltimore, MD: Johns Hopkins University Press.

Hansell, M. H. (1984). *Animal Architecture and Building Behaviour*. London: Longman.

Kohn, A. J. (1989). Natural history and the necessity of the organism. *American Zoologist* 29: 1095–1103.

Louw, G. N., and W. J. Hamilton. (1972). Physiological and behavioral ecology of the ultrapsammophilus Namib desert tenebrionid, *Lepidochora argentogrisea*. *Madoqua II* 54–62: 87–98.

Mayr, E. (1982). *The Growth of Biological Thought: Diversity, Evolution, and Inheritance*. Cambridge, MA: Belknap/Harvard University Press.

Reid, R. G. B. (1989). The unwhole organism. *American Zoologist* 29: 1133–1140.

Ruse, M. (1989). Do organisms exist? *American Zoologist* 29: 1061–1066.

Seely, M. K., and W. J. Hamilton. (1976). Fog catchment sand trenches constructed by tenebrionid beetles, *Lepidochora*, from the Namib Desert. *Science* 193: 484–486.

Seely, M. K., C. J. Lewis, et al. (1983). Fog response of tenebrionid beetles in the Namib Desert. *Journal of Arid Environments* 6: 135–143.

von Frisch, K., and O. von Frisch. (1974). *Animal Architecture*. New York: Harcourt Brace Jovanovich.

2. Physiology Beyond the Organism

Crick, R. E., ed. (1989). *Origin, Evolution and Modern Aspects of Biomineralization in Plants and Animals*. New York: Plenum Press.

Fermi, E. (1936). *Thermodynamics*. New York: Dover Press.

Hoar, W. S., and D. J. Randall. (1969). *Fish Physiology: Excretion, Ionic Regulation and Metabolism*. New York, Academic Press.

Keenan, J. H., G. N. Hatsopoulos, et al. (1994). Principles of thermodynamics. *Encyclopædia Britannica* 28: 619–644. Chicago: The Encyclopædia Britannica.

Koestler, A. (1967). *The Ghost in the Machine*. New York: Macmillan Company.

Lowenstam, H. A., and S. Weiner. (1989). *On Biomineralization*. Oxford: Oxford University Press.

Maloiy, G. M. O. (1979). *Comparative Physiology of Osmoregulation in Animals*. New York: Academic Press.

Marshall, A. T. (1996). Calcification in hermatypic and ahermatypic corals. *Science* 271(2 February 1996): 637–639.

McConnaughey, T. (1989). Biomineralization mechanisms. In *Origin, Evolution and Modern Aspects of Biomineralization in Plants and Animals*, ed. R. E. Crick, 57–73 New York: Plenum.

Pearse, V. B. (1970). Incorporation of metabolic CO_2 into coral skeleton. *Nature* 228: 383.

Pytkowicz, R. M. (1969). Chemical solution of calcium carbonate in sea water. *American Zoologist* 9: 673–679.

Roberts, L. (1990). Warm waters, bleached corals. *Science* 250: 213.

Shoemaker, V. H., and K. A. Nagy. (1977). Osmoregulation in amphibians and reptiles. *Annual Review of Physiology* 39: 449–471.

Shreeve, J. (1996). Are algae—not coral—reef's master builders? *Science* 271(2 February 1996): 597–598.

Tracy, C. R., and J. S. Turner. (1982). What is physiological ecology? *Bulletin of the Ecological Society of America* 63: 340–341.

Veron, J. E. N. (1995). *Corals in Space and Time: The Biogeography and Evolution of the Scleractina.* Ithaca, New York: Comstock/Cornell.

Vogel, K., and W. F. Gutmann. (1989). Organismic autonomy in biomineralization processes. In *Origin, Evolution and Modern Aspects of Biomineralization in Plants and Animals,* ed. R. E. Crick, 45–56. New York: Plenum.

3.   *Living Architecture*

Bohinski, R. C. (1979). *Modern Concepts in Biochemistry.* Boston, MA: Allyn and Bacon.

Calow, P. (1977). Conversion efficiencies in heterotrophic organisms. *Biological Reviews* 52: 385–409.

Handey, J. (1996). *Deep Thoughts.* London: Warner Books.

Ohmart, R. D., and R. C. Lasiewski. (1971). Road runners: Energy conservation by hypothermia and absorption of sunlight. *Science* 172: 67–69.

Phillipson, J. (1966). *Ecological Energetics.* London: Edward Arnold Ltd.

Pimm, S. L. (1982). *Food Webs.* London: Chapman and Hall.

Rosenberg, N. J. (1974). *Microclimate: The Biological Environment.* New York: Wiley and Sons.

Trimmer, J. D. (1950). *Response of Physical Systems.* New York: Wiley and Sons.

4.   *Broth and Taxis*

Frankel, R. B. (1984). Magnetic guidance of organisms. *Annual Review of Biophysics and Bioengineering* 13: 85–104.

Hemmersbach-Krause, R., W. Briegleb, W. Haeder, and H. Plattner. (1991). Gravity effects on *Paramecium* cells: An analysis of a possible sensory function of trichocysts and of simulated weightlessness on trichocyst exocytosis. *European Journal of Protistology* 27(1): 85–92.

Hennessey, T. M., Y. Sairni, and C. Kung. (1983). A heat induced depolarization of *Paramecium* and its relationship to thermal avoidance behavior. *Journal of Comparative Physiology* 153A: 39–46.

Hill, N. A., T. J. Pedley, and J. O. Kessler. (1989). Growth of bioconvection patterns in a suspension of gyrotactic micro-organisms in a layer of finite depth. *Journal of Fluid Mechanics* 208: 509–543.

Hillesdon, A. J., and T. J. Pedley. (1996). Bioconvection in suspensions of oxytactic bacteria: Linear theory. *Journal of Fluid Mechanics* 324: 223–259.

Kessler, J. O. (1985a). Co-operative and concentrative phenomena of swimming micro-organisms. *Contemporary Physics* 26(2): 147–166.

——— (1985b). Hydrodynamic focusing of motile algal cells. *Nature* 313(17 January 1985): 218–220.

Kils, U. (1993). Formation of micropatches by zooplankton-driven microturbulences. *Bulletin of Marine Science* 53(1): 160–169.

Kudo, R. R. (1966). *Protozoology.* 5th ed. Springfield, IL: Charles C Thomas.

Lewin, R. A. (1995). Bioconvection. *Archiv für Hydrobiologie, Supplement* 198(77): 67–73.

Mendelson, N. H., and J. Lega. (1998). A complex pattern of traveling stripes is produced by swimming cells of *Bacillus subtilis. Journal of Bacteriology* 180(13): 3285–3294.

Parkinson, J. S., and D. F. Blair. (1993). Does *E. coli* have a nose? *Science* 259: 1701–1702.

Pedley, T. J., N. A. Hill, and J. O. Kessler. (1988). The growth of bioconvection patterns in a uniform suspension of gyrotactic micro-organisms. *Journal of Fluid Mechanics* 195: 223–237.

Pedley, T. J., and J. O. Kessler. (1990). A new continuum model for suspensions of gyrotactic micro-organisms. *Journal of Fluid Mechanics* 212: 155–182.

Peterson, I. (1996). Shaken bead beds show pimples and dimples. *Science News* 150(31 August 1996): 135.

Vincent, R. V., and N. A. Hill. (1996). Bioconvection in a suspension of phototactic algae. *Journal of Fluid Mechanics* 327: 343–371.

Vogel, S. (1981). *Life in Moving Fluids.* Boston, MA: Willard Grant Press.

——— (1993). Life in a whirl. *Discover* 14(August 1993): 80–86.

5.   *Then a Miracle Occurs . . .*

Barthel, D. (1991). Influence of different current regimes on the growth form of *Halichondria panicea* Pallas. In *Fossil and Recent Sponges,* ed. J. Reitner and H. Keupp, 387–394. Berlin: Springer Verlag.

Bibby, C. (1978). *The Art of the Limerick.* Hamden, CT: Archon Books.

Bradbury, R. H., and R. E. Reichelt. (1983). Fractal dimension of a coral reef at ecological scales. *Marine Ecology— Progress Series* 10(3 January 1983): 169–171.

Bradbury, R. H., and D. C. Young. (1981). The effects of a major forcing function, wave energy, on a coral reef ecosystem. *Marine Ecology—Progress Series* 5: 229–241.

Bulloch, D. K. (1992). Close comfort. *Underwater Naturalist* 20(4): 13–16.

Dauget, J. M. (1991). Application of tree architectural models to reef-coral growth. *Marine Biology* 111: 157–165.

Davidson, E. H., K. J. Peterson, and R. A. Cameron. (1995). Origin of bilaterian body plans: Evolution of developmental regulatory mechanisms. *Science* 270(24 November 1995): 1319–1325.

Erwin, D., J. Valentine, and D. Jablonski. (1997). The origin of animal body plans. *American Scientist* 85(March–April 1997): 126–137.

Fry, W. G. (1979). Taxonomy, the individual and the sponge. In *Biology and Systematics of Colonial Organisms*, ed. G. Larwood and B. R. Rosen, 11: 49–80. London: Academic Press.

Gleick, J. (1987). *Chaos: Making a New Science.* New York: Penguin.

Harris, S. (1977). *What's So Funny about Science?* Los Altos, CA: William Kaufmann.

Hickman, C. P., and F. M. Hickman. (1992). *Laboratory Studies in Integrated Zoology.* 8th ed. Dubuque, IA: Wm. C. Brown Publishers.

Jackson, J. B. G. (1979). Morphological strategies of sessile animals. In *Biology and Systematics of Colonial Organisms*, ed. G. Larwood and B. R. Rosen, 11: 499–555. London: Academic Press.

Kaandorp, J. (1994). *Fractal Modelling: Growth and Form in Biology.* Berlin: Springer Verlag.

Kaandorp, J. A. (1991). Modelling growth forms of the sponge *Haliclona oculata* (Porifera, Demospongiae) using fractal techniques. *Marine Biology* 110: 203–215.

Kaandorp, J. A., and M. J. de Kluijver (1992). Verification of fractal growth models of the sponge *Haliclona oculata* (Porifera) with transplantation experiments. *Marine Biology* 113: 133–145.

Kerr, R. A. (1997). Life's winners keep their poise in tough times. *Science* 278: 1403.

Koehl, M. A. R. (1982). The interaction of moving water and sessile organisms. *Scientific American* 247(December 1982): 124–134.

Larwood, G., and B. R. Rosen. (1979). *Biology and Systematics of Colonial Organisms.* London: Academic Press.

McMenamin, M. A. S., and D. L. Schulte-McMenamin. (1990). *The Emergence of Animals: The Cambrian Breakthrough.* New York: Columbia University Press.

Meynell, F., and V. Meynell. (1938). *The Weekend Book.* Harmondsworth: Penguin.

Morell, V. (1997). Microbiology's scarred revolutionary. *Science* 276(2 May 1997): 699–702.

Palumbi, S. R. (1987). How body plans limit acclimation: Responses of a demisponge to wave force. *Ecology* 67: 208–214.

Pennisi, E., and W. Roush (1997). Developing a new view of evolution. *Science* 277(4 July 1997): 34–37.

Rodrigo, A. G., P. R. Bergquist, P. L. Bergquist, and R. A. Reeves (1994). Are sponges animals? An investigation into the vagaries of phylogenetic inference. In *Sponges in Time and Space: Biology, Chemistry, Paleontology*, ed. R. W. M. van Soest, T. M. G. van Kempen, and J. C. Braekman, 47–54. Rotterdam, A. A. Balkema.

Rosen, B. R. (1986). Modular growth and form of corals: A matter of metamers? *Philosophical Transactions of the Royal Society of London, Series B, Biological Sciences* 313: 115–142.

Storer, T. I., R. L. Usinger, R. C. Stebbins, and J. W. Nybakken. (1979). *General Zoology.* 6th ed. New York: McGraw-Hill.

Veron, J. E. N. (1995). *Corals in Space and Time: The Biogeography and Evolution of the Scleractina.* Ithaca, New York: Comstock/Cornell.

Vogel, S. (1974). Current-induced flow through the sponge, *Halichondria. Biological Bulletin* 147: 443–456.

——— (1978). Evidence for one-way valves in the water-flow system of sponges. *Journal of Experimental Biology* 76: 137–148.

——— (1981). *Life in Moving Fluids.* Boston, MA: Willard Grant Press.

Warburton, F. E. (1960). Influence of currents on the forms of sponges. *Science* 132: 89.

Williams, N. (1997). Fractal geometry gets the measure of life's scales. *Science* 276(4 April 1997): 34.

6. *Mud Power*

Aller, R. C. (1983). The importance of the diffusive permeability of animal burrow linings in determining marine sediment chemistry. *Journal of Marine Research* 41: 299–322.

Aller, R. C., and J. Y. Yingst. (1978). Biogeochemistry of tube dwellings: A study of the sedentary polychaete *Amphitrite ornata* (Leidy). *Journal of Marine Research* 36(2): 201–254.

Andersen, F. O., and E. Kristensen. (1991). Effects of burrowing macrofauna on organic matter decomposition in coastal marine sediments. In *The Environmental Impact of Burrowing Animals and Animal Burrows*, ed. P. S. Meadows and A. Meadows, 63: 69–88. Oxford: Clarendon.

Balavoine, G., and A. Adoutte. (1998). One or three Cambrian radiations? *Science* 280(17 April 1998): 397–398.

Bohinski, R. C. (1979). *Modern Concepts in Biochemistry*. Boston, MA: Allyn and Bacon.

Briggs, D. E. G. (1994). Giant predators from the Cambrian of China. *Science* 264(27 May 1994): 1283–1284.

Bromley, R. G. (1990). *Trace Fossils: Biology and Taphonomy*. London: Unwin Hyman Ltd.

Chapman, G. (1949). The thixotropy and dilatancy of a marine soil. *Journal of the Marine Biological Association of the United Kingdom* 28: 123–140.

Chapman, G. and G. E. Newell. (1947). The role of the body fluid in relation to movement in soft-bodied invertebrates I. The burrowing of *Arenicola*. *Proceedings of the Royal Society of London, Series B, Biological Sciences* 134: 432–455.

Clark, R. B., and J. B. Cowey. (1958). Factors controlling the change of shape of certain nemertean and turbellarian worms. *Journal of Experimental Biology* 35: 731–748.

Conway Morris, S. (1989). Burgess Shale faunas and the Cambrian explosion. *Science* 246: 339–346.

Conway Morris, S., and H. B. Whittington. (1979). The animals of the Burgess Shale. *Scientific American* 241(July 1979): 122–133.

Crimes, T. P. (1994). The period of early evolutionary failure and the dawn of evolutionary success: The record of biotic changes across the Precambrian-Cambrian boundary. In *The Paleobiology of Trace Fossils*, ed. S. K. Donovan. 105–133. Baltimore, MD: Johns Hopkins University Press.

Donovan, S. K., ed. (1994). *The Paleobiology of Trace Fossils*. Baltimore, MD: Johns Hopkins University Press.

Elder, H. Y. (1980). Peristaltic mechanisms. In *Aspects of Animal Movement*, ed. H. Y. Elder and E. R. Trueman. Cambridge, UK, Cambridge University Press. 5: 71–92.

Elder, H. Y., and E. R. Trueman, eds. (1980). *Aspects of Animal Movement*. Society for Experimental Biology Seminar Series. Cambridge: Cambridge University Press.

Ferry, J. G. (1997). Methane: Small molecule, big impact. *Science* 278: 1413–1414.

Gust, G., and J. T. Harrison. (1981). Biological pumps at the sediment-water interface: Mechanistic evaluation of the alpheid shrimp *Alpheus mackayi* and its irrigation pattern. *Marine Biology* 64: 71–78.

Haq, B. U., and F. W.B. van Eysinga. (1998). *Geological Time Scale*. 5th ed. Amsterdam: Elsevier Science B.V.

Heffernan, J. M., and S. A. Wainwright. (1974). Locomotion of the holothurian *Euapta lappa* and redefinition of peristalsis. *Biological Bulletin* 147: 95–104.

Hickman, C. P., L. S. Roberts, and A. Larson. (1993). *Integrated Principles of Zoology*. 9th ed. Dubuque, IA: Wm. C. Brown.

Hylleberg, J. (1975). Selective feeding by *Abarenicola pacifica* with notes on *Abarenicola vagabunda* and a concept of gardening in lugworms. *Ophelia* 14: 113–137.

Kerr, R. A. (1998). Tracks of billion-year-old animals? *Science* 282(2 October 1998): 19–21.

McMenamin, M. A. S., and D. L. Schulte-McMenamin. (1990). *The Emergence of Animals: The Cambrian Breakthrough*. New York: Columbia University Press.

Meadows, P. S., and A. Meadows, eds. (1991a). *The Environmental Impact of Burrowing Animals and Animal Burrows*. Symposia of the Zoological Society of London. Oxford: Clarendon Press.

——— (1991b). The geotechnical and geochemical implications of bioturbation in marine sedimentary ecosystems. In *The Environmental Impact of Burrowing Animals and Animal Burrows*, ed. P. S. Meadows and A. Meadows, 63: 157–181. Oxford: Clarendon.

Rhoads, D. C., and D. K. Young. (1971). Animal-sediment relations in Cape Cod Bay, Massachusetts. II. Reworking by *Molpadia oolitica* (Holothuroidea)." *Marine Biology* 11: 255–261.

Savazzi, E. (1994). Functional morphology of boring and burrowing invertebrates. In *The Paleobiology of Trace Fossils*, ed. S. K. Donovan, 43–82. Baltimore, MD: Johns Hopkins University Press.

Schopf, J. W. (1975). Precambrian paleobiology: Problems and perspectives. *Annual Review of Earth and Planetary Sciences* 3: 213–249.

Seilacher, A., P. K. Bose, and F. Pflüger. (1998). Triploblastic animals more than 1 billion years ago: Trace fossil evidence from India. *Science* 282(2 October 1998): 80–83.

Seymour, M. K. (1970). Skeletons of *Lumbricus terrestris* L. and *Arenicola marina* (L.). *Nature* 228(24 October 1970).

——— (1971). Burrowing behaviour in the European lugworm *Arenicola marina* (Polychaeta: Arenicolidae). *Journal of Zoology (London)* 164: 93–132.

——— (1973). Motion and the skeleton in small nematodes. *Nematologica* 19: 43–48.

Vidal, G. (1984). The oldest eukaryotic cells. *Scientific American* 250: 48–57.

Wells, G. P. (1948). Thixotropy, and the mechanics of burrowing in the lugworm (*Arenicola marina* L.). *Nature* 162: 652–653.

White, D. (1995). *The Physiology and Biochemistry of Prokaryotes.* Oxford and New York: Oxford University Press.

Wyatt, T. (1993). Submarine beetles. *Natural History* 102(7): 6, 8–9.

Ziebis, W., S. Forster, M. Huettel, and B. B. Jorgensen. (1996). Complex burrows of the mud shrimp *Callianassa truncata* and their geochemical impact in the sea bed. *Nature* 382(15 August 1996): 619–622.

7. As the Worm Turns

Alexander, R. M. (1983). *Animal Mechanics.* Oxford: Blackwell Scientific.

Boroffka, I. (1965). Elektrolyttransport im nephridium von *Lumbricus terrestris. Zeitschrift für Vergleichende Physiologie* 51: 25–48.

Campbell, G. S. (1977). *An Introduction to Environmental Biophysics.* New York: Springer Verlag.

Childs, E. C., and N. C. George. (1948). Soil geometry and soil-water equilibria. *Discussions of the Faraday Society* 3: 78–85.

Edwards, W. M., M. J. Shipitalo, S. J. Traina, C. A. Edwards, and L. B. Owens. (1992). Role of *Lumbricus terrestris* (L.) burrows on quality of infiltrating water. *Soil Biology and Biochemistry* 24: 1555–1561.

Ghilarov, M. S. (1983). Darwin's *Formation of Vegetable Mould.* Its philosophical basis. In *Earthworm Ecology from Darwin to Vermiculite,* ed. J. E. Satchell, 1–4. London: Chapman and Hall.

Goodrich, E. S. (1945). The study of nephridia and genital ducts since 1895. *Quarterly Journal of Microscopical Science* 86: 113–392.

Graff, O. (1983). Darwin on earthworms—the contemporary background and what the critics thought. In *Earthworm Ecology from Darwin to Vermiculite,* ed. J. E. Satchell, 5–18. London: Chapman and Hall.

Hoogerkamp, M., H. Rogaar, and H. J. P. Eijsackers. (1983). Effect of earthworms on grassland on recently reclaimed polder soils in the Netherlands. In *Earthworm Ecology: From Darwin to Vermiculite,* J. E. Satchell, 85–105. London: Chapman and Hall.

Joschko, M., W. Sochtig, and O. Larink. (1992). Functional relationship between earthworm burrows and soil water movement in column experiments. *Soil Biology and Biochemistry* 24: 1545–1547.

Kretzschmar, A. (1983). Soil transport as a homeostatic mechanism for stabilizing the earthworm environment. In *Earthworm Ecology: From Darwin to Vermiculite,* ed. J. E. Satchell, 59–83. London: Chapman and Hall.

Kretzschmar, A., and F. Aries. (1992). An analysis of the structure of the burrow system of the giant Gippsland earthworm *Megascolides australis* McCoy 1878 using 3-D images. *Exeter* 24: 1583–1586.

Kretzschmar, A., and P. Monestiez. (1992). Physical control of soil biological activity due to endogeic earthworm behaviour. *Soil Biology and Biochemistry* 24: 1609–1614.

Lee, K. E. (1983). Earthworms of tropical regions—some aspects of their ecology and relationships with soils. In *Earthworm Ecology: From Darwin to Vermiculite,* ed. J. E. Satchell, 179–194. London: Chapman and Hall.

——— (1985). *Earthworms: Their Ecology and Relationships with Soils and Land Use.* Sydney: Academic Press.

——— (1991). The diversity of soil organisms. In *Biodiversity of Microorganisms and Invertebrates: Its Role in Sustainable Agriculture,* ed. D. L. Hawksworth, 73–87. Wallingford: CAB International.

McKenzie, B. M., and A. R. Dexter. (1988a). Axial pressures generated by the earthworm, *Aporrectodea rosea. Biology and Fertility of Soils* 5: 323–327.

——— (1988b). Radial pressures generated by the earthworm, *Aporrectodea rosea. Biology and Fertility of Soils* 5: 328–332.

Meadows, P. S., and A. Meadows, eds. (1991). *The Environmental Impact of Burrowing Animals and Animal Burrows.* Symposia of the Zoological Society of London. Oxford: Clarendon Press.

Monteith, J. L., and M. H. Unsworth. (1990). *Principles of Environmental Biophysics.* London: Edward Arnold.

Oglesby, L. C. (1978). Salt and water balance. In *Physiology of*

Annelids, ed. P. Mill, 555–658. New York: Academic Press.

Satchell, J. E., ed. (1983). *Earthworm Ecology: From Darwin to Vermiculite*. London: Chapman and Hall.

Seymour, M. K. (1970). Skeletons of *Lumbricus terrestris* L. and *Arenicola marina* (L.). *Nature* 228(24 October 1970): 383–385.

——— (1971). Coelomic pressure and electromyogram in earthworm locomotion. *Comparative Biochemistry and Physiology* 40A: 859–864.

——— (1976). Pressure difference in adjacent segments and movement of septa in earthworm locomotion. *Journal of Experimental Biology* 64: 743–750.

Stehouwer, R. C., W. A. Dick, and S. J. Traina. (1992). Characteristics of earthworm burrow lining affecting atrazine sorption. *Journal of Environmental Quality* 22: 181–185.

——— (1994). Sorption and retention of herbicides in vertically oriented earthworm and artificial burrows. *Journal of Environmental Quality* 23: 286–292.

Trojan, M. D., and D. R. Linden. (1992). Microrelief and rainfall effects on water and solute movement in earthworm burrows. *Journal of the Soil Science Society of America* 56: 727–733.

Withers, P. C. (1992). *Comparative Animal Physiology*. Fort Worth, TX: Saunders College Publishing.

8.   Arachne's Aqualungs

Alexander, R. M. (1983). *Animal Mechanics*. Oxford: Blackwell Scientific.

Bristowe, W. S. (1930). Notes on the biology of spiders. II. Aquatic spiders. *The Annals and Magazine of Natural History; Zoology, Botany and Geology* 10(6): 343–347.

——— (1931a). A British semi-marine spider. *The Annals and Magazine of Natural History; Zoology, Botany and Geology* 9(12): 154–156.

——— (1931b). Notes on the biology of spiders. IV. Further notes on aquatic spiders, with a description of a new series of pseudoscorpion from Singapore. *The Annals and Magazine of Natural History; Zoology, Botany and Geology* 10(8): 457–464.

Clausen, C. P. (1931). Biological observations on *Agriotypus* (Hymenoptera). *Proceedings of the Entomological Society of Washington* 33(February 1931 (2)): 29–37.

Crisp, D. J. (1950). The stability of structures at a fluid interface. *Transactions of the Faraday Society* 46: 228–235.

Crisp, D. J., and W. H. Thorpe. (1948). The water-protecting properties of insect hairs. *Discussions of the Faraday Society* 3: 210–220.

Dawkins, R. (1982). *The Extended Phenotype*. San Francisco: W. H. Freeman.

Edwards, G. A. (1953). Respiratory mechanisms. In *Insect Physiology*, ed. K. D. Roeder, 55–95. New York: John Wiley and Sons.

Ege, R. (1915). On the respiratory function of the air stores carried by some aquatic insects (Corixidae, Dytiscidae and *Notonecta*). *Zeitschrift für allgemeine Physiologie* 17: 81–124.

Fabre, J. H. (1913). *The Life of the Spider*. New York: Dodd, Mead and Company.

Fish, D. (1977). An aquatic spittlebug (Homoptera: Cercopidae) from a *Heliconia* flower bract in southern Costa Rica. *Entomological News* 88: 10–12.

Fisher, K. (1932). *Agriotypus ornatus* (Walk) (Hymenoptera) and its relations with its hosts. *Proceedings of the Zoological Society of London*, pp. 451–461.

Foelix, R. F. (1996). *Biology of Spiders*. New York: Oxford University Press.

Foster, W. A. (1989). Zonation, behaviour and morphology of the intertidal coral-treader *Hermatobates* (Hemiptera: Hermatobatidae) in the south-west Pacific. *Zoological Journal of the Linnaean Society* 96(1): 87–105.

Guilbeau, B. (1908). The origin and formation of the froth in spittle insects. *American Naturalist* 42: 783–789.

Harvey, E. N. (1928). The oxygen consumption of luminous bacteria. *Journal of General Physiology* 11: 469–475.

Hinton, H. E. (1960). Plastron respiration in the eggs of blowflies. *Journal of Insect Physiology* 4: 176–183.

——— (1963). The respiratory system of the egg-shell of the blowfly, *Calliphora erythrocephala* Meig., as seen with the electron microscope. *Journal of Insect Physiology* 9: 121–129.

——— (1968). Spiracular gills. *Advances in Insect Physiology* 5: 65–162.

——— (1971). Plastron respiration in the mite *Platyseius italicus*. *Journal of Insect Physiology* 17: 1185–1199.

Hoffman, G. D., and P. B. McEvoy (1985). Mechanical limitations on feeding by meadow spittlebugs *Philaenus spumarius* (Homoptera: Cercopidae) on wild and cultivated host plants. *Ecological Entomology* 10: 415–426.

Horsfield, D. (1977). Relationships between feeding of *Philaenus spumarius* (L.) and the amino acid concentration in the xylem sap. *Ecological Entomology* 2: 259–266.

——— (1978). Evidence for xylem feeding by *Philaenus*

spumarius (L.) (Homoptera: Cercopidae). *Entomologia Experimentalis et Applicata* 24: 95–99.

Jackson, R., C. L. Craig, J. Henschel, P. J. Watson, S. Pollard, and N. Platnick. (1995). Arachnomania! *Natural History* 3/95: 28–53.

Jefferys, W. H., and J. O. Berger. (1992). Ockham's razor and Bayesian analysis. *American Scientist* 80(January–February 1992): 64–72.

Kastin, B. J. (1964). The evolution of spider webs. *American Zoologist* 4(2): 191–207.

Keller, C. (1979). *The Best of Rube Goldberg*. Englewood Cliffs, NJ: Prentice-Hall.

Kershaw, J. (1914). The alimentary canal of a cercopid. *Psyche* 21: 65–72.

King, P. E., and M. R. Fordy. (1984). Observations on *Aepophilus bonnairei* (Signoret) (Saldidae: Hemiptera), an intertidal insect of rocky shores. *Zoological Journal of the Linnaean Society* 80(2–3): 231–238.

Krantz, G. W. (1974). *Phaulodinychus mitis* (Leonardi 1899) (Acari, Uropodidae). An intertidal mite exhibiting plastron respiration. *Acaralogia* 16: 11–20.

Krogh, A. (1941). *The Comparative Physiology of Respiratory Mechanisms*. New York: Dover.

Kuenzi, F., and H. Coppel. (1985). The biology of *Clastoptera arborina* (Homoptera: Cercopidae) in Wisconsin [USA]. *Transactions of the Wisconsin Academy of Sciences Arts and Letters* 73: 144–153.

Langer, R. M. (1969). Elementary physics and spider webs. *American Zoologist* 9: 81–89.

Lounibos, L. P., D. Duzak, and J. R. Linley. (1997). Comparative egg morphology of six species of the *albimanus* section of *Anopheles* (Nyssorhynchus) (Diptera: Culicidae). *Journal of Medical Entomology* 34(2): 136–155.

Marshall, A. (1966). Histochemical studies on a mucocomplex in the Malpighian tubules of a cercopoid larvae. *Journal of Insect Physiology* 12: 925–932.

Marshall, A. (1966). Spittle production and tube-building by cercopoid larvae (Homoptera) 4. Mucopolysaccharide associated with spittle production. *Journal of Insect Physiology* 12: 635–644.

Mello, M., E. R. Pimentel, A. T. Yamada, and A. Storopoli-Neto. (1987). Composition and structure of the froth of the spittlebug, *Deois* sp. *Insect Biochemistry* 17: 493–502.

Mello, M. L. S. (1987). Effect of some enzymes, chemicals and insecticides on the macromolecular structure of the froth of the spittlebug, the cercopid *Deois* sp. *Entomologia Experimentalis et Applicata* 44: 139–144.

Messner, B., and J. Adis. (1992). Cuticular wax secretions as plastron retaining structures in larvae of spittlebugs and cicada (Auchenorhyncha: Cercopida). *Revue Suisse de Zoologie* 99: 713–720.

Prange, H. D. (1996). *Respiratory Physiology: Understanding Gas Exchange*. New York: Chapman and Hall.

Preston-Mafham, R., and K. Preston-Mafham. (1984). *Spiders of the World*. New York: Facts on File Inc.

Rahn, H. (1966). Aquatic gas exchange: Theory. *Respiration Physiology* 1: 1–12.

Rahn, H., and C. V. Paganelli. (1968). Gas exchange in gas gills of diving insects. *Respiration Physiology* 5: 145–164.

Ross, H. H. (1964). Evolution of caddisworm cases and nests. *American Zoologist* 4(2): 209–220.

Savory, T. H. (1926). *British Spiders: Their Haunts and Habitats*. Oxford: Oxford University Press.

Stride, G. O. (1955). On the respiration of an African beetle *Potamodytes tuberosus* Hinton. *Annals of the Entomological Society of America* 48: 345–351.

Thorpe, W. H. (1950). Plastron respiration in aquatic insects. *Biological Reviews* 25: 344–390.

Thorpe, W. H., and D. J. Crisp. (1947a). Studies on plastron respiration I. The biology of *Aphelocheirus* (Hemiptera, Aphelocheiridae (Naucoridae)) and the mechanism of plastron retention. *Journal of Experimental Biology* 24: 227–269.

——— (1947b). Studies on plastron respiration II. The respiratory efficiency of the plastron in *Aphelocheirus*. *Journal of Experimental Biology* 24: 270–303.

——— (1947c). Studies on plastron respiration III. The orientation responses of *Aphelocheirus* (Hemiptera, Aphelocheiridae (Naucoridae)) in relation to plastron respiration, together with an account of specialized pressure receptors in aquatic insects. *Journal of Experimental Biology* 24: 310–328.

——— (1947d). Studies on plastron respiration IV. Plastron respiration in the Coleoptera. *Journal of Experimental Biology* 26: 219–260.

Valerio, J. R., F. M. Wiendl, and O. Nakano. (1988). Injection of salivary secretion by the adult spittlebug *Zulia entreriana* (Berg 1879) (Homoptera, Cercopidae) in *Brachiaria decumbens* Stapf. *Revista Brasileira de Entomologia* 32: 487–492.

von Frisch, K., and O. von Frisch. (1974). *Animal Architecture*. New York: Harcourt Brace Jovanovich.

Walcott, C. (1969). A spider's vibration receptor: Its anatomy and physiology. *American Zoologist* 9: 133–144.

Weaver, C. R., and D. R. King. (1954). *Meadow Spittlebug*. Wooster, OH: Ohio Agricultural Station.

Wiegert, R. G. (1964a). The ingestion of xylem sap by meadow spittlebug, *Philaenus spumarius* (L.). *The American Midland Naturalist* 71: 422–428.

——— (1964b). Population energetics of meadow spittlebugs (*Philaenus spumarius* L.) as affected by migration and habitat. *Ecological Monographs* 34: 217–241.

Wigglesworth, V. B., and J. W. L. Beament. (1960). The respiratory structures in the eggs of higher Diptera. *Journal of Insect Physiology* 4: 184–189.

William, S. J., and K. S. Ananthasubramanian. (1989). Spittle of *Clovia punctata* Walker (Homoptera: Cercopidae) and its biological significance. *Journal of Ecobiology* 1: 278–282.

——— (1990). The resting behaviour of the adult spittlebug *Clovia quadridens* (Walker) (Homoptera: Cercopidae). *Uttar Pradesh Journal of Zoology* 10: 111–113.

Wilson, A. A., and C. K. Dorsey. (1957). Studies on the composition and microbiology of insect spittle. *Annals of the Entomological Society of America* 50: 399–406.

Woolley, T. A. (1972). Scanning electron microscopy of the respiratory apparatus of ticks. *Transactions of the American Microscopical Society* 91: 348–363.

9. Manipulative Midges and Mites

Acquaah, G., J. W. Saunders, and L. C. Ewart. (1992). Homeotic floral mutations. *Plant Breeding Reviews* 9: 63–99.

Ananthakrishnan, T. N. (1984a). Adaptive strategies in cecidogenous insects. In *The Biology of Gall Insects*, ed. T. N. Ananthakrishnan, 1–9. London: Edward Arnold.

——— ed. (1984b). *The Biology of Gall Insects*. London: Edward Arnold.

Bridgewater, E. J., ed. (1950). *The Columbia Encyclopedia*. Morningside Heights, NY: Columbia University Press.

Campbell, G. S. (1977). *An Introduction to Environmental Biophysics*. New York: Springer Verlag.

Channabasava, G. P., and N. Nangia. (1984). The biology of gall mites. In *The Biology of Gall Insects*, ed. T. N. Ananthakrishnan, 323–337. London: Edward Arnold.

Dreger-Jannf, F., and J. D. Shorthouse. (1992). Diversity of gall-inducing insects and their galls. In *Biology of Insect-Induced Galls*, ed. J. D. Shorthouse and O. Rohrfritsch, 8–33. Oxford: Oxford University Press.

Felt, E. P. (1917). *Key to American Insect Galls*. Albany, NY: State University of New York.

——— (1940). *Plant Galls and Gall Makers*. Ithaca, NY: Comstock Publishing Co.

Fitter, A. H., and R. K. M. Hay. (1987). *Environmental Physiology of Plants*. London: Academic Press.

Hodkinson, I. D. (1984). The biology and ecology of the gall-forming Psylloidea (Homoptera). In *The Biology of Gall Insects*, ed. T. N. Ananthakrishnan, 59–77. London: Edward Arnold.

Jacobs, W. P. (1979). *Plant Hormones and Plant Development*. Cambridge: Cambridge University Press.

Leopold, A. C., and P. E. Kriedemann. (1975). *Plant Growth and Development*. New York: McGraw-Hill.

Lewis, I. F., and L. Walton. (1920). Gall-formation on leaves of *Celtis occidentalis* L. resulting from material injected by *Pachypsylla* sp. *American Journal of Botany* 7: 62–78.

Llewellyn, M. (1982). The energy economy of fluid-feeding herbivores. In *Proceedings of the 5th International Symposium on Insect-Plant Relationships*, ed. J. H. Visser and A. K. Minks, 243–252. Wageningen, Netherlands: Centre for Agricultural Publishing and Documentation.

Meyer, J. (1987). *Plant Galls and Gall Inducers*. Stuttgart: Gebruder Borntraeger.

Miller, H. C., and R. A. Norton. (1980). The maple gall mites. *New York State Tree Pest Leaflet* F11.

Sattler, R. (1988). Homeosis in plants. *American Journal of Botany* 75(10): 1606–1617.

——— (1992). Process morphology: Structural dynamics in development and evolution. *Canadian Journal of Botany* 70: 708–714.

Turner, J. S. (1994). Anomalous water loss rates from spittle nests of spittle bugs (Homoptera: Cercopidae). *Comparative Biochemistry and Physiology* 107A: 679–683.

Vogel, S. (1968). "Sun leaves" and "shade leaves": Differences in convective heat dissipation. *Ecology* 49(6): 1203–1204.

——— (1970). Convective cooling at low airspeeds and the shapes of broad leaves. *Journal of Experimental Botany* 21(66): 91–101.

Wareing, P. F., and I. D. J. Phillips. (1981). *Growth and Differentiation in Plants*. Oxford: Pergamon Press.

Wells, B. W. (1920). Early stages in the development of certain *Pachypsylla* galls on *Celtis*. *American Journal of Botany* 7: 275–285.

Williams, M. A. J., ed. (1994). *Plant Galls: Organisms, Interactions, Populations.* Oxford: Clarendon Press.

10. Twist and Shout!

Alexander, R. M. (1983). *Animal Mechanics.* Oxford: Blackwell Scientific.

Barinaga, M. (1995). Focusing on the *eyeless* gene. *Science* 267(24 March 1995): 1766–1767.

Bennet-Clark, H. C. (1970). The mechanism and efficiency of sound production in mole crickets. *Journal of Experimental Biology* 52: 619–652.

——— (1971). Acoustics of insect song. *Nature* 234(3 December 1971): 255–259.

——— (1975). Sound production in insects. *Science Progress, Oxford* 62: 263–283.

——— (1987). The tuned singing burrow of mole crickets. *Journal of Experimental Biology* 128: 383–411.

Fowler, H. G. (1987). Predatory behavior of *Megacephala fulgida* (Coleoptera: Cicindelidae). *The Coleopterists Bulletin* 41(4): 407–408.

Guido, A. S., and H. G. Fowler. (1988). *Megacephala fulgida* (Coleoptera: Cicindelidae): A phonotactically orienting predator of *Scapteriscus* mole crickets (Orthoptera: Gryllotalpidae). *Cicindela* 20 (September–December 1988): 51–52.

Halder, G., P. Callaerts, and W. J. Gehring. (1995). Induction of ectopic eyes by targeted expression of the *eyeless* gene in *Drosophila. Science* 267(24 March 1995): 1788–1792.

Homer. (1944). *The Odyssey.* New York: Walter J. Black.

Michelsen, A., and H. Nocke. (1974). Biophysical aspects of sound communication in insects. *Advances in Insect Physiology* 10: 247–296.

Nickerson, J. C., D. E. Snyder, and C. C. Oliver. (1979). Acoustical burrows constructed by mole crickets. *Annals of the Entomological Society of America* 72: 438–440.

Prozesky-Schulze, L., O. P. M. Prozesky, F. Anderson, and G. J. J. van der Merwe. (1975). Use of a self-made sound baffle by a tree cricket. *Nature* 255(8 May 1975): 142–143.

Quiring, R., U. Walldorf, U. Kloter, and W. J. Gehring. (1994). Homology of the *eyeless* gene of *Drosophila* to the *Small eye* gene in mice and *Aniridia* in humans. *Science* 265(5 August 1996): 785–789.

Ulagaraj, S. M. (1976). Sound production in mole crickets (Orthoptera, Gryllotalpidae, *Scapteriscus*). *Annals of the Entomological Society of America* 69: 299–306.

Ulagaraj, S. M., and T. J. Walker. (1975). Response of flying mole crickets to three parameters of synthetic songs broadcast outdoors. *Nature* 253(13 February 1975): 530–532.

Vaughan, C. C., S. M. Glenn, and I. H. Butler. (1993). Characterization of prairie mole cricket chorusing sites in Oklahoma. *American Midland Naturalist* 130: 364–371.

Walker, T. J., and D. E. Figg. (1990). Song and acoustic burrow of the prairie mole cricket, *Gryllotalpa major* (Orthoptera: Gryllidae). *Journal of the Kansas Entomological Society* 63(2): 237–242.

Zuker, C. S. (1994). On the evolution of eyes: Would you like it simple or compound? *Science* 265(5 August 1994): 742–743.

11. The Soul of the Superorganism

Chapela, I. H., S. A. Rehner, T. R. Schultz, and U. G. Mueller. (1994). Evolutionary history of the symbiosis between fungus-growing ants and their fungi. *Science* 266(9 December 1994): 1691–1694.

Cherett, J. M., R. J. Powell, and D. J. Stradling. (1989). The mutualism between leaf-cutting ants and their fungus. In *Insect-Fungus Interactions,* ed. N. Wilding, N. M. Collins, P. M. Hammond, and J. F. Webber, 93–120. London: Academic Press.

Dangerfield, J. M., T. S. McCarthy, and W. N. Ellery. (1998). The mound-building termite *Macrotermes michaelseni* as an ecosystem engineer. *Journal of Tropical Ecology* 14: 507–520.

Darlington, J. (1987). How termites keep their cool. *The Entomological Society of Queensland News Bulletin* 15: 45–46.

Darlington, J. P. E. C. (1984). Two types of mounds built by the termite *Macrotermes subhyalinus* in Kenya. *Insect Science and Its Application* 5(6): 481–492.

Emerson, A. E. (1956). Regenerative behavior and social homeostasis in termites. *Ecology* 37: 248–258.

Harris, W. V. (1956). Termite mound building. *Insectes Sociaux* 3(2): 261–268.

Heller, H. C., L. I. Crawshaw, and H. T. Hammel. (1978). The thermostat of vertebrate animals. *Scientific American* 239 (August): 88–96.

Hinkle, G., J. K. Wetterer, T. R. Schultz, and M. L. Sogin. (1994). Phylogeny of the attine ant fungi based on analysis of small subunit ribosomal RNA gene sequences. *Science* 266(9 December 1994): 1695–1697.

Howse, P. E. (1984). Sociochemicals of termites. In *Chemical Ecology of Insects,* W. J. Bell and R. T. Cardé, 475–519. London: Chapman and Hall.

LaBarbera, M. (1990). Principles of design of fluid transport systems in zoology. *Science* 249: 992–1000.

LaBarbera, M., and S. Vogel. (1982). The design of fluid transport systems in organisms. *American Scientist* 70: 54–60.

Lofting, H. (1920). *The Story of Dr. Dolittle.* Philadelphia: J B Lippincott Co.

Loos, R. (1964). A sensitive anemometer and its use for the measurement of air currents in the nests of *Macrotermes natalensis* (Haviland). In *Etudes sur les Termites Africains,* ed. A. Bouillon, 364–372. Paris: Maisson.

Lüscher, M. (1956). Die Lufterneuerung im nest der termite *Macrotermes natalensis* (Haviland). *Insectes Sociaux* 3: 273–276.

——— (1961). Air conditioned termite nests. *Scientific American* 238(1): 138–145.

Maeterlinck, M. (1930). *The Life of the White Ant.* New York: Dodd, Mead.

Marais, E. N. (1939). *The Soul of the White Ant.* London: Methuen.

Neal, E. G., and T. J. Roper. (1991). The environmental impact of badgers (*Meles meles*) and their setts. In *The Environmental Impact of Burrowing Animals and Animal Burrows,* P. S. Meadows and A. Meadows, 63: 89–106. Oxford: Clarendon.

Nicolas, G., and D. Sillans. (1989). Immediate and latent effects of carbon dioxide on insects. *Annual Review of Entomology* 34: 97–116.

Peakin, G. J., and G. Josens. (1978). Respiration and energy flow. *Production Ecology of Ants and Termites,* 111–163.

Peters, R. H. (1983). *The Ecological Implications of Body Size.* Cambridge: Cambridge University Press.

Pomeroy, D. E. (1976). Studies on a population of large termite mounds in Uganda. *Ecological Entomology* 1: 49–61.

Ruelle, J. E. (1964). L'architecture du nid de *Macrotermes natalensis* et son sens fonctionnel. In *Etudes sur les termites Africains,* ed. A. Bouillon, 327–362. Paris: Maisson.

——— (1985). Order Isoptera (termites). In *Insects of Southern Africa,* ed. C. H. Scholtz and E. Holm, 502. Durban: Butterworth.

Scherba, G. (1957). Moisture regulation in mound nests of the ant *Formica ulkei* Emery. *American Midland Naturalist* 61(2): 499–507.

——— (1962). Mound temperatures of the ant *Formica ulkei* Emery. *American Midland Naturalist* 67(2): 373–385.

Schneirla, T. C. (1946). Problems in the biopsychology of social organization. *Journal of Abnormal and Social Psychology* 41(4): 385–402.

Seeley, T. D. (1974). Atmospheric carbon dioxide regulation in honey-bee (*Apis mellifera*) colonies. *Journal of Insect Physiology* 20: 2301–2305.

——— (1985). The information-center strategy of honeybee foraging. In *Experimental Behavioral Ecology and Sociobiology,* ed. B. Hölldobler and M. Lindauer, 75–90. Sunderland, MA: Sindauer Associates.

Southwick, E. E. (1983). The honey bee cluster as a homeothermic superorganism. *Comparative Biochemistry and Physiology* 75A(4): 641–645.

Southwick, E. E., and R. F. A. Moritz. (1987). Social control of ventilation in colonies of honey bees, *Apis mellifera. Journal of Insect Physiology* 33: 623–626.

Stuart, A. M. (1967). Alarms, defense and construction behavior relationships in termites (Isoptera). *Science* 156: 1123–1125.

——— (1972). Behavioral regulatory mechanisms in the social homeostasis of termites (Isoptera). *American Zoologist* 12: 589–594.

Turner, J. S. (1994). Ventilation and thermal constancy of a colony of a southern African termite (*Odontotermes transvaalensis:* Macrotermitinae). *Journal of Arid Environments* 28: 231–248.

Weir, J. S. (1973). Air flow, evaporation and mineral accumulation in mounds of *Macrotermes subhyalinus. Journal of Animal Ecology* 42: 509–520.

Wheeler, W. M. (1911). The ant colony as an organism. *Journal of Morphology* 22: 302–325.

Wilson, E. O. (1971). *The Insect Societies.* Cambridge, MA: Belknap, Harvard University Press.

Wood, T. G., and R. J. Thomas. (1989). The mutualistic association between Macrotermitinae and *Termitomyces.* In *Insect-Fungus Interactions,* ed. N. Wilding, N. M. Collins, P. M. Hammond, and J. F. Webber, 69–92. London: Academic Press.

12. *Love Your Mother*

Alper, J. (1998). Ecosystem "engineers" shape habitats for other species. *Science* 280(22 May 1998): 1195–1196.

Andreae, M. O., and P. J. Crutzen. (1997). Atmospheric aerosols: Biogeochemical sources and role in atmospheric chemistry. *Science* 276(16 May 1997): 1052–1058.

Aoki, I. (1989). Holological study of lakes from an entropy viewpoint—Lake Mendota. *Ecological Modelling* 45: 81–93.

Ayala, F. (1970). Teleological explanation in evolutionary biology. *Philosophy of Science* 27: 1–15.

Barlow, C., and T. Volk. (1990). Open systems living in a closed biosphere: A new paradox for the *Gaia* debate. *BioSystems* 23: 371–384.

Boston, P. J., and S. L. Thompson. (1991). Theoretical microbial and vegetation control of planetary environments. In *Scientists on Gaia,* ed. S. Schneider and P. Boston, 99–117. Cambridge, MA: MIT Press.

Broecker, W. S. (1997). Thermohaline circulation, the Achilles heel of our climate system: Will man-made CO_2 upset the current balance? *Science* 278(28 November 1997): 1582–1588.

Caldeira, K. (1991). Evolutionary pressures on planktonic dimethylsulfide production. In *Scientists on Gaia,* ed. S. Schneider and P. Boston, 153–158. Cambridge, MA: MIT Press.

Capra, F. (1996). *The Web of Life: A New Scientific Understanding of Living Systems.* New York: Anchor Books.

Chapin, F. S. (1993). The evolutionary basis of biogeochemical soil development. *Geoderma* 57: 223–227.

Dawkins, R. (1982). *The Extended Phenotype.* Oxford: W. H. Freeman.

Doolittle, W. F. (1981). Is nature really motherly? *The Co-Evolution Quarterly* (Spring 1981): 58–65.

Dorn, R. I. (1991). Rock varnish. *American Scientist* 79(November–December 1991): 542–553.

Dubos, R. (1979). Gaia and creative evolution. *Nature* 282(8 November 1979): 154–155.

Dyer, B. D. (1989). Symbiosis and organismal boundaries. *American Zoologist* 29: 1085–1093.

Fyfe, W. S. (1996). The biosphere is going deep. *Science* 273(26 July 1996): 448.

Gille, J.-C., M. J. Pelegrin, and P. Decaulne. (1959). *Feedback Control Systems: Analysis, Synthesis and Design.* New York: McGraw-Hill.

Goldsmith, E. (1993). *The Way: An Ecological World-View.* Boston: Shambhala Publications.

Goodwin, B. (1994). *How the Leopard Changed Its Spots.* New York: Charles Scribner's Sons.

Jones, C. G., J. H. Lawton, and M. Shachak. (1997). Positive and negative effects of organisms as physical ecosystem engineers. *Ecology* 78(7): 1946–1957.

Joseph, L. E. (1990). *Gaia: The Growth of an Idea.* New York: St. Martin's Press.

Keeling, R. (1991). Mechanisms for stabilization of a simple biosphere: Catastrophe on Daisyworld. In *Scientists on Gaia,* ed. S. Schneider and P. Boston, 118–120. Cambridge, MA: MIT Press.

Kerr, R. (1993). How ice age climate got the shakes. *Science* 260(14 May 1993): 890–892.

Kirschner, J. W. (1991). The Gaia hypotheses: Are they testable? Are they useful? In *Scientists on Gaia,* ed. S. Schneider and P. Boston, 38–46. Cambridge, MA: MIT Press.

Kump, L. R., and F. T. Mackenzie. (1996). Regulation of atmospheric O_2: Feedback in the microbial feedbag. *Science* 271(26 January 1996): 459–460.

Lewin, R. (1996). All for one, one for all. *New Scientist* 152(14 December 1996): 28–33.

Lovelock, J. E. (1987). *Gaia: A New Look at Life on Earth.* Oxford: Oxford University Press.

——— (1988). *The Ages of Gaia: A Biography of Our Living Earth.* New York: W. W. Norton.

——— (1991). Geophysiology—the science of Gaia. In *Scientists on Gaia,* ed. S. Schneider and P. Boston, 3–10. Cambridge, MA: MIT Press.

——— (1993). The soil as a model for the Earth. *Geoderma* 57: 213–215.

Machin, K. E. (1964). Feedback theory and its application to biological systems. In *Homeostasis and Feedback Mechanisms,* ed. G. M. Highes, 18: 421–455. London: Academic Press.

Margulis, L. (1997). Big trouble in biology: Physiological autopoiesis versus mechanistic neo-Darwinism. In *Slanted Truths: Essays on Gaia, Symbiosis and Evolution,* ed. L. Margulis and D. Sagan, 265–282. New York: Copernicus.

Margulis, L., and D. Sagan, eds. (1997). *Slanted Truths: Essays on Gaia, Symbiosis and Evolution.* New York: Copernicus.

Markos, A. (1995). The ontogeny of *Gaia:* The role of microorganisms in planetary information network. *Journal of Theoretical Biology* 176: 175–180.

Mellanby, K. (1979). Living with the Earth Mother. *New Scientist* (4 October 1979): 41.

Morrison, P. (1980). Books. *Scientific American* 242(March 1980): 44–46.

Novikoff, A. B. (1945). The concept of integrative levels and biology. *Science* 101(2 March 1945): 209–215.

Odum, H. T. (1995). Self organization and maximum empower. In *Maximum Power: The Ideas and Applications of H. T. Odum*, ed. C. A. S. Hall, 311–330. Boulder: University of Colorado Press.

Okubo, T., and J. Matsumoto. (1983). Biological clogging of sand and changes of organic constituents during artificial recharge. *Water Research* 12(7): 813–821.

Pierce, J. R. (1980). *An Introduction to Information Theory: Symbols, Signals and Noise*. New York: Dover.

Pimm, S. L. (1989). *The Balance of Nature?* Chicago: University of Chicago Press.

Schneider, S. H., and P. J. Boston, eds. (1991). *Scientists on Gaia*. Cambridge, MA: MIT Press.

Shearer, W. (1991). A selection of biogenetic influences relevant to the *Gaia* hypothesis. In *Scientists on Gaia*, ed. S. Schneider and P. Boston, 23–29. Cambridge, MA: MIT Press.

Stølum, H.-H. (1996). River meandering as a self-organizing process. *Science* 271(22 March 1996): 1710–1713.

Sudd, H. J. (1967). *An Introduction to the Behavior of Ants*. New York: St Martin's Press.

Thompson, W. I., ed. (1987). *Gaia: A Way of Knowing*. Great Barrington, MA: Lindisfarne Press.

van Cappellen, P., and E. D. Ingall. (1996). Redox stabilization of the atmosphere and oceans by phosphorus-limited marine productivity. *Science* 271(26 January 1996): 493–496.

Varela, F. G., H. R. Maturana, and R. Uribe. (1974). Autopoiesis: The organization of living systems, its characterization and a model. *BioSystems* 5: 187–196.

Visvader, J. (1991). *Gaia* and the myths of harmony: An exploration of ethical and practical implications. In *Scientists on Gaia*, ed. S. Schneider and P. Boston, 33–37. Cambridge, MA: MIT Press.

Volk, T. (1998). *Gaia's Body: Toward a Physiology of Earth*. New York: Copernicus/Springer-Verlag.

Watson, A., J. E. Lovelock, and L. Margulis. (1978). Methanogenesis, fires and the regulation of atmospheric oxygen. *BioSystems* 10: 293–298.

Watson, A. J., and L. Maddock. (1991). A geophysiological model for glacial-interglacial oscillations in the carbon and phosphorus cycles. In *Scientists on Gaia*, ed. S. Schneider and P. Boston, 240–246. Cambridge, MA: MIT Press.

Wicken, J. S. (1981). Evolutionary self-organization and the entropy principle: Teleology and mechanism. *Nature and System* 3: 129–141.

Williams, G. C. (1992). *Gaia*, nature worship and biocentric fallacies. *Quarterly Review of Biology* 76(4): 479–486.

Williams, N. (1997). Biologists cut reductionist approach down to size. *Science* 277(25 July 1997): 476–477.

Wilson, D. S. (1997). Biological communities as functionally organized units. *Ecology* 78(7): 2018–2024.

Wyatt, T. (1989). Do algal blooms play homeostatic roles? In *Toxic Marine Phytoplankton: Proceedings of the Fourth International Conference on Toxic Marine Phytoplankton*. New York: Elsevier.

Epilogue

Bowler, P. J. (1992). *The Norton History of the Environmental Sciences*. New York: W. W. Norton and Company.

Davies, P. (1999). *The Fifth Miracle*. New York: Simon and Schuster.

Golley, F. B. (1993). *A History of the Ecosystem Concept in Ecology*. New Haven, CT: Yale University Press.

Credits

Epigraph to Chapter 3: From Jack Handey, *Deep Thoughts* (London: Warner Books, 1996).

Figure 4.1: From R. R. Kudo, *Protozoology,* 5th ed. (1966); courtesy of Charles C. Thomas, Publisher, Ltd., Springfield, Illinois.

Figure 5.1: From Sidney Harris, *What's So Funny about Science?* (Los Altos, CA: William Kaufmann, 1977); reprinted courtesy of Sidney Harris.

Figures 5.3a and 5.4: From T. I. Storer, R. L. Usinger, R. C. Stebbins, and J. W. Nybakken, *General Zoology,* 6th ed. (New York: McGraw-Hill, 1979); courtesy of California Academy of Sciences.

Figures 5.3b and 5.3c: From C. P. and F. M. Hickman, *Laboratory Studies in Integrated Zoology,* 8th ed. (Dubuque, IA: Wm. C. Brown Publishers, 1992); reproduced with permission of The McGraw-Hill Companies.

Figure 6.1: From C. Hickman, L. S. Roberts, and A. Larson, *Integrated Principles of Zoology,* 9th ed. (Dubuque, IA: Wm. C. Brown, 1993); reproduced with permission of The McGraw-Hill Companies.

Figure 6.7: From R. G. Bromley, *Trace Fossils: Biology and Taphonomy* (London: Unwin Hyman Ltd., 1990), figs. 4.10, 4.10, 4.15; with kind permission from Kluwer Academic Publishers.

Figures 7.1a and 7.2: From E. S. Goodrich, "The study of nephridia and genital ducts since 1895," *Quarterly Journal of Microscopical Science* 86 (1945): 113–392.

Figure 8.1: From R. Preston-Mafham and K. Preston-Mafham, *Spiders of the World* (New York: Facts on File, Inc., 1984).

Figure 8.6: Reprinted from H. E. Hinton, "The respiratory system of the egg-shell of the blowfly, *Calliphora erythrocephala* Meig., as seen with the electron microscope," *Journal of Insect Physiology* 9 (1963): 121–129; copyright © 1963, with permission from Elsevier Science.

Figure 8.7: From G. O. Stride, "On the respiration of an African beetle *Potamodytes tuberosus* Hinton," *Annals of the Entomological Society of America* 48 (1955): 345–351; courtesy of the Entomological Society of America.

Figure 8.10: Illustration from *Animal Architecture* by Karl von Frisch, translated by Lisbeth Gombrich, illustrations copyright © 1974 by Turid Hölldobler, reproduced by permission of Harcourt, Inc.

Figure 8.11a: From W. H. Thorpe, "Plastron respiration in aquatic insects," *Biological Reviews* 25 (1950): 344–390; courtesy of Cambridge University Press.

Figure 8.11b: From C. P. Clausen, "Biological observations on *Agriotypus* (Hymenoptera)," *Proceedings of the Entomological Society of Washington* 33 (February 1931): 29–37.

Figure 8.14: Photograph courtesy of Deborah Goemans.

Figure 9.3: Republished with permission of the Columbia University Press, 562 W. 113th St., New York, NY 10025. *The Columbia Encyclopedia,* 2d ed. (Plate 34), W. Bridgwater and E. J. Sherwood, 1956. Reproduced by permission of the publisher via Copyright Clearance Center, Inc.

Figure 9.8b: From B. W. Wells, "Early stages in the development of certain *Pachypsylla* galls on *Celtis,*" *American Journal of Botany* 7 (1920): 275–285; courtesy of *American Journal of Botany.*

Figure 10.3: From L. Prozesky-Schulze, O. P. M. Prozesky, F. Anderson, and G. J. J. van der Merwe, "Use of a self-made sound baffle by a tree cricket," *Nature* 255 (May 1975): 142–143; courtesy of *Nature.*

Figure 10.7: From J. C. Nickerson, D. E. Snyder, and C. C. Oliver, "Acoustical burrows constructed by mole crickets," *Annals of the Entomological Society of America* 72 (1979): 438–440.

Figure 11.8: Photograph courtesy of Margaret Voss.

Index

acetic acid (acetate), 88–89, 94
acetogenic bacteria, 88–89, 90, 91, 94
Acropora spp. (coral), 73, 74
adaptation, 5–7, 33, 35, 37–39, 52, 89,
   100–105, 115, 118–119, 142, 151, 154,
   175, 188, 194, 197, 199–200, 211–212;
   process control, 5–6, 37, 39; of environment, 6–7, 115, 118–119; and natural
   selection, 33, 35, 52, 175; and anaerobic
   bacteria, 89; and kidney structure, 100–
   105; and terrestriality, 101–102, 103;
   and leaf temperature, 142, 151, 154;
   and structure of termite mounds, 188,
   194, 197, 199–200; and telesymbiosis,
   211–212
adenosine triphosphate (ATP), 16, 19, 23,
   28, 29, 31, 118–119; cycle, 16; conversion to blood pressure, 19; solute transport, 19; calciloblast and, 23; and physiological "choices" by earthworms, 118–
   119; in light production, 160; and control of climate, 202
alarm pheromone, 188–189
altruism, 181–182, 212
ammonia, 85, 89, 90, 96–97, 102–103,
   140–141; in early atmosphere, 85; ammonia-oxidizing bacteria, 89, 90; metabolic rectification in lugworm burrows,
   96–97; in air compared to aquatic habitats, 102–103; and volatilization from
   spittle bug nests, 140–141
anaerobic (anoxic), 88–90; metabolism,
   88–90; bacteria in muds, 90–92, 94, 96–
   98
Arachne, 120
Archaea, 55, 90
Arenicola. See lugworms

Argyroneta aquatica (diving bell spider),
   121; winter web, 135
Aristotle, 80
Atlantic conveyor, 204

baffles, 165–167
bioconvection, 42, 43, 45, 47–51, 54–55,
   69–70, 76
black boxes, 206–208, 209, 210, 211
body plans, 56–61, 76–78, 82, 119; and
   embryogenesis, 56–58; of animals, 57–
   58, 76–78; of sponges, 58–60;
   epigenetic influences on, 59–60; genetic
   influences on, 59–60; of coelenterates,
   60–61; of corals, 61; and homeostasis,
   76–78; and new methods of burrowing,
   82; and physiological "re-tooling," 119
book lungs, 122
Brownian motion, 50
bubbles, 126–130
bubble gills, 124–125, 132–135, 138–141;
   of *Notonecta*, 124–125; of *Potamodytes*,
   132–135; of spittlebugs, 138–141
buoyancy, 42, 44, 124, 192–194, 196, 199,
   203–204; and convection cells, 42; and
   bubbles carried by beetles, 124, 130;
   and pushmi-pullyu ventilation, 192–
   194; and ventilation of termite mounds,
   196, 199; and Atlantic conveyor, 203–
   204
burrows, 7, 81–84, 85, 90–97, 99–100,
   106, 116–118, 170–178, 195, 199; correlation with origin of macropredation,
   81; trace fossils of, 81–84; diversification in Precambrian, 82–84, 85; evolutionary "arms race" in, 82–84; methods
   of construction, 83–84; feeding, 91–94;

as metabolic rectifiers, 94–97; of earthworms, 99–100, 106, 116–118; as
   acoustical devices, 170–178; in termite
   mounds, 195, 199

calciloblast, 23
calcite (calcium carbonate), 3, 19–21, 23–
   24, 82, 84, 208, 210; deposition by
   coral, 19–21, 23–24; solubility of, 20; bicarbonate and, 21; hydrogen ions and,
   21; in earthworm casts, 103; in soil stabilization by earthworms, 116–118
Cambrian, 81–82; "explosion," 81–82;
   diversification of burrows in, 85
capacitative work, 164
capacitors, 39, 169
carbon dioxide, 13, 16, 21, 85, 87–88, 94,
   101–102, 103, 122, 129–130, 145, 148–
   149, 188, 190, 192, 196, 208–210; in
   photosynthesis, 13, 16, 20, 85, 87, 145,
   148–149; and carbonate, 21; and transport in bioconvection, 52, 74; in air
   compared to aquatic environments,
   101–102; and excretion from earthworms, 103; in bubble gills, 129–130; in
   social insect nests, 188, 190, 192, 196;
   and symbiosis, 208–210
Chlamydomonas, 41–46, 49–53;
   bioconvection and, 41–42, 45–46;
   gravitaxis in, 43, 49–50; swimming
   speed of, 43; centers of buoyancy and
   gravity in, 44–45; distribution of mass
   in, 44–45, 52; hydrodynamic focusing,
   45–46; "anti-bubbles" in culture, 50–51;
   energy flow in culture, 52–53
chromatography, 95
Clements, Frederic, 213

orderliness *(continued)*
    and adaptive structures, 178; and
    fitness, 209
oxidation, 85–88, 89, 96, 97, 212. *See also*
    reduction
oxygen, 7, 13, 16, 18, 21, 44, 47–49, 52–
    53, 55, 65, 74, 81, 84–92, 94, 97, 101,
    106–107, 114–115, 118, 122–123, 125–
    132, 135–138, 140–141, 145, 188, 190–
    194, 196, 199, 212; and photosynthesis,
    13, 85, 212; and respiration, 16, 87;
    partial pressures in microbial culture,
    44, 47–49, 52–53, 115; accumulation in
    atmosphere, 81, 84–85; and oxidation
    potential, 87–88, 89; threat to anaerobic
    microbes, 89–90; in muds, 90–92, 94,
    97; availability in air vs. water, 106–
    107; distribution in soils, 114–115, 118;
    effect on survival times of *Notonecta*,
    125; pressures in bubbles, 125–128; and
    performance of bubble gills, 129–130;
    and plastron gills, 130–132, 135, 136–
    138; in termite colonies, 188; concen-
    trations in honeybee hives, 190–194;
    consumption rate of termite colony,
    196, 199

parasite's dilemma, 156–157
photosynthesis, 11–13, 15–16, 20–21, 24,
    56, 72, 80, 85, 87, 89–90, 101, 106, 139,
    145, 148–150, 152, 154–158, 207; cre-
    ation of orderliness by, 11–13;
    zooxanthellae and, 20–21, 24, 72; and
    blue-green algae, 80, 85; and
    terrestriality, 101, 106; and leaf temper-
    ature, 148–150, 152, 154–158; and
    symbiosis, 207
physiological flux *(PF)*, 34, 182, 184, 191,
    192
plastron gills, 130–138, 140; and
    hydrofuge hairs, 132; in insect eggs,
    132; dynamic gill of *Potamodytes*, 132–
    135; and winter web of *Argyroneta*, 135;
    and *Agriotypus* cocoons, 135–138; two-
    dimensional flow in *Aphelocheirus*, 137–
    138; and spittle nests, 140
Plato, 27
Poiseuille's law, 35–36
polychaete worms, 83, 91, 97, 100, 104,
    105

Porifera. *See* sponges
positive feedback, 51, 70–74, 77, 188–189,
    206; in bioconvection, 51; and reso-
    nance, 70–71; in aggregation growth,
    74; in superorganismal physiology, 77;
    and social homeostasis, 188–189; and
    climate control, 206
Precambrian, 82–84, 85; trace fossils in,
    82–84; diversification of burrows dur-
    ing, 82–84, 85
pressure, 18–19, 24, 28, 33, 35, 48–49, 53,
    101, 104–105, 107–108, 112–113, 122,
    125–128, 130–135, 137–138, 162–165,
    188, 191, 198; blood, 18–19, 24, 28,
    101, 104, 105, 112; partial, 48–49, 53,
    122, 125–128, 130–132, 137–138, 188,
    191; potential (Ψ_p), 107–108, 112–113;
    osmotic, 108; in bubbles, 125–128, 130–
    131, 132, 135; dynamic, 133–135, 198;
    vapor, 151; and sound, 162, 163–165
propolis, 190
Protista, 55
protostomy, 56
pupfish, 6
pushmi-pullyu ventilation, 189–194
pyruvic acid (pyruvate), 88

reabsorption, 18–19, 102, 103–104
rectification, 38, 97; metabolic rectifica-
    tion in lugworm burrows, 94–97
reduction, 90–91; redox potentials in
    muds, 91; redox potential discontinuity
    (RPD layer), 90–91. *See also* oxidation
reflectivities of natural surfaces, 33
regeneration, 59
reproductive work, 33–35
resistance: electrical, 36–38; in electrical
    analogy, 36–38; hydraulic, 36
respiration, 88, 92, 97, 124, 131, 134; an-
    aerobic, 87, 88, 89, 91
roadrunners *(Geococcyx californianus)*, 32

Scapteriscus acletus (mole cricket), 170–173,
    175
scientific notation, 14–15
Second Law of Thermodynamics, 12, 15–
    17, 20, 22, 27, 29, 34, 41, 43, 48, 52,
    100–102, 116, 161; entropy and, 12;
    glucose metabolism and, 15–16; water
    balance and, 17, 100–102; calcite depo-

sition and, 20; chemical kinetics and,
    22; efficiency and, 27, 29; homeostasis
    and, 34; bioconvection and, 41, 43, 48,
    52; soil building by earthworms and,
    116; sound production and, 161
singing burrows, 170–171, 173–176, 177;
    as Klipsch horns, 170–171; of
    Scapteriscus acletus, 170–173, 175; of
    Gryllotalpa spp., 172–174, 175; as guid-
    ance vs. broadcast beacons, 173–175; as
    "organs of extreme perfection," 175–
    176; tuning during construction, 176–
    178
Sirens, 159–160
Socrates, 26
soil, 33, 83–84, 100, 106–107, 109–119,
    130, 173, 176; water in, 107, 109–111,
    114–115, 118–119; oxygen in, 107,
    114–115; and water balance of earth-
    worms, 112–114; horizons in, 114–115;
    building by earthworms, 115–119; and
    Charles Darwin, 115–116; weathering
    of, 116–117; reversal of weathering by
    earthworms, 116–118
sophistry, 27, 28–29, 40
Sophists, 26–27; Plato and, 26; Socrates
    and, 27
sound, 70–71, 160–167, 172–173; and
    positive feedback, 70–71; generation by
    crickets, 160–161, 163–164; energetics,
    161, 163–164; speed in various materi-
    als, 162; tone and intensity, 162; basic
    terminology, 162–163; and energy bal-
    ance, 164, 166; efficiency of production,
    165–167, 172–173
Southern Ocean raceway, 203–204
spittlebugs, 138–141; bubble gills of, 138–
    141; and nitrogen metabolism, 140–141
spittle nest, 138–141; and water loss from,
    140
sponges (Porifera), 55, 58–62, 64, 65, 68–
    71, 74, 76–77, 188; body plans of, 58–
    60, 77; epigenetic influences on shape,
    59–60; modular growth in, 61–62, 64,
    65; diffusion-limited aggregation in, 68–
    69; physiology and body shape in, 69–
    71, 74, 76, 77
stigmergy, 189, 190
sulfide, 97
sulfur bacteria, 89

sulphate-reducing bacteria, 90, 94
superorganism, 52, 179–180, 184, 195, 201, 210, 213
symbiosis, 20, 24, 56, 179, 182, 206–211; and homeostasis, 206, 208–210; and feedback, 207–208; and telesymbiosis, 210–211

Tansley, Arthur, 213
taxis, 43, 44, 45, 47, 49
telesymbiosis. *See* symbiosis
temperature, 6, 10, 12, 14, 16, 22–23, 33, 36, 39, 47, 50, 101, 108, 118, 119, 191, 205; homeostasis, 32, 33, 34, 35, 78, 180, 182, 183, 184–187, 196, 197, 203, 208; leaf, 148–153
termites. *See Macrotermes*
terrestriality, 100–102
thermodynamically favored flux *(TFF)*, 33, 100, 182, 191
thermodynamics, 2, 5, 11, 12, 15–16, 17, 27
thermoregulation. *See* temperature
torque, 44–46; and orientation of *Chlamydomonas*, 44–45; in hydrodynamic focusing of *Chlamydomonas*, 46–47

tracheal system, 122
triploblastic, 56
trophic interaction, 30–31
tunnel. *See* burrow
Twain, Mark, 5, 153

U-burrows, 91–93
units of measure, 14–15
urine, 18–19, 27

velocity profile: in hydrodynamic focusing, 47; shear rate and, 47; in *Chlamydomonas* anti-bubble, 51
Vernadsky, Vladimir, 214
viscosity: and instability in density inversions, 50; and modulation of bioconvection, 77
Volk, Tyler, 210

water, 4–6, 11, 13, 16–21, 23, 27, 29, 32, 34–36, 39, 43–44, 49–50, 58–61, 66–69, 71–72, 75, 77, 81, 85–87, 89, 90–91, 93–97, 100, 102–108, 110, 113–119, 121–141, 151–152, 162, 183, 188, 190, 192, 196, 203–206, 208, 209, 212; and glucose, 11, 13, 16, 20, 23, 29, 85, 87, 91, 208, 209, 212; water balance, 17–

19, 27, 77, 103–106, 113–115, 119; and reaction with carbon dioxide, 21; and hydric environments, 100, 102–103; availability of oxygen, 106–107; in soils, 107–112, 114–117; movement in porous media, 108–110; micropores vs. macropores in distribution in soils, 117–118; and insect respiration, 121–141; and evaporation, 151, 152, 183, 188, 190, 192, 196; as a conveyor of heat, 203–206
water potential (Ψ), 107–113, 117–118; pressure potential (Ψ_p), 107–108, 112–113; gravity potential (Ψ_g), 108, 117–118; osmotic potential (Ψ_o), 108, 112–113; matric potential (Ψ_m), 108–110; of soils, 108–110; and costs of water balance, 110–111, 113; of worms, 112–114; and water balance in earthworms, 112–114
Watson, James, 9
Whirlpool, The, 4–6
Whitehead, Alfred North, 213
Wilkins, Maurice, 9
Wood-Jones model for coral growth, 74

zooids, 61